*Translations of Mathematical Monographs* *Volume 7*

*Applications of*

# FUNCTIONAL ANALYSIS

*in Mathematical Physics*

*by*

S. L. Sobolev

AMERICAN MATHEMATICAL SOCIETY
PROVIDENCE, RHODE ISLAND
1963

НЕКОТОРЫЕ ПРИМЕНЕНИЯ
ФУНКЦИОНАЛЬНОГО АНАЛИЗА В
МАТЕМАТИЧЕСКОЙ ФИЗИКЕ

С. Л. СОБОЛЕВ

Издательство Ленинградского
Государственного
Университета

Ленинград 1950

Translated from the Russian by F. E. Browder

Publication aided by grant NSF-G12381
from the
NATIONAL SCIENCE FOUNDATION

Library of Congress Card Number 63-15658



*Second Printing*, 1964

Printed in the United States of America

## Contents

Author's Preface vii

Chapter I. Special problems of functional analysis

§1. Introduction 1
1. Summable functions (1). 2. The Hölder and Minkowski inequalities (3). 3. The reverse of the Hölder and Minkowski inequalities (7).

§2. Basic properties of the spaces $L_p$ 9
1. Norms. Definitions (9). 2. The Riesz-Fischer Theorem (11). 3. Continuity in the large of functions in $L_p$ (11). 4. Countable dense nets (13).

§3. Linear functionals on $L_p$ 16
1. Definitions. Boundedness of linear functionals (16). 2. Clarkson's inequalities (17). 3. Theorem on the general form of linear functionals (22). 4. Convergence of functionals (25).

§4. Compactness of spaces 28
1. Definition of compactness (28). 2. A theorem on weak compactness (29). 3. A theorem on strong compactness (30). 4. Proof of the theorem on strong compactness (31).

§5. Generalized derivatives 33
1. Basic definitions (33). 2. Derivatives of averaged functions (35). 3. Rules for differentiation (37). 4. Independence of the domain (39).

§6. Properties of integrals of potential type 42
1. Integrals of potential type. Continuity (42). 2. Membership in $L_{q^*}$ (43).

§7. The spaces $L_p^{(l)}$ and $W_p^{(l)}$ 45
1. Definitions (45). 2. The norms in $L_p^{(l)}$ (46). 3. Decompositions of $W_p^{(l)}$ and its norming (48). 4. Special decompositions of $W_p^{(l)}$ (50).

§8. Imbedding Theorems 56
1. The imbedding of $W_p^{(l)}$ in $C$ (56). 2. Imbedding of $W_p^{(l)}$ in $L_{q^*}$ (57). 3. Examples (58).

§9. GENERAL METHODS OF NORMING $W_p^{(l)}$ AND COROLLARIES OF THE IMBEDDING THEOREM 60
1. A theorem on equivalent norms (60). 2. The general form of norms equivalent to a given one (62). 3. Norms equivalent to the special norm (64). 4. Spherical projection operators (64). 5. Nonstar-like domains (66). 6. Examples (67).

§10. SOME CONSEQUENCES OF THE IMBEDDING THEOREM 68
1. Completeness of the space $W_p^{(l)}$ (68). 2. The imbedding of $W_p^{(l)}$ in $W_{p_k}^{(k)}$ (69). 3. Invariant norming of $W_p^{(l)}$ (72).

§11. THE COMPLETE CONTINUITY OF THE IMBEDDING OPERATOR (KONDRAŠEV'S THEOREM) 74
1. Formulation of the problem (74). 2. A lemma on the compactness of the special integrals in $C$ (75). 3. A lemma on the compactness of integrals in $L_{q^*}$ (77). 4. Complete continuity of the imbedding operator in $C$ (82). 5. Complete continuity of the operator of imbedding in $L_{q^*}$ (84).

CHAPTER II. VARIATIONAL METHODS IN MATHEMATICAL PHYSICS

§12. THE DIRICHLET PROBLEM 87
1. Introduction (87). 2. Solution of the variational problem (88). 3. Solution of the Dirichlet problem (91). 4. Uniqueness of the solution of the Dirichlet problem (94). 5. Hadamard's example (97).

§13. THE NEUMANN PROBLEM 99
1. Formulation of the problem (99). 2. Solution of the variational problem (100). 3. Solution of the Neumann problem (101).

§14. POLYHARMONIC EQUATIONS 103
1. The behaviour of functions from $W_2^{(m)}$ on boundary manifolds of various dimensions (103). 2. Formulation of the basic boundary value problem (105). 3. Solution of the variational problem (106). 4. Solution of the basic boundary value problem (108).

§15. UNIQUENESS OF THE SOLUTION OF THE BASIC BOUNDARY VALUE PROBLEM FOR THE POLYHARMONIC EQUATION 112
1. Formulation of the problem (112). 2. Lemma (112). 3. The structure of the domains $\Omega_h - \Omega_{3h}$ (115). 4. Proof of the lemma for $k < [s/2]$ (116). 5. Proof of the lemma for $k = [s/2]+1$ (118). 6. Proof of the lemma for $[s/2]+2 \leq k \leq m$ (120). 7. Remarks on the formulation of the boundary conditions (122).

§16. THE EIGENVALUE PROBLEM 123
1. Introduction (123). 2. Auxiliary inequalities (124). 3. Minimal sequences and the equation of variations (126). 4. Existence of the first eigenfunction (128). 5. Existence of the later eigenfunctions (131). 6. The infinite sequence of eigenvalues (134). 7. Closedness of the set of eigenfunctions (136).

## CHAPTER III. THE THEORY OF HYPERBOLIC PARTIAL DIFFERENTIAL EQUATIONS

§17. SOLUTION OF THE WAVE EQUATION WITH SMOOTH INITIAL CONDITIONS 139
1. Derivation of the basic inequality (139). 2. Estimates for the growth of the solution and its derivatives (142). 3. Solutions for special initial data (144). 4. Proof of the basic theorem (146).

§18. THE GENERALIZED CAUCHY PROBLEM FOR THE WAVE EQUATION 148
1. Twice differentiable solutions (148). 2. Example (150). 3. Generalized solutions (152). 4. Existence of initial data (153). 5. Solutions of the generalized Cauchy problem (155).

§19. LINEAR EQUATIONS OF NORMAL HYPERBOLIC TYPE WITH VARIABLE COEFFICIENTS (BASIC PROPERTIES) 157
1. Characteristics and bicharacteristics (157). 2. The characteristic conoid (164). 3. Equations in canonical coordinates (166). 4. The basic operators $M^{(0)}$ and $L^{(0)}$ in polar coordinates (168). 5. The system of basic relations on the cone (173).

§20. THE CAUCHY PROBLEM FOR LINEAR EQUATIONS WITH SMOOTH COEFFICIENTS 175
1. The operators adjoint to the operators of the basic system (175). 2. The construction of the functions $\sigma$ (177). 3. Investigation of the properties of the functions $\sigma$ (179). 4. Derivation of the basic integral identity $Bu=SF$ (181). 5. The inverse integral operator $B^{-1}$ and the method of successive approximations (183). 6. The adjoint integral operator $B^*$ (187). 7. The adjoint integral operator $S^*$ (191). 8. Solution of the Cauchy problem for an even number of variables (192). 9. The Cauchy problem for an odd number of variables (195).

§21. INVESTIGATION OF LINEAR HYPERBOLIC EQUATIONS WITH VARIABLE COEFFICIENTS 196
1. Simplification of the equation (196). 2. Formulation of the Cauchy problem for generalized solutions (198). 3. Basic inequalities (200). 4. A lemma on estimates for approximating solutions (204). 5. Solution of the generalized problem (209). 6. Formulation of the classical Cauchy problem (210). 7. A lemma on estimates for derivatives (213). 8. Solution of the classical Cauchy problem (215).

§22. QUASILINEAR EQUATIONS 217
1. Formulation of functions of functions (217). 2. Basic inequalities (221). 3. Petrovsky's functional equation (224). 4. The Cauchy problem with homogeneous initial conditions (229). 5. Properties of averaged functions (231). 6. Transformation of initial conditions (235). 7. The general case for the Cauchy problem for quasilinear equations (237).

## AUTHOR'S PREFACE

The present book arose as a result of revising a course of lectures given by the writer at the Leningrad State University. The notes for the lectures were taken and revised by H. L. Smolicki and I. A. Jakovlev, who contributed to them a series of valuable remarks and additions. Several additions, arising naturally during the lectures, were also made by the author himself.

In this fashion there came into being this monograph, a unifying treatment from a single point of view of a number of problems in the theory of partial differential equations. There are considered in it variational methods with applications to the Laplace equation and the polyharmonic equations as well as the Cauchy problem for linear and quasi-linear hyperbolic equations. The presentation of the problems of mathematical physics demands a suitable consideration of some new results and methods in functional analysis, which constitute in themselves the basis of all the later material. The first part is concerned with this basis. The material indicated above, the particular problems posed, and the methods for their investigation are not to be found in the ordinary course in mathematical physics and in particular, they are not in my book *Equations of mathematical physics*. The present book is of value for graduate students and research workers.

The author warmly thanks his assistants, H. L. Smolicki and I. A. Jakovlev, without whose assistance this book could not have been written in such a short time.

S. Sobolev

CHAPTER I

# SPECIAL PROBLEMS OF FUNCTIONAL ANALYSIS

§1. **Introduction.** For the consideration of all the problems handled in this book, it will be necessary to refer repeatedly to some of the simplest properties of the functions integrable in the sense of Lebesgue and to some of the simplest concepts and theorems of functional analysis, which have become well-known. For this reason we shall not for the most part go into the details of their proofs and merely present the necessary formulations and definitions.

For the understanding of the whole exposition below, it is sufficient to have such a knowledge of the theory of multiple integrals of functions of real variables as is to be found in Lecture VI of the author's *Equations of mathematical physics* or in *Course of higher mathematics* Volume V of V. I. Smirnov.

We recall now some properties of multiple integrals and summable functions.

1. SUMMABLE FUNCTIONS. For any function $f(x_1,x_2,\cdots,x_n)$ of $n$ variables on the bounded domain $\Omega$ we denote by $F$ any closed set on which $f$ is continuous.

By the inner integral of a positive function $f$, we denote the upper bound

$$(\text{in.})\int_{\Omega} f\,dx_1 \ldots dx_n = \sup_{F\subset\Omega}\int_{F} f(x_1, \ldots, x_n)\,dx_1 \ldots dx_n. \tag{1.1}$$

If for a positive function $f$, the inner integral exists and satisfies the condition

$$(\text{in.})\int_{\Omega} (f+1)\,dx_1 \ldots dx_n = \\ = (\text{in.})\int_{\Omega} f\,dx_1 \ldots dx_n + (\text{in.})\int_{\Omega} 1\cdot dx_1 \ldots dx_n, \tag{1.2}$$

then the function is called summable, and the integral (in.) $\int_\Omega f dx_1 \cdots dx_n$ is written simply

$$\int_\Omega f\,dx_1 \ldots dx_n \tag{1.3}$$

and is called the Lebesgue integral.

A function taking values of different signs is called summable if both of the following are summable

$$f^+ = \frac{1}{2}\{f + |f|\} \quad \text{and} \quad f^- = \frac{1}{2}\{|f| - f\}, \tag{1.4}$$

while the integral of the function $f$ is defined by the formula

$$\int_\Omega f\,dx_1 \ldots dx_n = \int_\Omega f^+\,dx_1 \ldots dx_n - \int_\Omega f^-\,dx_1 \ldots dx_n. \tag{1.5}$$

By the Lebesgue measure of a set $E$ is meant the integral

$$mE = \int_\Omega \varphi_E\,dx_1 \ldots dx_n, \tag{1.6}$$

where $\phi_E$ takes the value 1 at points of $E$ and the value 0 on the complement $(\Omega - E)$.

A function $f$ is said to be measurable on the domain $\Omega$ if the measure of the closed sets $F$ on which it is continuous may be taken as close as one pleases to the measure of $\Omega$.

Every summable function is measurable.

The Lebesgue integral has the same property as does the ordinary integral. In the following, in place of $dx_1 \cdots dx_n$, we shall write merely $dv$.

$$\left.\begin{aligned} &\int_\Omega (f_1 + f_2)\,dv = \int_\Omega f_1\,dv + \int_\Omega f_2\,dv, \\ &\int_\Omega af\,dv = a\int_\Omega f\,dv; \quad a = \text{const.} \end{aligned}\right\} \tag{1.7}$$

If the series $f_1 + f_2 + \cdots + f_k + \cdots = f_0$ converges uniformly, then

$$\int_\Omega (f_1 + f_2 + \ldots + f_k + \ldots)\,dv =$$

$$= \int_\Omega f_1\,dv + \int_\Omega f_2\,dv + \ldots + \int_\Omega f_k\,dv + \ldots. \tag{1.8}$$

In addition, formula (1.8) is valid whenever $|f_1 + f_2 + \cdots + f_N| \leqq \Psi$ for every $N$ and in addition $\Psi$ is a summable function.

If $f \geqq 0$ and $\int_\Omega f dv = 0$, then the set of points where $f \neq 0$ has measure 0 $(m\{f \neq 0\} = 0)$.

Two functions $f_1$ and $f_2$ are equivalent if $\int_\Omega |f_1-f_2|dv=0$.

If $\int_\Omega f\psi dv=0$, where $\psi$ is an arbitrary function continuous with all its derivatives in the interior of $\Omega$, then $f$ is equivalent to zero.

If $k<f<K$, then

$$k\cdot m\Omega<\int_\Omega f\,dv<K\cdot m\Omega. \tag{1.9}$$

The Lebesgue integral is absolutely continuous. In other words, for each $\epsilon>0$ and for each function $f$ summable on the domain $\Omega$, one can find $\delta(\epsilon)>0$ such that for any set $E\subset\Omega$ we have the inequality $\int_E |f|dv<\epsilon$ as soon as $mE<\delta(\epsilon)$.

We shall prove two important elementary inequalities.

2. THE HÖLDER AND MINKOWSKI INEQUALITIES. Let $p>1$; then if $p'=p/(p-1)$, then

$$\frac{1}{p}+\frac{1}{p'}=1 \text{ and } p'-1=\frac{1}{p-1}. \tag{1.10}$$

We consider the curve $y=x^{p-1}$ (Figure 1). On this curve

$$x=y^{\frac{1}{p-1}}=y^{p'-1}.$$

Let $x$ and $y$ be two arbitrary positive numbers.

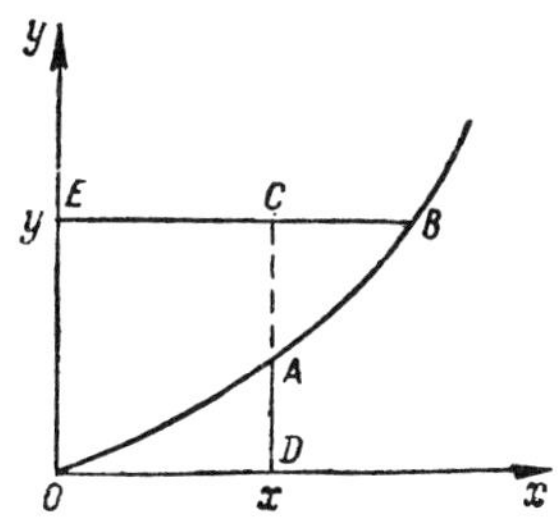

Figure 1

If we erect segments on the lines $AD{:}x=\text{const.}$ and $EC{:}y=\text{const.}$ up to their intersections with the curve, we see that the sum of the areas of the figures $OEB$ and $OAD$ is greater than the area of the rectangle $OECD$, whatever $x$ and $y$ may be. In other words

$$\int_0^x x^{p-1}\,dx+\int_0^y y^{p'-1}\,dy\geqslant xy \tag{1.11}$$

or

$$\frac{x^p}{p}+\frac{y^{p'}}{p'}\geqslant xy. \tag{1.12}$$

The equality will hold only in those cases where $y=x^{p-1}=x^{1/(p'-1)}$ or $x^p=y^{p'}$.

Let $\vec{Q}$ designate the variable point of the domain $\Omega$ of the $n$-dimensional space and $P(\vec{Q})>0$ be an arbitrary bounded function on $\Omega$. Let $x(\vec{Q})$ and $y(\vec{Q})$ be two positive functions on $\Omega$ satisfying the conditions

$$\int_\Omega |x(\vec{Q})|^p P(\vec{Q})\,dv=1;\quad \int_\Omega |y(\vec{Q})|^{p'} P(\vec{Q})\,dv=1. \tag{1.13}$$

Then, multiplying (1.12) by $P(\overrightarrow{Q})$, integrating over $\Omega$ and using (1.10), we obtain

$$\int_\Omega x(\vec{Q})\, y(\vec{Q})\, P(\vec{Q})\, dv \leqslant 1. \tag{1.14}$$

Let $X(\overrightarrow{Q})$ and $Y(\overrightarrow{Q})$ now be two arbitrary functions on $\Omega$ integrable respectively to the powers $p$ and $p'$. Then for the functions

$$x(\vec{Q}) = \frac{|X(\vec{Q})|}{\left[\int_\Omega |X|^p P\, dv\right]^{\frac{1}{p}}}; \quad y(\vec{Q}) = \frac{|Y(\vec{Q})|}{\left[\int_\Omega |Y|^{p'} P\, dv\right]^{\frac{1}{p'}}}$$

the inequality (1.13) is valid, and consequently, we have the inequality (1.14), which after simplification takes the form

$$\int_\Omega |X(\vec{Q})| \cdot |Y(\vec{Q})| P(\vec{Q})\, dv \leqslant \left[\int_\Omega |X|^p P\, dv\right]^{\frac{1}{p}} \cdot \left[\int_\Omega |Y|^{p'} P\, dv\right]^{\frac{1}{p'}},$$

from which follows the Hölder inequality

$$\left|\int_\Omega X(\vec{Q})\, Y(\vec{Q}) P(\vec{Q})\, dv\right| \leqslant \left[\int_\Omega |X|^p P\, dv\right]^{\frac{1}{p}} \times$$

$$\times \left[\int_\Omega |Y|^{p'} P\, dv\right]^{\frac{1}{p'}}. \tag{1.15}$$

It is obvious that the inequality sign can only hold in (1.14) if for almost all $\overrightarrow{Q}$ we have the equaltiy $x^p = y^{p'}$. Consequently, in the inequality (1.15) the equality sign holds only in the case in which

$$\frac{|X|^p}{\int_\Omega |X|^p P\, dv} = \frac{|Y|^{p'}}{\int_\Omega |Y|^{p'} P\, dv}, \quad \operatorname{sign} XY = \text{const.}$$

almost everywhere, i.e., if the functions $|X|^p$ and $|Y|^{p'}$ differ almost everywhere merely by a constant factor and $X$ and $Y$ have almost everywhere the same sign.

From (1.15) follows the generalized Hölder inequality for several functions.

Let $\lambda_1 + \lambda_2 + \cdots + \lambda_k = 1$, $\lambda_k > 0$ and let the functions $\phi_j$ $(j = 1, 2, \cdots, k)$ be integrable in their absolute values to the powers $1/\lambda_j$, i.e.,

$$\int_\Omega |\varphi_j|^{\frac{1}{\lambda_j}} P\, dv < \infty.$$

Then the product $\phi_1\phi_2\cdots\phi_k$ is summable, and we have the inequality

$$\left|\int_\Omega \varphi_1\varphi_2 \ldots \varphi_k P\,dv\right| \leqslant \left[\int_\Omega |\varphi_1|^{\frac{1}{\lambda_1}} P\,dv\right]^{\lambda_1} \times$$

$$\times \left[\int_\Omega |\varphi_2|^{\frac{1}{\lambda_2}} P\,dv\right]^{\lambda_2} \ldots \left[\int_\Omega |\varphi_k|^{\frac{1}{\lambda_k}} P\,dv\right]^{\lambda_k}, \tag{1.16}$$

where the inequality holds only if the $|\phi_j|^{1/\lambda_j}$ differ from one another merely by a constant factor (i.e., $|\phi_j|^{1/\lambda_j}=c_j\psi$) and the sign of $[\phi_1\phi_2\cdots\cdots\phi_k]$ is a constant almost everywhere.

We prove this inequality by induction from $k$ to $k+1$.

Let $\phi_1,\phi_2,\cdots,\phi_k,\phi_{k+1}$ be positive and suppose that for $k$ functions the inequality (1.16) has been proved. Then, if $\lambda_1+\lambda_2+\cdots+\lambda_k+\lambda_{k+1}=1$, we have, putting $p=1/\lambda_{k+1}$,

$$p'=\frac{1}{\lambda_1+\lambda_2+\ldots+\lambda_k}:$$

$$\left|\int_\Omega \varphi_1\varphi_2 \ldots \varphi_k\varphi_{k+1}P\,dv\right| \leqslant \left[\int_\Omega [\varphi_1\varphi_2 \ldots \varphi_k]^{\frac{1}{\lambda_1+\lambda_2+\ldots+\lambda_k}} \times\right.$$

$$\left.\times P\,dv\right]^{\lambda_1+\lambda_2+\ldots+\lambda_k}\left[\int_\Omega [\varphi_{k+1}]^{\frac{1}{\lambda_{k+1}}} P\,dv\right]^{\lambda_{k+1}}. \tag{1.17}$$

By virtue of the assumption of the correctness of (1.16) for $k$ functions, we obtain

$$\int_\Omega \varphi_1^{\frac{1}{\lambda_1+\lambda_2+\ldots+\lambda_k}} \varphi_2^{\frac{1}{\lambda_1+\lambda_2+\ldots+\lambda_k}} \ldots \varphi_k^{\frac{1}{\lambda_1+\lambda_2+\ldots+\lambda_k}} P\,dv \leqslant$$

$$\leqslant \left[\int_\Omega \left(\varphi_1^{\frac{1}{\lambda_1+\lambda_2+\ldots+\lambda_k}}\right)^{\frac{\lambda_1+\ldots+\lambda_k}{\lambda_1}} P\,dv\right]^{\frac{\lambda_1}{\lambda_1+\ldots+\lambda_k}} \ldots \times$$

$$\times \ldots \left[\int_\Omega \left(\varphi_k^{\frac{1}{\lambda_1+\ldots+\lambda_k}}\right)^{\frac{\lambda_1+\ldots+\lambda_k}{\lambda_k}} P\,dv\right]^{\frac{\lambda_k}{\lambda_1+\ldots+\lambda_k}}. \tag{1.18}$$

Substituting this expression in (1.17), we obtain the inequality (1.16) for $k+1$ functions. For $k=2$, the inequality was proved in (1.15). Consequently, the Hölder inequality is proved for all $k$.

We may verify, as a consequence of the result obtained earlier, that the equality sign holds only when all the functions

$$|\varphi_1|^{\frac{1}{\lambda_1}},\quad |\varphi_2|^{\frac{1}{\lambda_2}}, \ldots, |\varphi_k|^{\frac{1}{\lambda_k}}$$

differ from one another by constant factors (with the exception of a set of measure zero).

If the functions $\phi_1, \phi_2, \cdots, \phi_k$ take on only a finite number of values, the integrals may be written as sums, and we obtain

$$\sum_{i=1}^{N} a_1^{(i)} a_2^{(i)} \ldots a_k^{(i)} \leqslant [\sum_{i=1}^{N} [a_1^{(i)}]^{\frac{1}{\lambda_1}}]^{\lambda_1} \times$$
$$\times [\sum_{i=1}^{N} (a_2^{(i)})^{\frac{1}{\lambda_2}}]^{\lambda_2} \ldots [\sum_{i=1}^{N} (a_k^{(i)})^{\frac{1}{\lambda_k}}]^{\lambda_k} \qquad (1.19)$$

This inequality is also called the Hölder inequality. From (1.19) for the case $\lambda_1 = \lambda_2 = \frac{1}{2}$, we may obtain the useful inequality

$$[\sum_{i=1}^{N} a^{(i)}] = [\sum_{i=1}^{N} 1 \cdot a^{(i)}] \leqslant (\sum_{i=1}^{N} 1)^{\frac{1}{2}} [\sum_{i=1}^{N} (a^{(i)})^2]^{\frac{1}{2}} =$$
$$= \sqrt{N} [\sum_{i=1}^{N} (a^{(i)})^2]^{\frac{1}{2}}. \qquad (1.20)$$

Let $x(\overrightarrow{Q}) \geqq 0$, $y(\overrightarrow{Q}) \geqq 0$ on $\Omega$. We consider:

$$\int_\Omega (x+y)^p P\,dv = \int_\Omega x\,(x+y)^{p-1} P\,dv + \int_\Omega y\,(x+y)^{p-1} P\,dv.$$

To each term of the right hand side we apply the Hölder inequality. We obtain:

$$\int_\Omega (x+y)^p P\,dv \leqslant \left[\int_\Omega x^p P\,dv\right]^{\frac{1}{p}} \left[\int_\Omega (x+y)^{(p-1)p'} P\,dv\right]^{\frac{1}{p'}} +$$
$$+ \left[\int_\Omega y^p P\,dv\right]^{\frac{1}{p}} \left[\int_\Omega (x+y)^{(p-1)p'} P\,dv\right]^{\frac{1}{p'}} =$$
$$= \left[\int_\Omega (x+y)^p P\,dv\right]^{\frac{p-1}{p}} \left\{\left[\int_\Omega x^p P\,dv\right]^{\frac{1}{p}} + \left[\int_\Omega y^p P\,dv\right]^{\frac{1}{p}}\right\}.$$

Cancelling the first factor on the right, we obtain the Minkowski inequality:

$$\left[\int_\Omega (x+y)^p P\,dv\right]^{\frac{1}{p}} \leqslant \left[\int_\Omega x^p P\,dv\right]^{\frac{1}{p}} + \left[\int_\Omega y^p P\,dv\right]^{\frac{1}{p}}. \qquad (1.21)$$

It is obvious how to extend (1.21) to the sum of several functions of arbitrary sign on $\Omega$. Then we obtain the Minkowski inequality:

$$\left[\int_\Omega |x_1+\cdots+x_k|^p P\,dv\right]^{\frac{1}{p}} \leqslant \left[\int_\Omega |x_1|^p P\,dv\right]^{\frac{1}{p}} + \cdots + \left[\int_\Omega |x_k|^p P\,dv\right]^{\frac{1}{p}}. \tag{1.22}$$

The equality sign can hold only in case the functions $x_1, x_2, \cdots, x_k$ are all proportional.

If the functions $x$ and $y$ take on only a finite number of values, then the integrals may be taken as sums, and we obtain the Minkowski inequality for numerical series. From (1.21) we obtain:

$$[\sum a_i |x_i+y_i|^p]^{\frac{1}{p}} \leqslant [\sum a_i |x_i|^p]^{\frac{1}{p}} + [\sum a_i |y_i|^p]^{\frac{1}{p}}, \tag{1.23}$$

or for several numerical series:

$$[\sum_i a_i |\sum_j x_{ij}|^p]^{\frac{1}{p}} \leqslant \sum_j [\sum_i a_i |x_{ij}|^p]^{\frac{1}{p}}. \tag{1.24}$$

3. THE REVERSE OF THE HÖLDER AND MINKOWSKI INEQUALITIES: Let $0<p<1$, so that $p'=p/(p-1)<0$.

We consider the curve $y=x^{p-1}$ (Figure 2). On this curve, obviously,

$$x=y^{p'-1}.$$

Let $C$ be a point with coordinates $x$ and $y$ situated above the curve $y=x^{p-1}$.

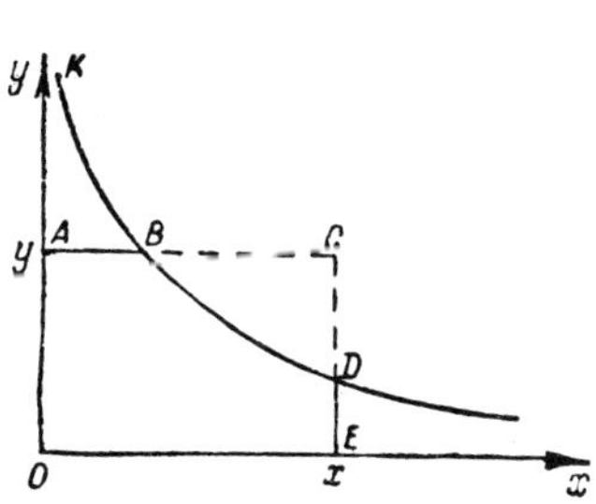


Figure 2

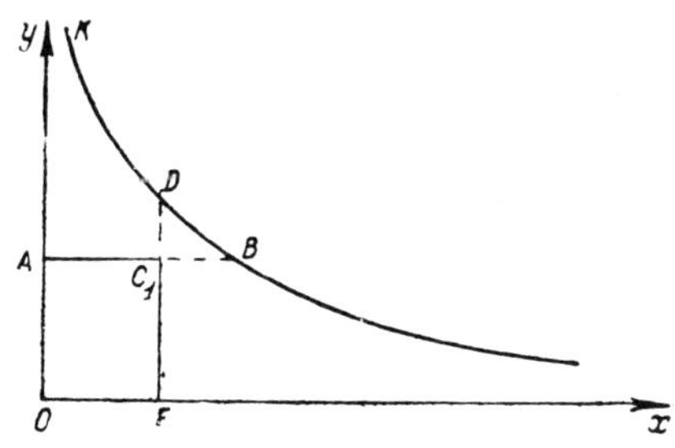


Figure 3

The area $q$ of the figure $OABDE$ is less than the area of the rectangle $OACE$. Let $K$ be an arbitrarily distant point on the $y$-axis. The area $q$ may be expressed as the difference between the area $OAKBDE=\int_0^x y\,dx$ and the area $AKB=\int_y^\infty x\,dy$. In other words:

$$\int_0^x x^{p-1}\,dx - \int_y^\infty y^{p'-1}\,dy \leqslant xy \quad \text{or} \quad \frac{x^p}{p}+\frac{y^{p'}}{p'} \leqslant xy.$$

Now we take a point $C_1$ below the curve $y=x^{p-1}$ (Figure 3). Consider the difference between the areas $KAOEC_1D=\int_0^x x^{p-1}dx$ and $KAC_1BD=\int_y^\infty y^{p'-1}dy$. This difference is equal to the difference between the areas $OAC_1E$ and $C_1DB$ and, consequently, is less than $OAC_1E$, i.e.,

$$\frac{x^p}{p}+\frac{y^{p'}}{p'} \leqslant xy. \tag{1.25}$$

In the cases where $y=x^{p-1}$, i.e., $x^p=y^{p'}$, the inequality (1.25) reduces to an equality.

From (1.25), as we have done for (1.15), we obtain for positive $X$ and $Y$ the reverse Hölder inequality:

$$\int_\Omega XYP\,dv \geqslant \left[\int_\Omega X^p P\,dv\right]^{\frac{1}{p}} \left[\int_\Omega Y^{p'} P\,dv\right]^{\frac{1}{p'}}. \tag{1.26}$$

From the reverse Hölder inequality (1.26), we may obtain the reverse Minkowski inequality:

$$\left[\int_\Omega (x_1+x_2+\ldots+x_k)^p P\,dv\right]^{\frac{1}{p}} \geqslant \left[\int_\Omega x_1^p P\,dv\right]^{\frac{1}{p}}+\ldots+ \left[\int_\Omega x_k^p P\,dv\right]^{\frac{1}{p}}, \tag{1.27}$$

valid for $0<p<1$ for positive $x_1,x_2,\cdots,x_k$.

As before, the equality sign can hold only in those cases where all the functions in the right hand side are proportional.

The proof is completely analogous to the preceding.

NOTE. Let $x(\overrightarrow{Q})$ be given on $\Omega$. If we put

$y=x$ on the set $E\subset\Omega$,
$y=0$ on the set $\Omega-E$,
$z=x-y$ on $\Omega$,

then we will have, if we apply to $y$ and $z$ the Minkowski inequality for $p\geqq 1$:

$$\left[\int_\Omega |x|^p\,dv\right]^{\frac{1}{p}} \leqslant \left[\int_E |x|^p\,dv\right]^{\frac{1}{p}} + \left[\int_{\Omega-E} |x|^p\,dv\right]^{\frac{1}{p}}.$$

For $p<1$, the inequality is reversed (if $x>0$).

## §2. Basic properties of the spaces $L_p$.

1. NORMS. DEFINITIONS. In the ordinary Euclidean space of $n$ dimensions, for the concepts of convergence and of passage to a limit one uses the distance between two points $\rho=(\sum(x_i-y_i)^2)^{1/2}$.

The function $\rho=(\sum\xi_i^2)^{1/2}$, expressing the length of the vector $\vec{x}$ with coordinates $\xi_i$, is a special case of the so-called norm of a vector. The distance between points is expressed in this way as the norm of the difference of the coordinate vectors of these points. The method of introducing the norms of vectors by means of this Euclidean length turns out not to be unique. We shall say that the function $\rho(x)$ is a permissible norm if it satisfies the three conditions:

(A) The function $\rho$ is a homogeneous function of the first degree with respect to $\xi_1,\xi_2,\cdots,\xi_n$, i.e.,

$$\rho(k\xi_1,\ k\xi_2,\ \ldots,\ k\xi_n)=k\rho(\xi_1,\ \xi_2,\ \ldots,\ \xi_n). \tag{2.1}$$

(B) The function $\rho(\xi_1,\cdots,\xi_n)$ is a convex function of its arguments. In other words, if we define the vector $\lambda\vec{x}+\mu\vec{y}$ as the vector with components $\lambda\xi_i+\mu\eta_i$, where $\xi_i$ and $\eta_i$ are the corresponding components of the vectors $\vec{x}$ and $\vec{y}$, then from the equalities $\rho(x)=a$, $\rho(y)=a$ follows

$$\rho(\lambda\vec{x}+\mu\vec{y})\leqslant a, \tag{2.2}$$

if $\lambda+\mu=1$, $0\leqq\lambda\leqq1$.

Geometrically the property of convexity means that if two points $\vec{x}$ and $\vec{y}$ lie on the surface $\rho=\text{const.}$, then an arbitrary point of the segment joining them will lie either on the same surface $\rho=\text{const.}$ or within it.

(C) From the equality $\rho(\xi_1,\xi_2,\cdots,\xi_n)=0$ follows

$$\xi_1=0,\ \xi_2=0,\ \ldots,\ \xi_n=0.$$

TRIANGLE INEQUALITY. The property of convexity is often formulated somewhat differently.

Let $\vec{\xi}$ and $\vec{\eta}$ be an arbitrary pair of vectors. We consider the quantity $\rho(\vec{\xi}+\vec{\eta})$ and attempt to estimate it. We have:

$$\vec{\xi}+\vec{\eta}=\left[\frac{\rho(\vec{\xi})}{\rho(\vec{\xi})+\rho(\vec{\eta})}\frac{\vec{\xi}}{\rho(\vec{\xi})}+\frac{\rho(\vec{\eta})}{\rho(\vec{\xi})+\rho(\vec{\eta})}\frac{\vec{\eta}}{\rho(\vec{\eta})}\right][\rho(\vec{\xi})+\rho(\vec{\eta})].$$

If we put

$$\frac{\vec{\xi}}{\rho(\vec{\xi})}=\vec{x};\quad \frac{\vec{\eta}}{\rho(\vec{\eta})}=\vec{y},$$

then we shall obtain, obviously,

$$\rho(\vec{x})=1;\quad \rho(\vec{y})=1.$$

Further, putting

$$\lambda = \frac{\rho(\vec{\xi})}{\rho(\vec{\xi}) + \rho(\vec{\eta})}; \quad \mu = \frac{\rho(\vec{\eta})}{\rho(\vec{\xi}) + \rho(\vec{\eta})}$$

and using the homogeneity of the norm, we have:

$$\rho(\vec{\xi} + \vec{\eta}) = [\rho(\vec{\xi}) + \rho(\vec{\eta})]\, \rho(\lambda\vec{x} + \mu\vec{y}).$$

Since the last factor is less than or equal to 1 by property (B), it follows that

$$\rho(\vec{\xi} + \vec{\eta}) \leqslant \rho(\vec{\xi}) + \rho(\vec{\eta}). \tag{2.3}$$

This inequality is called the triangle inequality. Thus the inequality (B) implies the triangle inequality for homogeneous functions of the first degree.

It is easy to see that, conversely, for homogeneous functions of degree one, the convexity of the function follows from the triangle inequality. Indeed, if $\rho(\vec{x}) = a$, $\rho(\vec{y}) = a$, then

$$\rho(\lambda\vec{x} + \mu\vec{y}) \leqslant \rho(\lambda\vec{x}) + \rho(\mu\vec{y}) = (\lambda + \mu)\, a = a,$$

which was the fact to be proved.

We shall call two norms $\rho_1(\vec{\xi})$ and $\rho_2(\vec{\xi})$ equivalent if for arbitrary $\vec{\xi}$ we have the inequality

$$m\rho_1(\vec{\xi}) \leqslant \rho_2(\vec{\xi}) \leqslant M\rho_1(\vec{\xi}),$$

where the constants $m$ and $M$ do not depend on $\vec{\xi}$.

On the Euclidean $n$-dimensional space all norms are equivalent. Indeed, on the surface $\rho_1(\vec{x}) = 1$ the function $\rho_2(\vec{x})$ assumes its maximum and minimum values, and the latter is positive by virtue of (C). If we denote these values by $M$ and $m$ respectively, we obtain:

$$\rho_2(\vec{\xi}) = \rho_2\left(\rho_1(\vec{\xi}) \frac{\vec{\xi}}{\rho_1(\vec{\xi})}\right) = \rho_1(\vec{\xi})\, \rho_2(\vec{x}),$$

from which the desired inequality immediately follows.

An arbitrary affine transformation of an $n$-dimensional space preserves the property of convexity for a surface. For this reason, if $\rho(y_1, \cdots, y_n)$ is a permissible norm on the space with coordinates $y_1, y_2, \cdots, y_n$, then

$$\rho\left(\sum a_{1j}x_j,\ \sum a_{2j}x_j,\ \ldots,\ \sum a_{nj}x_j\right)$$

will be again a permissible norm on the space of the $x_1, \cdots, x_n$ if the determinant of the transformation $|a_{ij}|$ is non-null. Indeed, if in the space $x_1, x_2, \cdots, x_n$, all the $y_1(x_1, \cdots, x_n), \cdots, y_n(x_1, \cdots, x_n)$ are zero, then all the coordinates $x_1,\cdots,x_n$ will be zero. Thus from $\rho=0$ will follow $x_1=x_2=\cdots=x_n=0$. The homogeneity of $\rho$ is also obvious.

The set of all functions $\phi$ integrable together with $|\phi|^p$ on a bounded domain we will denote by $L_p$. We introduce the notion of norm into the set of functions $L_p$. We set $\|\phi\|=[\int_\Omega|\phi|^p dv]^{1/p}$ and call $\|\phi\|$ the norm. The norm in a functional space serves as generalization of the geometrical concept of the length of a vector. Often, when it is necessary, we shall add as a subscript to the norm a special symbol indicating in which space this norm is defined. We shall write for example $\|\phi\|_{L_p}$.

For the norm, obviously, the following assertions are valid:

(a) $\|\phi+\psi\|\leqq\|\phi\|+\|\psi\|$ (Minkowski inequality),

(b) $\|a\phi\|=|a|\cdot\|\phi\|$, $a=$ const.,

(c) If $\|\phi\|=0$, then $\phi=0$ (with the exception of a set of measure zero).

In the following it will be necessary for us to use some of the notation of the theory of sets. We shall write $\phi\in L_p$ if the function $\phi$ is an element of $L_p$. A sequence $\phi_k$ of any sort of elements we shall denote by $\{\phi_k\}$. If the set $E_1$ is a part of $E_2$, we shall write $E_1\subset E_2$ or $E_1\subseteq E_2$, according to whether there are elements in $E_2$ not lying in $E_1$ or not. If the element $\phi$ does not lie in the set $E$, we shall write $\phi\bar{\in}E$.

We shall say that a sequence of functions $\{\phi_k\}$ converges strongly to the function $\phi_0$ in $L_p$ provided that $\|\phi_k-\phi_0\|\to 0$. We shall sometimes designate strong convergence by the notation

$$\varphi_k\Longrightarrow\varphi_0.$$

2. THE RIESZ-FISCHER THEOREM. Suppose that we have a sequence of functions $\{\phi_k\}$, $\phi_k\in L_p$, such that for an arbitrary $\epsilon>0$, $\|\phi_k-\phi_m\|<\epsilon$ for $k$, $m>N(\epsilon)$. Then there exists a function $\phi_0\in L_p$ such that $\phi_k\Rightarrow\phi_0$.

REMARK. This theorem asserts the completeness of the functional space $L_p$. We shall not give the proof of this theorem. It may be proved in the same way as it is proved for $p=2$.

3. CONTINUITY IN THE LARGE OF FUNCTIONS IN $L_p$. Let $\phi$ be given on the whole space, with $\phi\equiv 0$ outside of $\Omega$ and $\phi\in L_p$ on $\Omega$.

Let $\vec{P}(x_1,x_2,\cdots,x_n)$ be the coordinate vector in $n$-dimensional space, $|\vec{P}|$ its length.

DEFINITION. The function $\phi$ in $L_p$ on $\Omega$ is said to be continuous in the large in $L_p$ if for an arbitrary $\epsilon>0$ there exists $\delta(\epsilon)>0$ such that

$$\left[\int|\varphi(\vec{P}+\vec{Q})-\varphi(\vec{P})|^p dv\right]^{\frac{1}{p}}<\varepsilon, \tag{2.4}$$

if $|\vec{Q}|<\delta(\epsilon)$.

Theorem. *Every function $\phi \in L_p$ on a bounded domain $\Omega$ is continuous in the large on that domain.*

Proof. To simplify the argument, we extend the function $\phi$ outside of $\Omega$ by setting it equal to zero there. By virtue of the absolute continuity of Lebesgue integrals, for a given $\epsilon$ we can find $\delta_1(\epsilon)$ such that $\int_{\Omega'} |\phi|^p dv < \epsilon$ if $m\Omega' < \delta_1(\epsilon)$. By the definition of a summable function we can select a closed set $F_\delta$ such that $\phi$ is continuous on $F_\delta$ and $mF_\delta > m\Omega - (\delta_1/2)$, $F_\delta \subset \Omega$.

On the closed set $F_\delta$ the function $\phi$ is uniformly continuous by the theorem of Weierstrass, which means that if $\vec{P}$ and $\vec{P}+\vec{Q}$ both lie in $F_\delta$, then choosing $|\vec{Q}| < \delta_2(\epsilon)$ we have

$$|\varphi(\vec{P}+\vec{Q}) - \varphi(\vec{P})|^p < \frac{\varepsilon^p}{m\Omega}.$$

Let $F_\delta^{\vec{Q}}$ be the closed set obtained from $F_\delta$ by translating by $-\vec{Q}$.

We consider the sets $F_\delta$, $F_\delta^{\vec{Q}}$, and $\Omega$ as included in some sphere $K$ and form the sets

$$F_\delta^{(1)} = (K-\Omega) + F_\delta \quad \text{and} \quad F_\delta^{\vec{Q}(1)} = (K-\Omega) + F_\delta^{\vec{Q}}.$$

The measure of each of these is arbitrarily close to the measure of the sphere $K$. We have:

$$\left.\begin{aligned} m[(K-\Omega)+F_\delta] &= m[K-(\Omega-F_\delta)] > mK - \frac{\delta_1(\varepsilon)}{2}; \\ m[(K-\Omega)+F_\delta^{\vec{Q}}] &= m[K-(\Omega-F_\delta^{\vec{Q}})] > mK - \frac{\delta_1(\varepsilon)}{2}. \end{aligned}\right\} \tag{2.5}$$

We form the intersection $F_\delta^* = F_\delta^{(1)} \cdot F_\delta^{\vec{Q}(1)}$. Obviously,

$$mF_\delta^* > mF_\delta^{(1)} + mF_\delta^{\vec{Q}(1)} - mK > mK - \delta_1(\varepsilon).$$

Consequently,

$$m(K - F_\delta^*) < \delta_1(\varepsilon).$$

We have:

$$\left[\int_K |\varphi(\vec{P}+\vec{Q}) - \varphi(\vec{P})|^p dv\right]^{\frac{1}{p}} \leqslant \left[\int_{F_\delta^*} |\varphi(\vec{P}+\vec{Q}) - \varphi(\vec{P})|^p dv\right]^{\frac{1}{p}} +$$

$$+ \left[\int_{K-F_\delta^*} |\varphi(\vec{P}+\vec{Q})|^p dv\right]^{\frac{1}{p}} + \left[\int_{K-F_\delta^*} |\varphi(\vec{P})|^p dv\right]^{\frac{1}{p}}. \tag{2.6}$$

In the first of the integrals on the right hand side the integrated function is less than $\epsilon^p/m\Omega$, since in it both the points $\vec{P}$ and $\vec{P}+\vec{Q}$ lie in $F_\delta$. The second and third integrals are small by virtue of the absolute continuity of $\int_\Omega |\phi|^p dv$. From this our theorem follows.

4. COUNTABLE DENSE NETS.

THEOREM. *For every function $\phi \in L_p$ there exists a sequence $\{\phi_k\}$ of continuous functions with continuous derivatives of all orders converging strongly to $\phi$.*

PROOF. We shall assume that $\phi \equiv 0$ outside of the basic domain $\Omega$.

We consider the kernel:

$$\omega(\vec{Q}, h) = \begin{cases} e^{\frac{r^2}{r^2-h^2}} & \text{for} \quad r < h \\ 0 & \text{for} \quad r \geqslant h \end{cases} \quad (r = |\vec{P} - \vec{Q}|). \tag{2.7}$$

We note the obvious properties of the kernel:

(1) $\omega(\vec{Q},h)$ is continuous on the whole space together with all its derivatives.

(2) On the boundary of the sphere of radius $h$, the function $\omega(\vec{Q},h)$ and all its derivatives are equal to zero. (The second property follows from the fact that $e^{r^2/(r^2-h^2)}$ and any of its derivatives $\to 0$ as $r \to h$ with $r < h$).

In fact, by induction we may show that any derivative

$$\frac{\partial^\alpha e^{\frac{r^2}{r^2-h^2}}}{\partial x_1^{\alpha_1} \dots \partial x_n^{\alpha_n}}$$

has the form

$$\frac{P_{\alpha_1, \dots, \alpha_n}(x_1, \dots, x_n)}{(r^2 - h^2)^{2\alpha}} e^{\frac{r^2}{r^2-h^2}}, \tag{2.8}$$

where in the numerator there appears a polynomial in $(x_1, x_2, \cdots, x_n)$.

(3) We set $\kappa = \int_{r<1} e^{r^2/(r^2-1)} dv$; then $\int \omega(\vec{Q},h) dv = \kappa h^n$, where $n$ is the dimension of the space.

Thus

$$\frac{1}{\varkappa h^n} \int \omega(\vec{Q}, h)\, dv = 1.$$

We form the averaged function for $\phi$ on the sphere of radius $h$ with center at $\vec{P}$:

$$\varphi_h(\vec{P}) = \frac{1}{\varkappa h^n} \int \omega(\vec{P} - \vec{P}_1; h)\, \varphi(\vec{P}_1)\, dv_{\vec{P}_1}. \tag{2.9}$$

PROPERTIES OF THE AVERAGED FUNCTIONS.

(1) *The functions* $\phi_h(\vec{P})$ *have derivatives of arbitrary order.*

PROOF. Let $(x_1,\cdots,x_n)$ and $(y_1,\cdots,y_n)$ be the coordinates of the points $\vec{P}$ and $\vec{P}_1$. We consider the expression

$$H=\frac{\varphi_h(x_1+k, x_2, \ldots, x_n)-\varphi_h(x_1, x_2, \ldots, x_n)}{k}=$$

$$=\frac{1}{\varkappa h^n}\int\frac{\omega(x_1+k-y_1, x_2-y_2,\ldots,x_n-y_n)-\omega(x_1-y_1, x_2-y_2,\ldots,x_n-y_n)}{k}\times$$

$$\times\varphi(y_1,\ldots,y_n)\,dv_{\vec{P}_1}\to\frac{1}{\varkappa h^n}\int_\Omega\frac{\partial}{\partial x_1}\left[\omega(\vec{P}-\vec{P}_1;h)\right]\varphi(\vec{P}_1)\,dv_{\vec{P}_1}$$

$$\text{for } k\to 0. \qquad (2.10)$$

Taking the limit under the integral sign, since

$$\frac{\omega(x_1+k-y_1, x_2-y_2,\ldots,x_n-y_n)-\omega(x_1-y_1, x_2+y_2,\ldots,x_n-y_n)}{k}\to$$

$$\to\frac{\partial}{\partial x_1}\left[\omega(\vec{P}-\vec{P}_1;h)\right]$$

uniformly, as follows from the equality

$$\left|\frac{\omega(x+k)-\omega(x)}{k}-\omega'(x)\right|=\left|\frac{k}{2!}\omega''(x+\theta k)\right|<\eta$$

and thus

$$\left|H-\frac{1}{\varkappa h^n}\int\frac{\partial\omega}{\partial x_1}\varphi\,dv\right|\leqslant\frac{1}{\varkappa h^n}\left[\int\left|\frac{\omega(x_1+k)-\omega(x_1)}{k}-\right.\right.$$

$$\left.\left.-\frac{\partial\omega}{\partial x_1}\right|^{p'}dv\right]^{\frac{1}{p'}}\left[\int|\varphi|^p\,dv\right]^{\frac{1}{p}}\leqslant$$

$$\leqslant\frac{1}{\varkappa h^n}\left[\int\eta^{p'}\,dv\right]^{\frac{1}{p'}}\left[\int|\varphi|^p\,dv\right]^{\frac{1}{p}}\to 0$$

for $k\to 0$ (for then $\eta\to 0$).

In a completely analogous way, we may carry through the proof of the existence and continuity of an arbitrary derivative of arbitrary order.

(2) *The functions* $\phi_h$ *converge strongly to* $\phi$ *for* $h\to 0$, *i.e.*,

$$\int_\Omega|\varphi_h-\varphi|^p\,dv\to 0 \quad \textit{for} \quad h\to 0.$$

PROOF. Consider $\delta=\|\phi-\phi_h\|$.

$$\delta^p = \int \left| \varphi(\vec{P}) - \frac{1}{\varkappa h^n} \int \omega(\vec{P} - \vec{P}_1, h) \varphi(\vec{P}_1) \, dv_{\vec{P}_1} \right|^p dv \leqslant$$
$$\leqslant \int_{\Omega} \left\{ \frac{1}{\varkappa h^n} \int \omega(\vec{P} - \vec{P}_1, h) \left| \varphi(\vec{P}) - \varphi(\vec{P}_1) \right| dv_{\vec{P}_1} \right\}^p dv.$$

We set $\vec{P} - \vec{P}_1 = \vec{Q}$ and for the estimation of the interior integral we apply Hölder's inequality:

$$\delta^p \leqslant \int \left\{ \frac{1}{\varkappa h^n} \left[ \int_{|\vec{Q}| \leqslant h} |\omega(\vec{Q}, h)|^{p'} dv_{\vec{Q}} \right]^{\frac{1}{p'}} \times \right.$$
$$\left. \times \left[ \int_{|\vec{Q}| \leqslant h} |\varphi(\vec{P} + \vec{Q}) - \varphi(\vec{P})|^p dv_{\vec{Q}} \right]^{\frac{1}{p}} \right\}^p dv; \quad \frac{1}{p} + \frac{1}{p'} = 1.$$

From the boundedness of $\omega$ it follows that:

$$\left| \int_{|\vec{Q}| \leqslant h} |\omega(\vec{Q}, h)|^{p'} dv_{\vec{Q}} \right| = Ch^n,$$

so that

$$\delta^p \leqslant C_2 (h^n)^{-p\left(1 - \frac{1}{p'}\right)} \int_{|\vec{Q}| \leqslant h} \int |\varphi(\vec{P} + \vec{Q}) - \varphi(\vec{P})|^p \, dv_{\vec{Q}} \, dv =$$
$$= C_2 h^{-n} \int_{|\vec{Q}| \leqslant h} \int |\varphi(\vec{P} + \vec{Q}) - \varphi(\vec{P})|^p \, dv \, dv_{\vec{Q}};$$

in consequence of the continuity of the function $\phi$ in the large it follows that for $h$ sufficiently small $\int |\phi(\vec{P} + \vec{Q}) - \phi(\vec{P})|^p dv < \eta$, and then

$$\delta^p \leqslant C_2 h^{-n} \int_{|\vec{Q}| \leqslant h} \eta \, dv_{\vec{Q}} \leqslant C_3 \eta \to 0. \tag{2.11}$$

The averaged functions form, obviously, a sequence of functions which are continuous together with their derivatives of arbitrary order and which converge strongly to the function $\phi \in L_p$, and our theorem is proved.

THEOREM. *In the space $L_p$ of functions $\phi$ given on a bounded domain $\Omega$ there exists a countable everywhere dense set of functions $\{\phi_k\}$ $(k = 1, 2, \cdots)$, $\phi_k \in L_p$, i.e., a countable set such that for an arbitrary element $\psi \in L_p$ and for an arbitrary $\epsilon > 0$, one can find an element $\phi_s$ for which $\|\psi - \phi_s\| < \epsilon$.*

For the proof it suffices to show the existence of such $\phi_s$ for all continuous functions. By the theorem of Weierstrass an arbitrary continuous function $f$ may be approximated arbitrarily closely by a polynomial $P$. An arbitrary polynomial $P$ may be approximated by polynomials $P_r$ with rational coefficients. The polynomials $P_r$ form a countable set.

Let $\psi \in L_p$. We construct $\psi_h$, an averaged function for $\psi$ such that

$$\|\psi-\psi_h\|<\frac{\varepsilon}{3}.$$

Further, for $\psi_h$ by the Weierstrass theorem we may find a polynomial $P(\vec{Q})$ such that

$$\|\psi_h-P\|<\frac{\varepsilon}{3}.$$

Finally, there exists a polynomial $P_r$ with rational coefficients such that

$$\|P-P_r\|<\frac{\varepsilon}{3}.$$

From these three inequalities, it follows that

$$\|\psi-P_r\|<\varepsilon,$$

and since the functions $P_r$ form a countable set, the theorem is proved.

REMARK. The property of a space having a countable everywhere dense set (expressed as "countable everywhere dense net") is called separability. Thus, the space $L_p$ is separable.

§3. **Linear functionals on** $L_p$.

1. DEFINITIONS. BOUNDEDNESS OF LINEAR FUNCTIONALS. If for each $\phi \in L_p$, there is given a corresponding number $l\phi$, then we say that $l\phi$ is a functional of the function $\phi$. Functionals may be added, subtracted, and multiplied by constants. We shall set

$$(al)\varphi=a(l\varphi),$$
$$(l_1+l_2)\varphi=l_1\varphi+l_2\varphi.$$

A functional $l\phi$ having the properties:

(1) Distributivity:

$$l(af_1+bf_2)=alf_1+blf_2, \tag{3.1}$$

where $a$ and $b$ are arbitrary constants;

(2) Continuity:

$$\text{if}\quad \varphi_k \Longrightarrow \varphi, \text{ then } l\varphi_k\to l\varphi, \tag{3.2}$$

is called a linear functional on $L_p$.

The functional $l\phi$ is called bounded if there exists a constant $M>0$ such that for any $\phi$

$$|l\varphi|\leqslant M\|\varphi\|, \tag{3.3}$$

where $\|\phi\| = [\int |\phi|^p dv]^{1/p}$ is the norm of the function $\phi$ in $L_p$.

THEOREM. *Every linear functional on $L_p$ is bounded.*

We prove this theorem by reductio ad absurdum. If the assertion were false, there would exist a sequence $\{\phi_k\} \subset L_p$ such that $l\phi_k/\|\phi_k\| \to \infty$, i.e., $l\phi_k/\|\phi_k\| > k$ ($k$ an arbitrary integer).

We consider $\{\psi_k\} = \{\phi_k/(\sqrt{k}\|\phi_k\|)\}$. Obviously,

$$\psi_k \in L_p \text{ and } \|\psi_k\| = (1/\sqrt{k}) \to 0;$$

as a result $\psi_k \Rightarrow 0$, and by the continuity of the functional $l\psi_k \to 0$ also. On the other hand, $l\psi_k > \sqrt{k}$, and as a consequence $l\psi_k \to \infty$. We have arrived at a contradiction. The theorem is proved.

Of all the numbers $M$ which satisfy (3.3) we can find a least, called the norm of the functional $l\phi$ and denoted by $\|l\|$. It is worth remarking that the norm of the functional satisfies the conditions:

$$\|al\| = |a| \|l\|$$

and

$$\|l_1 + l_2\| \leqslant \|l_1\| + \|l_2\| \qquad \text{(triangle inequality)}.$$

These properties follow immediately from the definition of the sum of functionals and of the functional $(al)$.

Our problem appears as that of establishing the general form of linear functionals on $L_p$. As a preliminary, we establish two auxiliary inequalities.

2. CLARKSON'S INEQUALITIES.

LEMMA 1.

$$\Phi(x) = (1+x)^\lambda + (1-x)^\lambda - 2^\lambda \leqslant 0, \tag{3.4}$$

*if* $\lambda \geqq 1$ *and* $0 < x < 1$.

PROOF. $\Phi'(x) = \lambda(1+x)^{\lambda-1} - \lambda(1-x)^{\lambda-1} \geqq 0$, and $\Phi(1) = 0$, i.e., $\Phi(x)$ is a nondecreasing function on $[0,1]$ which is equal to zero at the right hand end of the interval and consequently, $\Phi(x) \leqq 0$ for $x < 1$ (q.e.d.).

LEMMA 2.

$$\left(\frac{1+x}{2}\right)^p + \left(\frac{1-x}{2}\right)^p \leqslant \frac{1}{2}(1+x^p)$$
$$\textit{for } p \geqslant 2, \quad 0 < x < 1 \tag{3.5}$$

PROOF. We consider

$$F(x)=\left(\frac{1+x}{2}\right)^p+\left(\frac{1-x}{2}\right)^p-\frac{1}{2}(1+x^p)$$

and

$$\Phi(x)=\frac{2^p}{x^p}F(x)=\left(\frac{1}{x}+1\right)^p+\left(\frac{1}{x}-1\right)^p-2^{p-1}\left(\frac{1}{x^p}+1\right).$$

We have

$$\Phi(1)=0;$$

$$\Phi'(x)=-\frac{p}{x^2}\left[\left(\frac{1}{x}+1\right)^{p-1}+\left(\frac{1}{x}-1\right)^{p-1}\right]+\frac{p2^{p-1}}{x^{p+1}}=$$

$$=-\frac{p}{x^{p+1}}[(1+x)^{p-1}+(1-x)^{p-1}-2^{p-1}].$$

By virtue of Lemma 1, $\Phi'(x)\geqq 0$, and as a result, $\Phi(x)\leqq 0$, which means that $F(x)\leqq 0$, as was to be proved.

CLARKSON'S FIRST INEQUALITY. We consider $|(\phi+\psi)/2|^p+|(\phi-\psi)/2|^p$. Suppose that at the given point, for example, $|\psi|\leqq|\phi|$; we put $|\psi/\phi|=x$. Then by Lemma 2, we obtain

$$\left|\frac{\varphi+\psi}{2}\right|^p+\left|\frac{\varphi-\psi}{2}\right|^p=|\varphi|^p\left\{\left(\frac{1+x}{2}\right)^p+\left(\frac{1-x}{2}\right)^p\right\}\leqslant$$

$$\leqslant\frac{|\varphi|^p}{2}(1+x^p)=\frac{1}{2}(|\varphi|^p+|\psi|^p). \qquad (3.6)$$

Integrating, we obtain Clarkson's first inequality:

$$\int\left|\frac{\varphi+\psi}{2}\right|^p dv+\int\left|\frac{\varphi-\psi}{2}\right|^p dv\leqslant\frac{1}{2}\int|\varphi|^p dv+$$

$$+\frac{1}{2}\int|\psi|^p dv,\quad p\geqslant 2. \qquad (3.7)$$

LEMMA 3.

$$\omega(p)=[z^p+(1-z)^p]^{\frac{1}{p-1}}, \qquad (3.8)$$

*is an increasing function of* $p$ *for* $p>1$, $\frac{1}{2}<z<1$.

PROOF. We consider the auxiliary function

$$\lambda(p)=\log[z^p+(1-z)^p]. \text{ then } \log\omega(p)=\frac{\lambda(p)}{p-1}.$$

We have

$$\lambda'(p)=\frac{z^p\log z+(1-z)^p\log(1-z)}{z^p+(1-z)^p}.$$

We shall show that $\lambda'(p)$ is an increasing function, i.e., that $\lambda(p)$ is a convex function.

In fact, $z/(1-z)>1$, and consequently,

$$\log\frac{z}{1-z}>0.$$

In addition,

$$\mu(p)=\frac{z^p}{z^p+(1-z)^p}=\frac{1}{1+\left(\frac{1-z}{z}\right)^p}$$

is an increasing function of $p$, since $(1-z)/z<1$. For this reason,

$$\lambda'(p)=\mu(p)\log z+[1-\mu(p)]\log(1-z)=\mu(p)\log\frac{z}{1-z}+\log(1-z)$$

is also an increasing function of $p$, as was to be proved. As a result, the curve of $\lambda(p)$ $(p>1)$ is convex. From the monotonicity of $\lambda'(p)$ follows obviously that $\lambda''(p)>0$.

We consider the derivative of $\lambda(p)/(p-1)=y(p)=\log\omega(p)$. We obtain

$$y'(p)=\frac{1}{(p-1)^2}[\lambda'(p)(p-1)-\lambda(p)]. \tag{3.9}$$

The expression in square brackets is always positive. Indeed,

$$[\lambda'(p)(p-1)-\lambda(p)]'=(p-1)\lambda''(p)$$

and as we have observed, is always positive. It follows that the bracket $[\lambda'(p)(p-1)-\lambda(p)]$ is always increasing. For $p=1$, it is zero since $\lambda(1)=0$. Thus, it is always positive and so is the function $y'(p)$. This means that $y(p)$ is an increasing function, as was to be proved.

REMARK. We note that the function $\omega(p,\alpha)=[z^p+(1-z)^p]^{1/p-\alpha}$ will be increasing for $\alpha>1$ (and in particular for $p=\alpha$) and decreasing for $\alpha<0$. For $0<\alpha<1$, it will have a single minimum. It is easy to prove this assertion geometrically.

We have

$$\log\omega(p,\alpha)=\frac{\lambda(p)}{p-\alpha}.$$

The curve $\lambda(p)$; $(p>1)$ has the form pictured in Figure 4. For $p>1$ it is convex since $\lambda''(p)>0$. It intersects the $p$-axis at $p=1$. The curve has the asymptote $\lambda=p\log z$ since

$$\lambda(p)=p\log z+\log\left[1+\left(\frac{1-z}{z}\right)^p\right],$$

and consequently the difference $\lambda(p)-p\log z$ tends to zero as $p\to\infty$. An arbitrary line can meet this curve in at most two points. It is obvious that $\lambda(p)/(p-\alpha)$ represents geometrically the tangent of the angle $\phi$ made with the $p$-axis by the line running through the given point on the curve and the point $p=\alpha$, $\lambda=0$ on the $p$-axis. If $\alpha>1$, this angle grows from $-\pi$ to arc tg log $z$. Analogously if $\alpha\leqq 0$, this angle $\phi$ decreases from 0 to arc tg log $z$.

Finally, if $0<\alpha<1$, then from the point $p=\alpha$, $\lambda=0$, one may draw one tangent to our curve. As a result, the angle $\phi$ to begin with decreases to our minimum and then increases to the value arc tg log $z$. From this, since $\omega(p,\alpha)=e^{\mathrm{tg}\,\phi}$ there follows the correctness of our assertion about $\omega(p,\alpha)$.

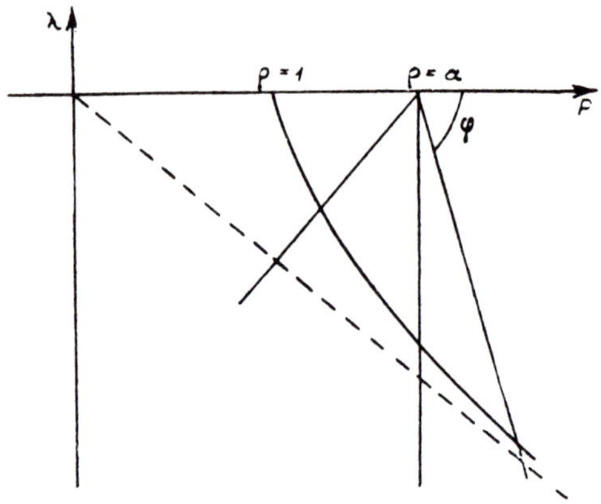


Figure 4

Clarkson's second inequality. We assume for the sake of definiteness that $\phi>\psi>0$ and consider the function

$$F(\varphi)=\left[\left(\frac{\varphi+\psi}{2}\right)^{\frac{p}{p-1}}+\left(\frac{\varphi-\psi}{2}\right)^{\frac{p}{p-1}}\right]^{p-1}-\frac{1}{2}\left[\varphi^p+\psi^p\right];\quad 1<p\leqslant 2. \tag{3.10}$$

Obviously if $\phi=\psi$, then $F(\phi)=0$. Furthermore, we have

$$\frac{\partial F}{\partial\varphi}=\left[\left(\frac{\varphi+\psi}{2}\right)^{\frac{p}{p-1}}+\left(\frac{\varphi-\psi}{2}\right)^{\frac{p}{p-1}}\right]^{p-2}\cdot\frac{1}{2}p\left[\left(\frac{\varphi+\psi}{2}\right)^{\frac{p}{p-1}-1}+\left(\frac{\varphi-\psi}{2}\right)^{\frac{p}{p-1}-1}\right]-\frac{p}{2}\varphi^{p-1}=\frac{p}{2}\varphi^{p-1}\left\{\left[\left(\frac{\varphi+\psi}{2\varphi}\right)^{\frac{p}{p-1}}+\left(\frac{\varphi-\psi}{2\varphi}\right)^{\frac{p}{p-1}}\right]^{p-2}\left[\left(\frac{\varphi+\psi}{2\varphi}\right)^{\frac{1}{p-1}}+\left(\frac{\varphi-\psi}{2\varphi}\right)^{\frac{1}{p-1}}\right]-1\right\}. \tag{3.11}$$

We put

$$\frac{\varphi+\psi}{2\varphi}=z>\frac{1}{2},\text{ and }\frac{p}{p-1}=p'\geqslant 2,$$

Then

$$\frac{\varphi-\psi}{2\varphi}=1-z, \quad \frac{1}{p-1}=p'-1, \quad p-2=-\frac{p'-2}{p'-1}.$$

Therefore

$$\frac{\partial F}{\partial \varphi}=\frac{p}{2}\varphi^{p-1}\left\{\left(\frac{[z^{p'-1}+(1-z)^{p'-1}]^{\frac{1}{p'-2}}}{[z^{p'}+(1-z)^{p'}]^{\frac{1}{p'-1}}}\right)^{p'-2}-1\right\}=$$

$$=\frac{p}{2}\varphi^{p-1}\left\{\left[\frac{\omega(p'-1)}{\omega(p')}\right]^{p'-2}-1\right\}\leqslant 0, \tag{3.12}$$

as in Lemma 3:

$$\frac{\omega(p'-1)}{\omega(p')}<1, \qquad p'-2\geqslant 0.$$

As a consequence, $F(\phi)$ is a decreasing function, and since for $\phi=\psi$ $F(\phi)=0$, it follows that $F(\phi)\leqq 0$ for $\phi>\psi$.

Obviously, we have the inequality

$$\left[\left|\frac{\varphi+\psi}{2}\right|^{\frac{p}{p-1}}+\left|\frac{\varphi-\psi}{2}\right|^{\frac{p}{p-1}}\right]^{p-1}-\frac{1}{2}[|\varphi|^p+|\psi|^p]\leqslant 0 \tag{3.13}$$

for $1<p\leqq 2$ and arbitrary $\phi$ and $\psi$.

For the proof of the second inequality of Clarkson, we apply the reverse Minkowski inequality

$$\left[\int(|x|+|y|)^q\,dv\right]^{\frac{1}{q}}\geqslant\left[\int|x|^q\,dv\right]^{\frac{1}{q}}+\left[\int|y|^q\,dv\right]^{\frac{1}{q}};$$
$$0<q<1. \tag{3.14}$$

We set in (3.14) $x=\left|\frac{\varphi+\psi}{2}\right|^{\frac{p}{p-1}}$, $y=\left|\frac{\varphi-\psi}{2}\right|^{\frac{p}{p-1}}$, $q=p-1$. We obtain

$$\left[\int_\Omega\left(\left|\frac{\varphi+\psi}{2}\right|^{\frac{p}{p-1}}+\left|\frac{\varphi-\psi}{2}\right|^{\frac{p}{p-1}}\right)^{p-1}dv\right]^{\frac{1}{p-1}}\geqslant\left[\int_\Omega\left|\frac{\varphi+\psi}{2}\right|^p dv\right]^{\frac{1}{p-1}}+$$

$$+\left[\int_\Omega\left|\frac{\varphi-\psi}{2}\right|^p dv\right]^{\frac{1}{p-1}}$$

and using (3.13), we obtain the second Clarkson inequality:

$$\left[\int_\Omega \left|\frac{\varphi+\psi}{2}\right|^p dv\right]^{\frac{1}{p-1}} + \left[\int_\Omega \left|\frac{\varphi-\psi}{2}\right|^p dv\right]^{\frac{1}{p-1}} \leqslant$$

$$\leqslant \left[\frac{1}{2}\int_\Omega |\varphi|^p dv + \frac{1}{2}\int_\Omega |\psi|^p dv\right]^{\frac{1}{p-1}} \quad 1 < p \leqslant 2. \tag{3.15}$$

Let $\phi$ and $\psi$ be two unit vectors. The sum $(\phi+\psi)/2$ represents the midpoint of their chord, the length of the chord being equal to $\|\phi-\psi\|$.

Clarkson's inequality enables one to assert that the midpoint of each chord having a given definite length $\delta$ will have its norm strictly less than some number $\eta(\delta)<1$, i.e., will actually lie in the interior of the unit sphere. This property may be called the uniform convexity of the unit sphere.

3. THEOREM ON THE GENERAL FORM OF LINEAR FUNCTIONALS.

THEOREM. *Every linear functional on $L_p$ may be represented in the form*

$$l\varphi = \int_\Omega \varphi\psi_0\, dv, \tag{3.16}$$

*where*

$$\psi_0 \in L_{p'} \qquad \left(\frac{1}{p}+\frac{1}{p'}=1\right).$$

PROOF. Set $\sup\limits_{\|\phi\|=1} |l\phi| = g$. This means that there exists a sequence $\{\phi_k\}$, $\|\phi_k\|=1$, such that $\lim\limits_{k\to\infty} l\phi_k = g$. We shall show that $\{\phi_k\}$ converges strongly.

Assume the contrary. Then there exists an $\epsilon_0>0$ such that we may find pairs of numbers $n_k$ and $m_k$ ($n_k\to\infty$, $m_k\to\infty$, as $k\to\infty$) such that

$$\|\phi_{m_k}-\phi_{n_k}\| > \epsilon_0.$$

Applying Clarkson's inequality (3.7) to the functions $\phi_{m_k}$ and $\phi_{n_k}$ if $p\geqq 2$, and (3.15) if $1<p\leqq 2$, we obtain

$$\left\|\frac{\varphi_{m_k}+\varphi_{n_k}}{2}\right\|^p + \left\|\frac{\varphi_{m_k}-\varphi_{n_k}}{2}\right\|^p \leqslant 1; \qquad (p \geqslant 2)$$

$$\left\|\frac{\varphi_{m_k}+\varphi_{n_k}}{2}\right\|^{\frac{p}{p-1}} + \left\|\frac{\varphi_{m_k}-\varphi_{n_k}}{2}\right\|^{\frac{p}{p-1}} \leqslant 1. \qquad (1<p\leqslant 2)$$

It follows from this that

$$\left\|\frac{\varphi_{m_k}+\varphi_{n_k}}{2}\right\| < 1-\eta, \tag{3.17}$$

where $\eta>0$ and does not depend on $k$.

We consider $\chi_k=(\phi_{m_k}+\phi_{n_k})/\|\phi_{m_k}+\phi_{n_k}\|$. We have $\|\chi_k\|=1$. From the distributivity of linear functionals, we have:

$$l\chi_k=\frac{1}{2}\,[l\varphi_{m_k}+l\varphi_{n_k}]\frac{1}{\left\|\dfrac{\varphi_{m_k}+\varphi_{n_k}}{2}\right\|}>\frac{\frac{1}{2}(l\varphi_{m_k}+l\varphi_{n_k})}{1-\eta}.$$

However, $l\phi_{m_k}\to g$, $l\phi_{n_k}\to g$, from which it follows that for sufficiently large $k$, we will have:

$$l\chi_k>\frac{g-\varepsilon_k}{1-\eta}>g,$$

which contradicts the fact that $\sup\limits_{\|\phi\|=1} l\phi=g$.

The sequence $\{\phi_m\}$ converges strongly and by virtue of the completeness of $L_p$ it has a limit element $\phi_0\in L_p$.

Obviously $\|\phi_0\|=1$.

REMARK. From this argument follows the uniqueness of the function $\phi_0\in L_p$ such that $\|\phi_0\|=1$ and $l\phi_0=g$, since otherwise we could construct a divergent sequence for which $\lim l\phi_k=g$, which is impossible.

We shall show that

$$l\varphi=g\int\limits_{\Omega}[|\varphi_0|^{p-1}\operatorname{sign}\varphi_0]\,\varphi\,dv \tag{3.18}$$

or, putting $g|\phi_0|^{p-1}\operatorname{sign}\phi_0=\psi_0$:

$$l\varphi=\int\limits_{\Omega}\psi_0\varphi\,dv.$$

LEMMA 4. *If for an arbitrary function $\psi\in L_p$*

$$\int\limits_{\Omega}[|\varphi_0|^{p-1}\operatorname{sign}\varphi_0]\,\psi\,dv=0, \tag{3.19}$$

*then $l\psi=0$.*

PROOF. We consider $y(\lambda)=l((\phi_0+\lambda\psi)/\|\phi_0+\lambda\psi\|)$, where $\psi\neq c\phi_0$.

Since $\|(\phi_0+\lambda\psi)/\|\phi_0+\lambda\psi\|\|=1$ and since for $\lambda\neq 0$, $(\phi_0+\lambda\psi)/\|\phi_0+\lambda\psi\|\neq\phi_0$, therefore $y(\lambda)\leq g$, while the equality holds only for $\lambda=0$. As a result $y(\lambda)$ has a maximum for $\lambda=0$. If $y'(0)$ exists, then $y'(0)=0$. It is not hard to convince oneself of the existence of this derivative.

Differentiating formally, we have

$$\frac{d}{d\lambda}\|\varphi_0+\lambda\psi\|=\frac{d}{d\lambda}\Big[\int_\Omega|\varphi_0+\lambda\psi|^p dv\Big]^{\frac{1}{p}}=\frac{1}{p}\Big[\int_\Omega|\varphi_0+\lambda\psi|^p dv\Big]^{\frac{1}{p}-1}\times$$

$$\times p\int_\Omega|\varphi_0+\lambda\psi|^{p-1}\operatorname{sign}(\varphi_0+\lambda\psi)\,\psi\,dv,$$

and since the integral $\int_\Omega|\phi_0+\lambda\psi|^{p-1}\operatorname{sign}(\phi_0+\lambda\psi)\psi dv$ converges uniformly, the formal differentiation with respect to $\lambda$ is valid. Further,

$$\Big(\frac{d}{d\lambda}\|\varphi_0+\lambda\psi\|\Big)_{\lambda=0}=\Big[\int_\Omega|\varphi_0|^p dv\Big]^{\frac{1}{p}-1}\int_\Omega[|\varphi_0|^{p-1}\operatorname{sign}\varphi_0]\,\psi\,dv,$$

and therefore,

$$y'(0)=\frac{l\psi}{\|\varphi_0\|}-\frac{l\varphi_0}{\|\varphi_0\|^2}\|\varphi_0\|^{1-p}\int_\Omega[|\varphi_0|^{p-1}\operatorname{sign}\varphi_0]\,\psi\,dv,$$

from which by virtue of (3.19) it follows that $l\psi=0$, and the lemma is proved.

We now establish (3.18). Let $\phi\in L_p$ be an arbitrary function. We put

$$\alpha=\int_\Omega[|\varphi_0|^{p-1}\operatorname{sign}\varphi_0]\,\varphi\,dv.$$

Then $\psi=\phi-\alpha\phi_0$ satisfies (3.19), since

$$\int_\Omega[|\varphi_0|^{p-1}\operatorname{sign}\varphi_0]\,(\varphi-\alpha\varphi_0)\,dv=\alpha-\alpha\int_\Omega|\varphi_0|^{p-1}|\varphi_0|\,dv=\alpha-\alpha=0.$$

Consequently, $l\psi=l\phi-\alpha l\phi_0=0$, i.e.,

$$l\varphi=\alpha l\varphi_0=g\int_\Omega[|\varphi_0|^{p-1}\operatorname{sign}\varphi_0]\,\varphi\,dv,$$

and (3.18) is established.

Since $\phi_0\in L_p$, it follows that

$$||\varphi_0|^{p-1}\operatorname{sign}\varphi_0|^{p'}=[|\varphi_0|^{p-1}]^{\frac{p}{p-1}}=|\varphi_0|^p,$$

and as a result,

$$\psi=g|\varphi_0|^{p-1}\operatorname{sign}\varphi_0\in L_{p'},\quad\Big(\frac{1}{p}+\frac{1}{p'}=1\Big),$$

from which $\|\psi\|_{L_{p'}}=g$. In this way, we obtain from (3.18)

$$l\varphi=\int_\Omega\psi\varphi\,dv,\quad\psi\in L_{p'},$$

and the theorem is proved. In addition, obviously $\|l\|=g$.

REMARK. It follows from (3.16) that every linear functional on $L_p$ corresponds to a function $\psi \in L_{p'}$, and conversely, every $\psi \in L_{p'}$ generates a linear functional on $L_p$. The spaces $L_p$ and $L_{p'}$ are said to be mutually conjugate functional spaces.

4. CONVERGENCE OF FUNCTIONALS. The concept of convergence is easily extended to functional spaces.

We shall say that a sequence of functionals $\{l_k\}$ is weakly convergent if for each element $\phi$ from $L_p$ there exists the limit

$$\lim_{k \to \infty} l_k \varphi = l_0 \varphi.$$

The limit is obviously an additive functional, since

$$l_0 (a_1 \varphi_1 + a_2 \varphi_2) = \lim_{k \to \infty} l_k (a_1 \varphi_1 + a_2 \varphi_2) = \lim_{k \to \infty} (a_1 l_k \varphi_1 + a_2 l_k \varphi_2) =$$
$$= a_1 \lim_{k \to \infty} l_k \varphi_1 + a_2 \lim_{k \to \infty} l_k \varphi_2 = a_1 l_0 \varphi_1 + a_2 l_0 \varphi_2.$$

If the limit function $l_0$ is bounded, then it will also be continuous, i.e., linear.

We establish the theorem.

THEOREM 1. *The space $L_p^*$, i.e., the space of functionals for $L_p$, is complete in the sense of weak convergence. In other words, every sequence which is weakly convergent has its limit a linear functional.*

Obviously it suffices to show the boundedness of every weakly convergent sequence of functionals. There will follow from this the boundedness of the limit functional.

We show the correctness of a proposition from which this will follow.

THEOREM 2. *If a sequence of linear functionals*

$$l_1, l_2, \ldots, l_k, \ldots \tag{3.20}$$

*is unbounded, i.e., if it takes on the unit sphere in $L_p$ arbitrarily large values, then one can find an element $\omega_0$ of $L_p$ on which this sequence diverges.*

The idea of the proof consists in choosing from the sequence of functionals (3.20) a subsequence

$$m_1, m_2, \ldots, m_s, \ldots, \tag{3.21}$$

where

$$m_s = l_{k_s} \quad (s = 1, 2, \ldots), \ k_s \to \infty \qquad \text{for } s \to \infty$$

and

$$m_s\varphi = \int_\Omega \psi_s\varphi\, dv, \quad \psi_s \in L_{p'}, \tag{3.22}$$

Corresponding to the $m_s$, we introduce the system of $\omega_s \in L_p$:

$$\omega_s = \frac{|\psi_s|^{p'-1}\operatorname{sign}\psi_s}{\|\psi_s\|^{p'-1}}.$$

Obviously,

$$\omega_s \in L_p, \quad \|\omega_s\| = 1, \quad (m_s\omega_s) = \|\psi_s\|_{L_{p'}}.$$

We form a series with the functions $\omega_s$:

$$\omega_0 = \sum_{s=1}^{\infty} \alpha_s\omega_s, \tag{3.23}$$

where

$$\alpha_s = \|\psi_s\|^{-\frac{1}{2}}.$$

The series will converge in $L_p$ if we have convergence for the series $\sum\|\psi_s\|^{-1/2}$. As we show for a suitable choice of the $\{m_s\}$, the series (3.23) turns out to be strongly convergent to the element $\omega_0$, and the sequence $l_k\omega_0$ will be convergent.

The sequence $m_s$ will be constructed inductively. Suppose $m_s$ has been chosen: we shall show how $m_{s+1}$ must be chosen.

We consider the sequence $(l_1\omega_s), (l_2\omega_s), \cdots, (l_j\omega_s), \cdots$. If it is unbounded, then the theorem has been proved. If the sequence is bounded, we put

$$A_s = \sup_{\substack{k=1,2,\ldots,s \\ j=1,2,\ldots}} |l_j\omega_k|.$$

We have

$$A_1 \leqslant A_2 \leqslant \ldots \leqslant A_s < \infty.$$

Each functional $l_k$ of the sequence (3.20) corresponds to $\bar{\psi}_k \in L_{p'}$ such that the norm of $l_k$ is equal to $\|\bar{\psi}_k\|_{L_{p'}}$. Since the sequence (3.20) is unbounded and a fortiori these $\{\|\psi_k\|\}$ will be unbounded, therefore as $m_{s+1}$ we can always choose an $l_k$ which satisfies the inequalities.

$$\|\psi_s\|_{L_{p'}} > 1 \quad (s = 1, 2, \ldots), \tag{3.24}$$

$$\|\psi_{s+1}\|_{L_{p'}}^{\frac{1}{2}} > 4(A_1 + A_2 + \ldots + A_s), \tag{3.25}$$

$$\|\psi_{s+1}\|_{L_{p'}} > 3^{2(s+1-j)}\|\psi_j\|_{L_{p'}}; \quad (j \leqslant s). \tag{3.26}$$

By virtue of (3.24) and (3.26), we have $\|\psi_s\|_{L_{p'}} > 3^{2(s-1)}$, from which it follows that the series $\sum\alpha_s = \sum\|\psi_s\|^{-1/2}$ converges and thereby the series (3.23) converges on $L_p$. Therefore

$$|m_s\omega_0| = \left|\sum_{j=1}^{s-1} \alpha_j m_s\omega_j + \alpha_s m_s\omega_s + \sum_{j=s+1}^{\infty} \alpha_j m_s\omega_j\right| \geqslant \alpha_s m_s\omega_s -$$

$$- \left|\sum_{j=1}^{s-1} \alpha_j m_s\omega_j\right| - \left|\sum_{j=s+1}^{\infty} \alpha_j m_s\omega_j\right|. \tag{3.27}$$

We have

$$\alpha_s m_s\omega_s = \|\psi_s\|^{-\frac{1}{2}} \|\psi_s\| = \|\psi_s\|^{\frac{1}{2}}. \tag{3.28}$$

By virtue of (3.24) and (3.25), we have:

$$\left|\sum_{j=1}^{s-1} \alpha_j m_s\omega_j\right| < \sum_{j=1}^{s-1} |m_s\omega_j| \leqslant \sum_{j=1}^{s-1} A_j \leqslant \frac{1}{4}\|\psi_s\|^{\frac{1}{2}}. \tag{3.29}$$

From (3.26), changing notation, we easily find:

$$\alpha_j = \|\psi_j\|^{-\frac{1}{2}} \leqslant \frac{1}{3^{j-s}}\|\psi_s\|^{-\frac{1}{2}}; \quad j > s,$$

from which

$$\left|\sum_{j=s+1}^{\infty} \alpha_j m_s\omega_j\right| \leqslant \sum_{j=s+1}^{\infty} \frac{1}{3^{j-s}}\|\psi_s\|^{-\frac{1}{2}} \|\psi_s\| =$$

$$= \|\psi_s\|^{\frac{1}{2}} \sum_{j=1}^{\infty} \frac{1}{3^j} = \frac{\|\psi_s\|^{\frac{1}{2}}}{2}. \tag{3.30}$$

Then from (3.27), (3.28), (3.29), (3.30), we obtain

$$|m_s\omega_0| > \frac{1}{4}\|\psi_s\|^{\frac{1}{2}} > \frac{3^{s-1}}{4},$$

whence clearly it follows that $l_k\omega_0$ cannot converge to any limit with increasing $k$. The theorem is proved. Hence, if a sequence $\{l_k\}$ converges weakly, it cannot be unbounded and therefore for all $\phi \in L_p$ we have

$$|l_0\varphi| = |\lim_{k\to\infty} l_k\varphi| = \lim_{k\to\infty} |l_k\varphi| \leqslant A\|\psi\|, \tag{3.31}$$

where $A$ is the common bound on the norm of all the $l_k$. Thus the functional $l_0$ is bounded and therefore continuous. The theorem on the completeness of $L_p^*$ is thereby proved.

WEAK CONVERGENCE IN $L_p$. A sequence of functions $\{\phi_k\}$ is said to be weakly convergent to the function $\phi_0$ if for an arbitrary function $l \in L_p^*$ we have

$$\lim_{k\to\infty} l\varphi_k = l\varphi_0. \tag{3.32}$$

In the space $L_p$ the set of functionals coincides with the space $L_{p'}$. Therefore the formula (3.32) will hold whenever the functionals $l_k \in L^*_{p'}$, corresponding to the $\phi_k$ converge to the functional $l_0 \in L^*_{p'}$, corresponding to $\phi_0$. We have finally on the space $L_p$ two forms of convergence: strong and weak, the latter being written

$$\varphi_k \to \varphi_0.$$

Obviously from the strong convergence $\phi_k \Rightarrow \phi_0$ there follows the weak convergence $\phi_k \to \phi_0$. The converse is not always true. We give an example of a weakly convergent sequence which does not converge strongly.

EXAMPLE. Let $\Omega = [0,2\pi]$, $p=2$, $\phi_k(x) = \sin kx$. Then $\sin kx \to 0$ since for an arbitrary function $\psi \in L_2$,

$$\int_0^{2\pi} \psi(x) \sin kx \, dx = \pi b_k \to 0,$$

in view of the fact that

$$\pi \sum_{k=1}^{\infty} b_k^2 = \int_0^{2\pi} \psi^2 \, dx,$$

i.e., the series $\sum_{k=1}^{\infty} b_k^2$ is convergent. But $\sin kx \not\Rightarrow 0$, since

$$\int_0^{2\pi} \sin^2 kx \, dx = \pi \not\to 0.$$

REMARK. Theorem 2 may be formulated in terms of the weak convergence of functions as: an unbounded sequence of functions cannot be weakly convergent.

**§4. Compactness of spaces.**

1. DEFINITION OF COMPACTNESS. A set $M$ is called compact if from each infinite subset one can choose a convergent sequence.

EXAMPLES. 1. Every bounded set of points of the plane is compact (the Bolzano-Weierstrass principle).

2. Compactness in the space of continuous functions is established by Arzela's theorem: if a family of functions $\{\phi\}$ is uniformly bounded and equicontinuous, then it is compact (i.e., if $|\phi| < A$ and if for $\epsilon > 0$ one can find $\delta(\epsilon) > 0$ such that for all $\Phi$ in the family, $|\phi(\vec{p} + \vec{Q}) - \phi(\vec{p})| < \epsilon$ whenever $|\vec{Q}| < \delta(\epsilon)$, then the family $\{\phi\}$ is compact).

COMPACTNESS IN $L_p$. Corresponding to the two forms of convergence in $L_p$, we distinguish between strong and weak compactness.

1. A set of functions $\{\phi\} \subset L_p$ is called *weakly compact* if any infinite subset of it contains a weakly convergent sequence.

2. A set of functions $\{\phi\} \subset L_p$ is called *strongly compact* if any infinite subset of it contains a strongly convergent sequence.

2. A THEOREM ON WEAK COMPACTNESS. *In order that a set $X \subset L_p$ should be weakly compact it is necessary and sufficient that it should be bounded.*

PROOF OF NECESSITY. The necessity of the condition follows simply from the theorem on the weak completeness of $L_p$.

If the set $X$ is unbounded, then from it we may choose a sequence $\{\phi_k\}$ such that $\|\phi_k\| \to \infty$, and from which on the basis of the remark in §3, item 4, we cannot choose a weakly convergent sequence, and consequently, the set $X$ cannot be compact.

PROOF OF SUFFICIENCY. Suppose that for all $\phi \in X \subset L_p$,

$$\|\varphi\| < A. \tag{4.1}$$

We consider the conjugate space $L_{p'}$. It is separable (§2, item 4). Let us denote a net which is everywhere dense in $L_{p'}$ by

$$\psi_1, \psi_2, \ldots, \psi_k, \ldots. \tag{4.2}$$

Let $\{\phi_k\}$ be an arbitrary infinite sequence $\subset X$. We shall show that from it we can choose a weakly convergent subsequence. Take $\psi_1$ and consider the sequence of numbers $(\phi_k \psi_1) = \int_\Omega \phi_k \psi_1 dv = a_k^{(1)}$. This sequence is bounded: $|a_k^{(1)}| \leq A \|\psi_1\|$, and from it we may choose a convergent subsequence, corresponding to a subsequence

$$\varphi_1^{(1)}, \varphi_2^{(1)}, \ldots, \varphi_k^{(1)}, \ldots; \int_\Omega \varphi_k^{(1)} \psi_1 \, dv \to a^{(1)} \quad \text{for } k \to \infty.$$

Consider $\psi_2$. The sequence of numbers $a_k^{(2)} = (\phi_k^{(1)}, \psi_2)$ is bounded and from it we may choose a convergent subsequence, corresponding to the subsequence $\phi_1^{(2)}, \phi_2^{(2)}, \cdots, \phi_k^{(2)}, \cdots, \int_\Omega \phi_k^{(2)} \psi_2 dv \to a^{(2)}$ for $k \to \infty$.

Continuing this process, we obtain a series of convergent sequences of numbers $a_k^{(s)}$ and corresponding sequences of functionals, all obtained from the sequence $\{\phi_k\}$ and such that each of them is a subsequence of the preceding.

$$\{\varphi_k^{(s)}\} \subset \{\varphi_k^{(s-1)}\} \subset \{\varphi_k\} \text{ and } a_k^{(s)} = \int_\Omega \varphi_k^{(s)} \psi_s \, dv \to a^{(s)}$$

for $k \to \infty$ and $s = 1, 2, 3, \cdots$.

By the diagonal process we may select the sequence $\{\phi_k^{(k)}\}$, which is weakly convergent on the whole countable dense set $\{\psi_s\}$, i.e.,

$$l_k \psi_s = \int_\Omega \varphi_k^{(k)} \psi_s \, dv \to a^{(s)} = l \psi_s \quad \text{for} \quad k \to \infty \text{ and } s = 1, 2, 3, \ldots;$$

$$|l \psi_s| < A \|\psi_s\|.$$

The convergence of $\{l_k\}$ on the set $\{\psi_s\}$ implies its convergence everywhere.

For in fact, every element $\psi \in L_{p'}$ may be represented in the form

$$\psi = \psi_s + \psi',$$

where $\|\psi'\| < \epsilon$, and $\psi_s$ is an element from our net (4.2). Then by virtue of the boundedness of all the $l_k$, we have

$$|l_k\psi - l_k\psi_s| < A\varepsilon,$$
$$|l_m\psi - l_m\psi_s| < A\varepsilon$$

for arbitrary $k$ and $m$.

Furthermore,

$$|l_m\psi - l_k\psi| \leqslant |l_k\psi_s - l_m\psi_s| + 2A\epsilon,$$

and choosing sufficiently large $m$ and $k$, we obtain:

$$|l_m\psi - l_k\psi| < 3A\,\varepsilon,$$

from which follows the convergence of the functionals $l_k$ on $\psi$. As shown earlier, the limit $l\psi$ will be a linear functional on $L_{p'}$ and may therefore be written in the form $l\psi = \int \phi_0\psi dv$. Thus, for every $\psi \in L_{p'}$, we have $\int_\Omega \phi_k^{(k)}\psi dv \to \int_\Omega \phi_0\psi dv$, and consequently, for every functional from $L_p^*$ it follows that $(l\phi_k^{(k)}) \to (l\phi_0)$. Therefore the set $X$ is weakly compact, as was to be proved.

3. A THEOREM ON STRONG COMPACTNESS. *In order that a set $X \subset L_p$ should be strongly compact, it is necessary and sufficient that*

(1) $\|\phi\| < A$ *for all* $\phi \in X$. (4.3)

(2) *The set $X$ should be equicontinuous in the large, i.e., for an arbitrary $\epsilon > 0$, one can find $\delta(\epsilon) > 0$ such that for all $\phi \in X$*

$$J(\vec{Q}) = \int_\Omega |\varphi(\vec{P} + \vec{Q}) - \varphi(\vec{P})|^p\, dv < \varepsilon, \tag{4.4}$$

*if only $|\vec{Q}| < \delta(\epsilon)$.*

LEMMA. *The set $X \subset L_p$ will be strongly compact if and only if for every $\epsilon > 0$ one can find a finite $\epsilon$-net in $X$, i.e., a finite number of functions $\phi_1, \phi_2, \cdots, \phi_{N(\epsilon)}$, the functions of the net, such that for arbitrary $\phi \in X$, one can find among the functions of the net a function $\phi_s$ for which $\|\phi - \phi_s\| < \epsilon$.*

PROOF OF NECESSITY. Suppose that there existed an $\epsilon_0 > 0$ such that one could not construct a finite $\epsilon$-net for it. This means that for any element $\phi_1 \in X$, one can find $\phi_2 \in X$ such that $\|\phi_1 - \phi_2\| > \epsilon_0$. The elements $\phi_1$ and $\phi_2$ do not form an $\epsilon_0$-net, consequently one can find $\phi_3 \in X$ such that

$$\|\varphi_1-\varphi_3\|>\varepsilon_0 \text{ and } \|\varphi_2-\varphi_3\|>\varepsilon_0.$$

Continuing this process indefinitely, we construct an infinite sequence $\{\phi_k\}\subset X$ such that $\|\phi_i-\phi_k\|>\epsilon_0$ for $i\neq k$. Obviously from this sequence one cannot choose a strongly convergent subsequence, which contradicts the assumption of the strong compactness of $X$. The necessity of the condition is proved.

PROOF OF SUFFICIENCY. Suppose that $X$ is such that in it one can construct finite $\epsilon$-nets for arbitrary $\epsilon>0$ (and for which moreover this is possible for each of its subsets). Let $Y_1\subset X$ be an arbitrary infinite subset. We construct in it a finite $(\frac{1}{2})$-net, $\phi_1^{(1)},\phi_2^{(1)},\cdots,\phi_{N_1}^{(1)}$. For arbitrary $\phi\in Y_1$ we can find $\phi_k^{(1)}$ such that $\|\phi-\phi_k^{(1)}\|<\frac{1}{2}$. Let $Y_1^{(s)}$ be the set of functions $\in Y_1$ and at distance from $\phi_s^{(1)}$ less than $1/2^2$. At least one of the $Y_1^{(s)}$ is infinite since $Y_1$ is an infinite set. We denote it by $Y_2$ and construct on it a $1/2^2$-net. Repeating this process, we obtain on the $k$th step $Y_k\subset Y_{k-1}\subset\cdots\subset Y_2\subset Y_1$ and on it a $(1/2^k)$-net $\phi_1^{(k)},\phi_2^{(k)},\cdots,\phi_{N_k}^{(k)}$. Let $\phi'$ and $\phi''\subset Y_k$. Then $\|\phi'-\phi''\|\leqq\|\phi'-\phi_k\|+\|\phi''-\phi_k\|\leqq(1/2^k)+(1/2^k)=1/2^{k-1}$, i.e., an arbitrary pair of functions from $Y_k$ differ from one another by not more than $1/2^{k-1}$.

The sequence of functions $\{\phi_k\}$, where $\phi_k\in Y_k$, will be strongly convergent since $\|\phi_{k+m}-\phi_k\|<1/2^{k-1}<\epsilon$ for sufficiently large $k$ ($\phi_{k+m}\in Y_{k+m}\subset Y_k$ for arbitrary $m$). Thus from $Y_1\subset X$ we have extracted a strongly convergent sequence, and consequently $X$ is compact.

4. PROOF OF THE THEOREM ON STRONG COMPACTNESS.

1. SUFFICIENCY. We consider a family $\{\phi\}=X$ satisfying the conditions (4.3) and (4.4). For convenience we introduce a domain $\Omega_1\supset\Omega$ such that any sphere of radius $h<\delta$ with center in $\Omega$ will lie in the interior $\Omega_1$. Let $\phi=0$ be outside of $\Omega$. We construct a family of averaged functions for each $\phi\in X$ by means of the family of kernels $\omega(\vec{Q},h)$ (§2, item 4).

$$\varphi_h(\vec{P})=\frac{1}{h^n}\int_{\Omega_1}\varphi(\vec{P}_1)\,\omega(\vec{P}-\vec{P}_1,h)\,dv_1. \tag{4.5}$$

By the estimates of §2, item 4: $\|\phi_h-\phi\|<KJ_\phi(h)$, where $K$ does not depend upon the particular function of the family or upon the parameter $h$ of the kernel, and $J_\phi(h)=\sup_{|\vec{Q}|\leq h}J_\phi(\vec{Q})$. By condition (4.4), for a given $\epsilon>0$ one can find $\delta(\epsilon)>0$ such that for $h\leqq\delta(\epsilon)$

$$\|\varphi_h-\varphi\|<\frac{\varepsilon}{2} \quad\text{for all}\quad \varphi\in X. \tag{4.6}$$

Then an $\epsilon/2$ net for $\{\phi_h\}$ will be an $\epsilon$-net for $\{\phi\}$.

If we now establish the compactness of the family of averaged functions, by the basic lemma there will follow from this existence for the averaged functions of finite $\epsilon/2$ nets. This means that in the set $X$, we may construct finite $\epsilon$-nets for arbitrary $\epsilon$. By the basic lemma once more, $X$ will be compact. In order to show the compactness of $\{\phi_h\}$, we show that $\{\phi_h\}$ for given $h$ are uniformly bounded and equicontinuous. Then by Arzela's theorem, this set of functions will be compact in the sense of uniform convergence and a fortiori in the sense of convergence in $L_p$. But for sufficiently small $|\vec{Q}|$, by virtue of the continuity of the kernel:

$$|\omega_h(\vec{P}_1-\vec{P}-\vec{Q})-\omega_h(\vec{P}_1-\vec{P})|<\eta,$$

from which

$$|\varphi_h(\vec{P}+\vec{Q})-\varphi_h(\vec{P})|=\left|\frac{1}{\varkappa h^n}\int\limits_{\Omega}\varphi(\vec{P}_1)[\omega_h(\vec{P}_1-\vec{P}-\vec{Q})-\omega_h(\vec{P}_1-\vec{P})]\,dv_1\right|\leqslant$$

$$\leqslant\frac{\eta}{\varkappa h^n}\left\{\int\limits_{\Omega}|\varphi(\vec{P})|^p\,dv_1\right\}^{\frac{1}{p}}=\frac{\eta}{\varkappa h^n}\cdot\|\varphi\|\leqslant A\cdot\frac{\eta}{\varkappa h^n}$$

and for sufficiently small $|\vec{Q}|$ and fixed $h$

$$|\varphi_h(\vec{P}+\vec{Q})-\varphi_h(\vec{P})|<\varepsilon,$$

i.e., $\{\phi_h\}$ is equicontinuous.

The uniform boundedness follows from the estimate:

$$|\varphi_h(\vec{P})|\leqslant\left\{\frac{1}{\varkappa h^n}\left[\int\limits_{\Omega_1}|\omega(\vec{P}_1-\vec{P})|^{p'}\,dv_1\right]^{\frac{1}{p'}}\left[\int\limits_{\Omega_1}|\varphi(\vec{P}_1)|^p\,dv_1\right]^{\frac{1}{p}}\right\}<\frac{C}{\varkappa}A.$$

Thus, the family $\{\phi_h\}$ is compact, and by the lemma, we may construct on it a finite $(\epsilon/2)$-net. This will be a finite $\epsilon$-net for $X$ and, as a result, by the lemma, $X$ is strongly compact as was to be proved.

2. NECESSITY. Let $X$ be strongly compact. Then it is weakly compact and by the theorem on weak compactness for all $\phi\in X$ we have $\|\phi\|<A$, i.e., the necessity of condition (4.3) is established.

By the lemma there exists a finite $(\epsilon/3)$-net $\{\phi_k\}$ $(k=1,2,\cdots,N)$. Each one of the functions $\phi_k$ is continuous in the large, and since there a finite number, for a given $\epsilon$ we may find $\delta(\epsilon/3)>0$ such that

$$\left[\int\limits_{\Omega}|\varphi_k(\vec{P}+\vec{Q})-\varphi_k(\vec{P})|^p\,dv\right]^{\frac{1}{p}}<\frac{\varepsilon}{3},\qquad(k=1,2,\ldots,N),$$

if $|\vec{Q}|<\delta$.

Take an arbitrary function $\phi \in X$ and choose the nearest $\phi_k$ to it from the $(\epsilon/3)$-net. Using the Minkowski inequality, we obtain:

$$\left[\int_\Omega |\varphi(\vec{P}+\vec{Q})-\varphi(\vec{P})|^p \, dv\right]^{\frac{1}{p}} =$$

$$=\left[\int_\Omega |\varphi(\vec{P}+\vec{Q})-\varphi_k(\vec{P}+\vec{Q})+\varphi_k(\vec{P}+\vec{Q})-\right.$$

$$\left.-\varphi_k(\vec{P})+\varphi_k(\vec{P})-\varphi(\vec{P})|^p \, dv\right]^{\frac{1}{p}} \leqslant$$

$$\leqslant\left[\int_\Omega |\varphi(\vec{P}+\vec{Q})-\varphi_k(\vec{P}+\vec{Q})|^p \, dv\right]^{\frac{1}{p}}+$$

$$+\left[\int_\Omega |\varphi_k(\vec{P}+\vec{Q})-\varphi_k(\vec{P})|^p \, dv\right]^{\frac{1}{p}}+\left[\int_\Omega |\varphi_k(\vec{P})-\varphi(\vec{P})|^p \, dv\right]^{\frac{1}{p}}<$$

$$<\frac{\varepsilon}{3}+\frac{\varepsilon}{3}+\frac{\varepsilon}{3}=\varepsilon \qquad \text{for } |\vec{Q}|<\delta,$$

i.e., for all $\phi \in X$ for the given $\epsilon$, we may give $\delta$ such that $J_\phi(\vec{Q})<\epsilon$ whenever $|\vec{Q}|<\delta$, and consequently, the necessity of condition (4.4) has been shown.

## §5. Generalized derivatives.

1. BASIC DEFINITIONS. Suppose that $\phi$ is given on the whole space and summable on every bounded domain.

We consider a function $\psi$ continuous together with all its derivatives up to order $l$ inclusive and equal to zero outside some bounded domain $V_\psi$.

If $\phi$ has continuous derivatives, then:

$$\int_\Omega \left[\varphi \frac{\partial^l \psi}{\partial x_1^{\alpha_1} \ldots \partial x_n^{\alpha_n}}+(-1)^{l+1} \psi \frac{\partial^l \varphi}{\partial x_1^{\alpha_1} \ldots \partial x_n^{\alpha_n}}\right] dv=0. \tag{5.1}$$

Suppose now that we know nothing about the existence of derivatives of $\phi$ and suppose that there exists a function $\omega_{\alpha_1 \alpha_2 \cdots \alpha_n}$ which is summable and satisfies the equation

$$\int_\Omega \left[\varphi \frac{\partial^l \psi}{\partial x_1^{\alpha_1} \ldots \partial x_n^{\alpha_n}}+(-1)^l \psi \omega_{\alpha_1 \alpha_2 \ldots \alpha_n}\right] dv=0 \tag{5.2}$$

for all functions $\psi$ from the class considered. Then we set

$$\omega_{\alpha_1 \alpha_2 \ldots \alpha_n}=\frac{\partial^l \varphi}{\partial x_1^{\alpha_1} \partial x_2^{\alpha_2} \ldots \partial x_n^{\alpha_n}} \tag{5.3}$$

and call $\omega_{\alpha_1\alpha_2\cdots\alpha_n}$ a "generalized derivative" of the function $\phi$.

REMARK. The equation (5.3), by the basic lemma of the calculus of variations, is true in the ordinary sense almost everywhere if $\phi$ admits the corresponding derivatives and the integration by parts is justified.

The new definition of derivatives does not coincide with the definition of derivatives almost everywhere, as is shown by examples.

EXAMPLE 1. Let $\phi(x)$ be continuous on $[0, 1]$ but not absolutely continuous. If $\phi(x)$ has a generalized derivative $\omega(x)$, then for every $\psi(x)$, continuous together with its first derivative and vanishing outside $(\epsilon, 1-\epsilon)$ $(0<\epsilon<\frac{1}{2})$, we have the equation:

$$\int_0^1 \varphi \frac{d\psi}{dx}\, dx = -\int_0^1 \omega(x)\psi(x)\, dx.$$

Let $\Omega(x) = \int_0^x \omega(\xi)\, d\xi$. Then

$$-\int_0^1 \omega(x)\psi(x)\, dx = \int_0^1 \Omega(x)\frac{d\psi}{dx}\, dx,$$

and consequently, we have

$$\int_0^1 [\varphi - \Omega(x)]\frac{d\psi}{dx}\, dx = 0,$$

from which, by virtue of the arbitrariness of $d\psi/dx$, satisfying only the condition $\int_0^1 d\psi dx/dx = 0$, will follow $\phi - \Omega(x) = C$, i.e., $\phi = \Omega(x) + C$, and consequently $\phi(x)$ is absolutely continuous, which contradicts the assumption. Thus, $\phi(x)$ does not have a generalized derivative. For this reason, if $\phi(x)$ is continuous and has a derivative $d\phi/dx$ almost everywhere but is not absolutely continuous, then it does not have a generalized derivative. Thus, from the existence of the derivative almost everywhere, does not follow the existence of the generalized derivative.

EXAMPLE 2. We consider the function of two variables $\phi(x, y) = f_1(x) + f_2(y)$, where neither $f_1(x)$ nor $f_2(y)$ is differentiable: then $\phi(x, y)$ does not have derivatives in the ordinary sense, but the generalized derivative $\partial^2\phi/\partial x\partial y$ exists and is equal to zero. Indeed:

$$\int_\Omega \varphi \frac{\partial^2\psi}{\partial x\,\partial y}\, dv = \int_\Omega f_1(x)\frac{\partial^2\psi}{\partial x\,\partial y}\, dv + \int_\Omega f_2(y)\frac{\partial^2\psi}{\partial x\,\partial y}\, dv.$$

But on the boundary $\partial\psi/\partial x$ and $\partial\psi/\partial y = 0$, and therefore

$$\int_\Omega f_1(x)\frac{\partial^2\psi}{\partial x\,\partial y}\,dv=\int_a^b f_1(x)\int_{\varphi_1(x)}^{\varphi_2(x)}\frac{\partial^2\psi}{\partial x\,\partial y}\,dy\,dx=$$

$$=\int_a^b f_1(x)\left[\frac{\partial\psi}{\partial x}\right]_{y=\varphi_1(x)}^{y=\varphi_2(x)}dx=0,$$

and analogously

$$\int_\Omega f_2(y)\frac{\partial^2\psi}{\partial x\,\partial y}\,dv=0.$$

Consequently,

$$\int_\Omega \varphi\frac{\partial^2\psi}{\partial x\,\partial y}\,dv=0,$$

as was to be proved.[1)]

2. DERIVATIVES OF AVERAGED FUNCTIONS.

EXISTENCE OF GENERALIZED DERIVATIVES. *Derivatives of averaged functions are equal to the averaged functions of the generalized derivatives*

$$\frac{\partial^\alpha\varphi_h}{\partial x_1^{\alpha_1}\partial x_2^{\alpha_2}\dots\partial x_n^{\alpha_n}}=\left[\frac{\partial^\alpha\varphi}{\partial x_1^{\alpha_1}\partial x_2^{\alpha_2}\dots\partial x_n^{\alpha_n}}\right]_h. \tag{5.4}$$

PROOF.

$$\varphi_h(\vec{P})=\frac{1}{\varkappa h^n}\int_\Omega\varphi(\vec{P}_1)\,\omega_h(\vec{P}_1-\vec{P})\,dv_1,$$

where

$$\vec{P}=(x_1,x_2,\dots,x_n);\quad \vec{P}_1=(y_1,y_2,\dots,y_n)$$

$$\frac{\partial^\alpha\varphi_h}{\partial x_1^{\alpha_1}\partial x_2^{\alpha_2}\dots\partial x_n^{\alpha_n}}=\frac{1}{\varkappa h^n}\int\varphi(\vec{P}_1)\frac{\partial^\alpha\omega_h(\vec{P}_1-\vec{P})}{\partial x_1^{\alpha_1}\partial x_2^{\alpha_2}\dots\partial x_n^{\alpha_n}}\,dv_1=$$

$$=\frac{(-1)^\alpha}{\varkappa h^n}\int\varphi(\vec{P}_1)\frac{\partial^\alpha\omega_h(\vec{P}_1-\vec{P})}{\partial y_1^{\alpha_1}\partial y_2^{\alpha_2}\dots\partial y_n^{\alpha_n}}\,dv_1.$$

By the definition of the generalized derivative, for every $\psi$ we have equa-

---

[1)] There can exist at most one generalized derivative $\omega_{\alpha_1\cdots\alpha_n}$ of the function $\phi$. If there were two: $\omega^{(1)}_{\alpha_1\cdots\alpha_n}$ and $\omega^{(2)}_{\alpha_1\cdots\alpha_n}$, then on the basis of (5.2) we would have

$$\int_\Omega\left(\omega^{(1)}_{\alpha_1\dots\alpha_n}-\omega^{(2)}_{\alpha_1\dots\alpha_n}\right)\psi\,dv=0,$$

from which by the arbitrariness of $\psi$ we would have $\omega^{(1)}_{\alpha_1\cdots\alpha_n}=\omega^{(2)}_{\alpha_1\cdots\alpha_n}$ almost everywhere. Thus generalized derivatives are uniquely defined (up to a set of measure zero).

tion (5.2). Taking $\psi = \omega_h(\vec{P}_1 - \vec{P})$, we obtain

$$\frac{\partial^\alpha \varphi_h}{\partial x_1^{\alpha_1} \partial x_2^{\alpha_2} \ldots \partial x_n^{\alpha_n}} = \frac{1}{\varkappa h^n} \int \omega_h (\vec{P}_1 - \vec{P}) \frac{\partial^\alpha \varphi}{\partial y_1^{\alpha_1} \partial y_2^{\alpha_2} \ldots \partial y_n^{\alpha_n}} dv_1 = \left[ \frac{\partial^\alpha \varphi^*}{\partial x_1^{\alpha_1} \partial x_2^{\alpha_2} \ldots \partial x_n^{\alpha_n}} \right]_h,$$

as was to be proved.

COROLLARY. *If $\phi$ is a summable function and*

$$\frac{\partial^\alpha \varphi}{\partial x_1^{\alpha_1} \ldots \partial x_n^{\alpha_n}} \in L_p,$$

*then*

$$\frac{\partial^\alpha \varphi_h}{\partial x_1^{\alpha_1} \partial x_2^{\alpha_2} \ldots \partial x_n^{\alpha_n}} \Longrightarrow \frac{\partial^\alpha \varphi}{\partial x_1^{\alpha_1} \partial x_2^{\alpha_2} \ldots \partial x_n^{\alpha_n}} \text{ in } L_p,$$

*which follows from the properties of averaged functions.*

THEOREM. *If a given summable function $\phi$ can be approximated by a sequence of continuously differentiable functions $\phi_k(\vec{P})$ $(k=0,1,2,\cdots)$ in the sense that for every function $\psi$, continuous and such that $\psi \equiv 0$ outside of $V_\psi \subset \Omega$, we have*

$$\lim \int_\Omega \varphi_k(\vec{P}) \psi(\vec{P}) dv = \int_\Omega \varphi(\vec{P}) \psi(\vec{P}) dv, \tag{5.5}$$

*and if, in addition,*

$$\int_\Omega \left| \frac{\partial^\alpha \varphi_k}{\partial x_1^{\alpha_1} \partial x_2^{\alpha_2} \ldots \partial x_n^{\alpha_n}} \right|^p dv < A(\Omega), \tag{5.6}$$

*then the generalized derivative exists*

$$\frac{\partial^\alpha \varphi}{\partial x_1^{\alpha_1} \partial x_2^{\alpha_2} \ldots \partial x_n^{\alpha_n}}.$$

PROOF. From the boundedness of the integrals (5.6) follows the weak compactness in $L_p$ of the sequence $\{\partial^\alpha \phi_k / (\partial x_1^{\alpha_1} \partial x_2^{\alpha_2} \cdots \partial x_n^{\alpha_n})\}$, and consequently there exists a weakly convergent subsequence

$$\frac{\partial^\alpha \varphi_{k_i}}{\partial x_1^{\alpha_1} \partial x_2^{\alpha_2} \ldots \partial x_n^{\alpha_n}} \to \omega_{\alpha_1 \alpha_2 \ldots \alpha_n}$$

(where $\omega_{\alpha_1 \alpha_2 \cdots \alpha_n} \in L_p$). We have the equation:

$$\int_{\Omega}\left\{\varphi_{k_i}\frac{\partial^\alpha\psi}{\partial x_1^{\alpha_1}\dots\partial x_n^{\alpha_n}}+(-1)^{\alpha+1}\psi\frac{\partial^\alpha\varphi_{k_i}}{\partial x_1^{\alpha_1}\dots\partial x_n^{\alpha_n}}\right\}dv=0$$

for all $\psi$ continuous with all derivatives up to order $\alpha$ and vanishing outside $V_\psi\subset\Omega$.

Passing to the limit for $k_i\to\infty$, we obtain:

$$\int_{\Omega}\left\{\varphi\frac{\partial^\alpha\psi}{\partial x_1^{\alpha_1}\dots\partial x_n^{\alpha_n}}+(-1)^{\alpha+1}\psi\omega_{\alpha_1\dots\alpha_n}\right\}dv=0,$$

which establishes the theorem.

REMARK. By virtue of (3.31), $\int_\Omega|\omega_{\alpha_1\cdots\alpha_n}|^p dv<A(\Omega)$, since $\psi_{\alpha_1\cdots\alpha_n}\in L_p$ and is the weak limit of $\partial^\alpha\phi_k/(\partial x_1^{\alpha_1}\cdots\partial x_n^{\alpha_n})$.

COROLLARY. *If the set of derivatives of $\alpha$th order of the averaged functions is weakly compact, then the given function has generalized derivatives of $\alpha$th order.*

3. RULES OF DIFFERENTIATION. From the definition of generalized derivatives follow the assertions:

(1) *If $\phi_1$ and $\phi_2$ have generalized derivatives of a given order*

$$\frac{\partial^\alpha\varphi_1}{\partial x_1^{\alpha_1}\dots\partial x_n^{\alpha_n}}\quad \textit{and}\quad \frac{\partial^\alpha\varphi_2}{\partial x_1^{\alpha_1}\dots\partial x_n^{\alpha_n}},$$

*and if $c_1$ and $c_2$ are constants, then $c_1\phi_1+c_2\phi_2$ has generalized derivatives of the same order.*

$$\frac{\partial^\alpha(c_1\varphi_1+c_2\varphi_2)}{\partial x_1^{\alpha_1}\partial x_2^{\alpha_2}\dots\partial x_n^{\alpha_n}}=c_1\frac{\partial^\alpha\varphi_1}{\partial x_1^{\alpha_1}\partial x_2^{\alpha_2}\dots\partial x_n^{\alpha_n}}+c_2\frac{\partial^\alpha\varphi_2}{\partial x_1^{\alpha_1}\dots\partial x_n^{\alpha_n}}.$$

The proof is obvious.

(2) *If $\phi$ has the generalized derivative*

$$\frac{\partial^\alpha\varphi}{\partial x_1^{\alpha_1}\partial x_2^{\alpha_2}\dots\partial x_n^{\alpha_n}}=\omega(x_1,x_2,\dots,x_n)\quad \textit{and}\quad \omega(x_1,x_2,\dots,x_n)$$

*has the generalized derivative*

$$\frac{\partial^\beta\omega}{\partial x_1^{\beta_1}\partial x_2^{\beta_2}\dots\partial x_n^{\beta_n}}=\lambda(x_1,\dots,x_n),\quad \textit{then}\quad \lambda(x_1,x_2,\dots,x_n)$$

*is the generalized derivative of $\phi$ of order $\alpha+\beta$:*

$$\lambda = \frac{\partial^{\alpha+\beta}\varphi}{\partial x_1^{\alpha_1+\beta_1}\,\partial x_2^{\alpha_2+\beta_2}\ldots\partial x_n^{\alpha_n+\beta_n}}.$$

PROOF. Let $\psi$ have continuous derivatives up to order $\alpha+\beta$ on the whole space and be equal to zero outside some domain $V_\psi \subset \Omega$. Then $\partial^\beta\psi/(\partial x_1^{\beta_1}\cdots\partial x_n^{\beta_n}) = \psi_{\beta_1\cdots\beta_n}$ has continuous derivatives up to order $\alpha$ and is equal to zero outside of $V_\psi$.

From the definition of generalized derivatives there follows:

$$\int_\Omega \lambda\psi\,dv = (-1)^\beta \int_\Omega \omega \frac{\partial^\beta\psi}{\partial x_1^{\beta_1}\ldots\partial x_n^{\beta_n}}\,dv = (-1)^\beta \int_\Omega \omega\psi_{\beta_1\ldots\beta_n}\,dv =$$

$$= (-1)^{\alpha+\beta} \int_\Omega \varphi \frac{\partial^\alpha\psi_{\beta_1\ldots\beta_n}}{\partial x_1^{\alpha_1}\partial x_2^{\alpha_2}\ldots\partial x_n^{\alpha_n}}\,dv =$$

$$= (-1)^{\alpha+\beta} \int_\Omega \varphi \frac{\partial^{\alpha+\beta}\psi}{\partial x_1^{\alpha_1+\beta_1}\ldots\partial x_n^{\alpha_n+\beta_n}}\,dv,$$

which in view of the arbitrariness of $\psi$ establishes the assertion.

(3) *If $\phi_1$ and $\partial\phi_1/\partial x_1 \in L_p$, $\phi_2$ and $\partial\phi_2/\partial x_1 \in L_{p'}$, then the product $\phi_1\phi_2$ has the generalized derivative*

$$\frac{\partial(\varphi_1\varphi_2)}{\partial x_1} = \varphi_1\frac{\partial\varphi_2}{\partial x_1} + \varphi_2\frac{\partial\varphi_1}{\partial x_1}.$$

Indeed, $\phi_1 \in L_p$ and $\phi_2 \in L_{p'}$, and therefore $\phi_1\phi_2$ is summable. Analogously, $\phi_1(\partial\Phi_2/\partial x_1)$, $\phi_2(\partial\Phi_1/\partial x_1)$, and $\phi_1(\partial\phi_2/\partial x_1)+\phi_2(\partial\phi_1/\partial x_1)$ are summable. Let $\phi_{1h}$ and $\phi_{2h}$ be the averaged functions for $\phi_1$ and $\phi_2$. By the properties of the averaged functions, we have:

$$\varphi_{1h} \Rightarrow \varphi_1,\ \frac{\partial\varphi_{1h}}{\partial x_1} \Rightarrow \frac{\partial\varphi_1}{\partial x_1} \text{ in } L_p,\ \varphi_{2h} \Rightarrow \varphi_2,\ \frac{\partial\varphi_{2h}}{\partial x_1} \Rightarrow \frac{\partial\varphi_2}{\partial x_1} \text{ in } L_{p'}.$$

Suppose that $\psi$ is continuous with its derivatives of first order on the whole space and is equal to zero outside $V_\psi \subset \Omega$. Suppose $|\psi|, |\partial\psi/\partial x_1| < A$. Then

$$\left|\int_\Omega (\varphi_{1h}\varphi_{2h} - \varphi_1\varphi_2)\frac{\partial\psi}{\partial x_1}\,dv\right| < A\int_\Omega |\varphi_{1h}\varphi_{2h} - \varphi_1\varphi_2|\,dv =$$

$$= A\int_\Omega |\varphi_{1h}(\varphi_{2h}-\varphi_2) + \varphi_2(\varphi_{1h}-\varphi_1)|\,dv \leqslant A\int_\Omega |\varphi_{1h}|\cdot|\varphi_{2h}-\varphi_2|\,dv +$$

$$+ A\int_\Omega |\varphi_2|\cdot|\varphi_{1h}-\varphi_1|\,dv \leqslant A[\|\varphi_{1h}\|_{L_p}\|\varphi_{2h}-\varphi_2\|_{L_{p'}} +$$

$$+ \|\varphi_2\|_{L_{p'}}\|\varphi_{1h}-\varphi_1\|_{L_p}] \to 0 \text{ for } h \to 0,$$

i.e.,

$$\int_\Omega \varphi_{1h}\varphi_{2h}\frac{\partial\psi}{\partial x_1}\,dv \to \int_\Omega \varphi_1\varphi_2\frac{\partial\psi}{\partial x_1}\,dv, \text{ for } h \to 0.$$

Analogously, one shows

$$\int_\Omega\left(\varphi_{1h}\frac{\partial\varphi_{2h}}{\partial x_1}+\varphi_{2h}\frac{\partial\varphi_{1h}}{\partial x_1}\right)\psi\,dv \to \int_\Omega \psi\left(\varphi_1\frac{\partial\varphi_2}{\partial x_1}+\varphi_2\frac{\partial\varphi_1}{\partial x_1}\right)dv.$$

Then from the obvious equation

$$\int_\Omega \varphi_{1h}\varphi_{2h}\frac{\partial\psi}{\partial x_1}\,dv = -\int_\Omega \psi\left(\varphi_{1h}\frac{\partial\varphi_{2h}}{\partial x_1}+\varphi_{2h}\frac{\partial\varphi_{1h}}{\partial x_1}\right)dv$$

for $h \to 0$ follows

$$\int_\Omega \varphi_1\varphi_2\frac{\partial\psi}{\partial x_1}\,dv = -\int_\Omega \psi\left(\varphi_1\frac{\partial\varphi_2}{\partial x_1}+\varphi_2\frac{\partial\varphi_1}{\partial x_1}\right)dv,$$

and this means that

$$\varphi_1\frac{\partial\varphi_2}{\partial x_1}+\varphi_2\frac{\partial\varphi_1}{\partial x_1}=\frac{\partial(\varphi_1\varphi_2)}{\partial x_1}.$$

REMARK. If $\phi_1$ and $\phi_2$ have continuous derivatives up to order $\alpha$, then

$$\frac{\partial^\alpha(\varphi_1\varphi_2)}{\partial x_1^{\alpha_1}\partial x_2^{\alpha_2}\ldots\partial x_n^{\alpha_n}} =$$

$$= \sum_{\beta=0}^{\alpha}\sum_{\Sigma\beta_i=\beta} C_{\beta_1\beta_2\ldots\beta_n}\frac{\partial^\beta\varphi_1}{\partial x_1^{\beta_1}\ldots\partial x_n^{\beta_n}}\frac{\partial^{\alpha-\beta}\varphi_2}{\partial x_1^{\beta_1}\ldots\partial x_n^{\beta_n}},$$

where $C_{\beta_1\beta_2\ldots\beta_n}$ are the binomial coefficients, and just the ones which appear in the corresponding formula for continuously differentiable functions.

This same formula is correct if $\phi_1$ and $\phi_2$ lie in $L_p$ and $L_{p'}$, respectively, and have all the generalized derivatives occurring on the right hand side, with the derivatives of $\phi_1$ lying in $L_p$ and the derivatives of $\phi_2 \in L_{p'}$. The proof is analogous to the one just given.

4. INDEPENDENCE OF THE DOMAIN. Let $\phi$ and $\lambda$ be two summable functions on $\Omega$. If $\lambda$ is the generalized derivative of $\phi$

$$\lambda = \frac{\partial^\alpha\varphi}{\partial x_1^{\alpha_1}\ldots\partial x_n^{\alpha_n}}$$

on $\Omega$, then $\lambda$ is the generalized derivative of $\phi$ on an arbitrary subset of $\Omega$, as easily follows from the definition of generalized derivatives. Thus, although the definition of the generalized derivative is in the large, its character in an arbitrary neighborhood of any point is defined by the local properties of the function $\phi$. We consider the possibility of extending the domain of the initial definition of the generalized derivative.

As a preliminary, we establish a lemma.

LEMMA. *Let $\phi$ and $\lambda$ be summable functions on $\Omega$, for which on every sphere of radius less than some $\delta>0$ lying within $\Omega$,$\lambda$ is the generalized derivative of $\phi$:*

$$\lambda = \frac{\partial^{\alpha}\varphi}{\partial x_1^{\alpha_1} \dots \partial x_n^{\alpha_n}} .$$

*Then $\lambda$ is the generalized derivative of $\phi$ on $\Omega$.*

We denote by $\Omega_\delta$ the set of all the points of $\Omega$ whose distance from the boundary is greater than $\delta$. Then the sphere $C_h(\vec{P})$ with center at an arbitrary point of $\Omega_\delta$ with radius $h$ $(h<\delta)$ lies in $\Omega$. Let $\lambda_h$ be the averaged function of $\lambda$ with respect to the kernel $\omega_h(\vec{P}_1-\vec{P})$. Using the definition of the generalized derivative, we have for an arbitrary point $\vec{P}\in\Omega_h$

$$\lambda_h = \int\limits_{C_h(\vec{P})} \lambda\omega_h\, dv = (-1)^{\alpha} \int\limits_{C_h(\vec{P})} \varphi \frac{\partial^{\alpha}\omega_h}{\partial y_1^{\alpha_1} \dots \partial y_n^{\alpha_n}} dv =$$

$$= (-1)^{\alpha} \int\limits_{C_h(\vec{P})} (-1)^{\alpha} \varphi \frac{\partial^{\alpha}\omega_h}{\partial x_1^{\alpha_1} \dots \partial x_n^{\alpha_n}} dv =$$

$$= \frac{\partial^{\alpha}}{\partial x_1^{\alpha_1} \dots \partial x_n^{\alpha_n}} \int\limits_{C_h(\vec{P})} \varphi\omega_h\, dv = \frac{\partial^{\alpha}\varphi_h}{\partial x_1^{\alpha_1} \dots \partial x_n^{\alpha_n}} .$$

Thus the averaged function $\lambda_h$ is the derivative of the averaged function $\phi_h$. For $h\to 0$, $\lambda_h \Rightarrow \lambda$ on an arbitrary $\Omega_\delta$ $(\delta>0)$ and consequently

$$\frac{\partial^{\alpha}\varphi}{\partial x_1^{\alpha_1} \partial x_2^{\alpha_2} \dots \partial x_n^{\alpha_n}} \Rightarrow \lambda$$

on an arbitrary $\Omega_\delta$. Similarly

$$\varphi_h \Rightarrow \varphi.$$

From this, taking the limit as $h\to 0$ of the obvious equality

$$\int\limits_{\Omega_\delta} \psi\lambda_h\, dv = (-1)^{\alpha} \int\limits_{\Omega_\delta} \varphi_h \frac{\partial^{\alpha}\psi}{\partial x_1^{\alpha_1} \dots \partial x_n^{\alpha_n}} dv$$

($\psi\equiv 0$ outside of $\Omega_\delta$ and has continuous derivatives up to order $\alpha$ on the whole space), we arrive at the equality:

$$\int_{\Omega_\delta} \psi\lambda\, dv = (-1)^\alpha \int_{\Omega} \varphi \frac{\partial^\alpha \psi}{\partial x_1^{\alpha_1} \ldots \partial x_n^{\alpha_n}} dv,$$

and the lemma is proved.

Let $\Omega^{(1)}$ and $\Omega^{(2)}$ be two arbitrary domains. Obviously

$$\Omega_\delta^{(1)} + \Omega_\delta^{(2)} \subset (\Omega^{(1)} + \Omega^{(2)})_\delta.$$

If for some $\delta > 0$, we can find $\delta'$ $(0 < \delta' < \delta)$ such that

$$(\Omega^{(1)} + \Omega^{(2)})_\delta \subset \Omega_{\delta'}^{(1)} + \Omega_{\delta'}^{(2)},$$

then the pair of domains $\Omega^{(1)}$ and $\Omega^{(2)}$ will be called summable.

THEOREM. *Let $\Omega^{(1)}$ and $\Omega^{(2)}$ be a summable pair of domains having the intersection $\Omega^{(1)}\Omega^{(2)}$. Let $\phi_1$ be defined on $\Omega^{(1)}$, $\phi_2$ on $\Omega^{(2)}$, and $\phi_1 = \phi_2$ on $\Omega^{(1)} \cdot \Omega^{(2)}$ (with the exception of a set of measure zero).*

*Suppose that $\phi_1$ has on $\Omega^{(1)}$ the generalized derivative $\partial^\alpha \phi_1 / \partial x_1^{\alpha 1} \cdots \partial x_n^{\alpha n}$ and $\phi_2$ on $\Omega^{(2)}$ has the generalized derivative $\partial^\alpha \phi_2 / \partial x_1^{\alpha 1} \cdots \partial x_n^{\alpha n}$. Then the function $\phi$ defined on $\Omega^{(1)} + \Omega^{(2)}$ by*

$$\varphi = \begin{cases} \varphi_1 & \text{for } \Omega^{(1)} \\ \varphi_2 & \text{for } \Omega^{(2)} \end{cases}$$

*has on $\Omega^{(1)} + \Omega^{(2)}$ the generalized derivative equal to*

$$\frac{\partial^\alpha \varphi}{\partial x_1^{\alpha_1} \ldots \partial x_n^{\alpha_n}} = \lambda = \begin{cases} \dfrac{\partial^\alpha \varphi_1}{\partial x_1^{\alpha_1} \ldots \partial x_n^{\alpha_n}} \text{ for } \Omega_1 \\[2ex] \dfrac{\partial^\alpha \varphi_2}{\partial x_1^{\alpha_1} \ldots \partial x_n^{\alpha_n}} \text{ for } \Omega_2. \end{cases} \tag{5.7}$$

The proof of this theorem presents no difficulty. By virtue of the fact that $\phi_1 = \phi_2$ on $\Omega^{(1)}\Omega^{(2)}$, we have $\partial^\alpha \phi_1 / \partial x_1^{\alpha 1} \cdots \partial x_n^{\alpha n} = \partial^\alpha \phi_2 / \partial x_1^{\alpha 1} \cdots \partial x_n^{\alpha n}$ on $\Omega^{(1)}\Omega^{(2)}$, which follows from the local character of the generalized derivative. Thereby the right hand side of the formula (5.7) is consistent on $\Omega^{(1)}\Omega^{(2)}$.

Obviously it suffices to show that $\lambda$ is the generalized derivative of $\phi$ on an arbitrary closed domain lying within $(\Omega^{(1)} + \Omega^{(2)})$. We shall prove it for $(\Omega^{(1)} + \Omega^{(2)})_\delta$, where $\delta > 0$ is arbitrarily small, since an arbitrary closed domain lies in $(\Omega^{(1)} + \Omega^{(2)})_\delta$ for some $\delta > 0$.

Since the pair of domains $\Omega^{(1)}$ and $\Omega^{(2)}$ is summable, $(\Omega^{(1)} + \Omega^{(2)})_\delta \subset \Omega_{\delta'}^{(1)} + \Omega_{\delta'}^{(2)}$ for some $\delta' > 0$, and consequently each point $\vec{P} \in (\Omega^{(1)} + \Omega^{(2)})_\delta$ lies in at least one of the domains $\Omega_{\delta'}^{(1)}$, $\Omega_{\delta'}^{(2)}$. Suppose $\vec{P} \in \Omega_\delta^{(1)}$.

Then the sphere $C_{\delta'}(\vec{P}) \subset \Omega^{(1)}$, where $\lambda$ coincides with $\partial^\alpha \phi_1 / \partial x_1^{\alpha 1} \cdots \partial x_n^{\alpha n}$ and

$\phi$ coincides with $\phi_1$. Therefore $\lambda$ is the generalized derivative of $\phi$ on $C_{\delta'}(\vec{P})$ for any $\vec{P}\in(\Omega^{(1)}+\Omega^{(2)})_\delta$. By the basic lemma, $\lambda$ is the generalized derivative of $\phi$ on $(\Omega^{(1)}+\Omega^{(2)})_\delta$. By virtue of the arbitrariness of $\delta$, $\lambda$ is the generalized derivative of $\phi$ on $\Omega^{(1)}+\Omega^{(2)}$.

**§6. Properties of integrals of potential type.**

1. INTEGRALS OF POTENTIAL TYPE. CONTINUITY. Suppose $f\in L_p$ $(p>1)$ on the unbounded space of $n$ variables, while $f=0$ outside of some bounded domain $\Omega$.

We construct the function

$$U(\vec{Q})=\int_{r\leqslant R} r^{-\lambda}f(\vec{P})\,dv_{\vec{P}}, \tag{6.1}$$

where $\lambda$ is a number,

$$0<\lambda<n, \quad r=|\vec{P}-\vec{Q}|=\sqrt{\sum_{i=1}^{n}(x_i-y_i)^2}, \quad \vec{P}(x_i),\ \vec{Q}(y_i);$$

$r\leqq R$ is a sphere including the domain $\Omega$ in its interior.

THEOREM 1. *If* $\lambda<n/p'$, $((1/p)+(1/p')=1)$, *then* $U(\vec{Q})$ *is unbounded and continuous and satisfies the inequality*

$$|U(\vec{Q})|<K\|f\|_{L_p}. \tag{6.2}$$

PROOF. We have $\lambda p'<n$. Therefore the integral $\int_{r\leqq h}r^{-\lambda p'}dv_{\vec{P}}$ converges and tends to zero as $h\to 0$. Using Hölder's inequality, we obtain:

$$\left|\int_{r\leqslant h} r^{-\lambda}f(\vec{P})\,dv_{\vec{P}}\right|\leqslant\left\{\int_{r\leqslant h}|f|^p dv\right\}^{\frac{1}{p}}\left\{\int_{r\leqslant h} r^{-\lambda p'}dv\right\}^{\frac{1}{p'}}\leqslant$$

$$\leqslant\|f\|_{L_p}\left\{\int_{r\leqslant h} r^{-\lambda p'}dv\right\}^{\frac{1}{p'}}<\varepsilon, \tag{6.3}$$

if $h$ is sufficiently small, independently of the position of the point $\vec{Q}$.

Since

$$U(\vec{Q})=\int_{h<r<R} r^{-\lambda}f(\vec{P})\,dv+\int_{r\leqslant h} r^{-\lambda}f(\vec{P})\,dv$$

and the first term on the right side is a continuous function of $\vec{Q}$, $U(\vec{Q})$ is continuous as the uniform limit of continuous functions. Applying the Hölder inequality to (6.1), just as in (6.3), we find

$$\left|U(\vec{Q})\right| \leqslant \|f\|_{L_p}\left\{\int_{r\leqslant R} r^{-\lambda p'} dv_{\vec{P}}\right\}^{\frac{1}{p'}} = K\|f\|_{L_p},$$

i.e., the estimate (6.2). The theorem is proved.

2. MEMBERSHIP IN $L_{q^*}$.

THEOREM 2. *Consider the hyperplane of s variables* $y_{s+1}=y_{s+2}=\cdots=y_n=0$ *and suppose that* $\vec{Q}^{(s)}(y_1,y_2,\cdots,y_s)$ *is a point of that hyperplane. Then, if* $\lambda \geqq (n/p')$ *and* $(s/p)>\lambda-(n/p')$, *i.e.,* $s>n-(n-\lambda)p$, *then* $U(\vec{Q}^{(s)})$ *is summable (on an arbitrary finite domain of the hyperplane) to the power* $q^*$, *where* $q^*<q$, $(s/q)=\lambda-(n/p')$, *i.e.,* $q=sp/(n-(n-\lambda)p)$, *and the inequality holds*

$$\|U(\vec{Q}^{(s)})\|_{L_{q^*}} \leqslant K_1\|f\|_{L_p}. \tag{6.4}$$

(In those cases where we shall simultaneously meet $L_p$ spaces on Euclidean spaces of a different number of variables, we shall often designate these spaces by subscripts in the form $L_{q^*,s}$.)[2)]

REMARK. It is probable that the number $q^*$ may be taken equal to $q$, but this has not yet been proved.

PROOF. From the definition of $q$, it follows that $q>p$.

Suppose that $q^*$ is some number satisfying the inequality

$$p<q^*<q.$$

We put $\lambda=(n/p')+(s/q^*)-2\epsilon$, where $\epsilon=(s/2)((1/q^*)-(1/q))>0$. From (6.1), we have

$$|U(\vec{Q})| \leqslant \int_{r\leqslant R} \left(|f|^{\frac{p}{q^*}} r^{-\frac{s}{q^*}+\varepsilon}\right)\left(|f|^{p\left(\frac{1}{p}-\frac{1}{q^*}\right)}\right)\left(r^{-\frac{n}{p'}+\varepsilon}\right) dv_{\vec{P}},$$

and applying the Hölder inequality for three factors and putting $\lambda_1=1/q^*$, $\lambda_2=(q^*-p)/q^*p=(1/p)-(1/q^*)$, $\lambda_3=1/p'$ (obviously $\lambda_1+\lambda_2+\lambda_3=1$), we obtain

$$|U(\vec{Q})| \leqslant \left\{\int_{r\leqslant R} |f|^p r^{-s+q^*\varepsilon} dv_{\vec{P}}\right\}^{\frac{1}{q^*}} \left\{\int_{r\leqslant R} |f|^p dv_{\vec{P}}\right\}^{\frac{1}{p}-\frac{1}{q^*}} \times$$
$$\times\left\{\int_{r\leqslant R} r^{-n+p'\varepsilon} dv_{\vec{P}}\right\}^{\frac{1}{p'}}.$$

Taking into account the fact that the integral $\int_{r\leqq R} r^{-n+p'\epsilon}dv$ converges

[2)]This result for the case $s=n$ was proved by S. L. Sobolev in somewhat stronger form for $q^*=q$. For $s<n$, S. L. Sobolev proved this theorem for $p=2$, $q^*=2$. The theorem proved in the text is due to V. I. Kondrašov.

(since $n-\epsilon p'<n$) and that $\int_{r\leq R}|f|^p dv=\|f\|^p_{L_p}$, we get

$$|U(\vec{Q})|\leqslant K_1\|f\|^{1-\frac{p}{q^*}}\left\{\int\limits_{r\leqslant R}|f|^p r^{-s+\varepsilon q^*}dv_{\vec{P}}\right\}^{\frac{1}{q^*}}. \tag{6.5}$$

Raising (6.5) to the $q^*$th power, integrating on the plane

$$y_{s+1}=y_{s+2}=\ \dots\ =y_n=0$$

and interchanging the order of integration, we find

$$\overbrace{\int\cdots\int}^{s}|U(\vec{Q}^{(s)})|^{q^*}dv_{\vec{Q}^{(s)}}\leqslant$$
$$\leqslant K_1^{q^*}\|f\|^{q^*-p}\overbrace{\int\limits_{r\leqslant R}\cdots\int}^{n}|f|^p\left[\overbrace{\int\cdots\int}^{s}r^{-s+\varepsilon q^*}dv_{\vec{Q}^{(s)}}\right]dv_{\vec{P}}. \tag{6.6}$$

We shall show that the integral

$$\overbrace{\int\cdots\int}^{s}r^{-s+\varepsilon q^*}dv_{\vec{Q}^{(s)}}$$

is bounded.

Indeed in polar coordinates in the plane of $\vec{Q}^{(s)}$, we will have:

$$r=\sqrt{\rho^2+h^2},\ dv_{\vec{Q}^{(s)}}=\rho^{s-1}d\rho\, d\omega^{(s)}, \tag{6.7}$$

where $h$ is the distance of the point $\vec{P}$ to the plane, $d\omega^{(s)}$ is the element of solid angle in the plane around the foot of the perpendicular dropped from $\vec{P}$ to that plane.

If $\kappa_s$ is the area of the sphere in $s$-dimensional space, then

$$\overbrace{\int\limits_{r\leqslant R}\cdots\int}^{s}r^{-s+\varepsilon q^*}dv_{\vec{Q}^{(s)}}=$$
$$=\varkappa_s\int\limits_0^{\sqrt{R^2-h^2}}(\sqrt{\rho^2+h^2})^{-s+\varepsilon q^*}\rho^{s-1}d\rho\leqslant\varkappa_s\int\limits_0^R\rho^{\varepsilon q^*-1}d\rho=K_2.$$

Substituting in (6.6) and denoting $K_1K_2^{1/q^*}$ by $K$, we find:

$$\|U(\vec{Q}^{(s)})\|_{L_{q^*}}=\left\{\overbrace{\int\cdots\int}^{s'}|U(\vec{Q}^{(s)})|^{q^*}dv_{\vec{Q}^{(s)}}\right\}^{\frac{1}{q^*}}<K\|f\|_{L_p},$$

which was to be proved.

Suppose $q_1<q^*$. Then:

$$\overbrace{\int \cdots \int}^{s} |U(\vec{Q}^{(s)})|^{q_1} dv_{\vec{Q}^{(s)}} \leqslant$$

$$\leqslant \Big[ \overbrace{\int \cdots \int}^{s} |U(\vec{Q}^{(s)})|^{q_1 \cdot \frac{q^*}{q_1}} dv_{\vec{Q}^{(s)}} \Big]^{\frac{q_1}{q^*}} \Big[ \overbrace{\int \cdots \int}^{s} dv_{\vec{Q}^{(s)}} \Big]^{\frac{q^*-q_1}{q^*}},$$

from which it follows that

$$\| U(\vec{Q}) \|_{L_{q_1}} \leqslant C \| U(\vec{Q}) \|_{L_{q^*}},$$

and consequently that the inequality (6.4) is valid for arbitrary $q_1$ with $1<q_1<q^*$. Consequently the condition that $q^*>p$ may be dropped and the theorem is completely proved.

REMARK 1. The constants $K$ in Theorems 1 and 2 depend exclusively upon the form of the domain and the constants $\lambda, p, s, n$, and $q^*$ but do not depend upon the function $f(\vec{P})$.

If the $s$-dimensional manifold is not a plane, then the theorem may be reduced to the preceding by a change of variables. One must assume that there exists a coordinate transformation which introduces only a finite distortion of distance (i.e., such that on bounded parts of the space, one can find constants $M>m>0$ such that $m<\rho/r<M$, where $r$ is the distance in the old system and $\rho$ in the new) and which carries the manifold under consideration into a plane.

REMARK 2. For the case $s=n$ and $R=\infty$, a much stronger theorem may be proved:

$$U(\vec{Q}) \in L_q, \quad \| U(\vec{Q}) \|_{L_q} \leqslant K \| f \|_{L_p}.$$ [3]

**§7. The spaces $L_p^{(l)}$ and $W_p^{(l)}$.**

1. DEFINITIONS. 1. The linear manifold of all summable functions $\phi(x_1,x_2,\cdots,x_n)$ having on a finite domain $\Omega$ all generalized derivatives of order $l$ summable to power $p>1$, will be called $W_p^{(l)}$:

$$\frac{\partial^l \varphi}{\partial x_1^{\alpha_1} \partial x_2^{\alpha_2} \cdots \partial x_n^{\alpha_n}} \in L_p \text{ in } \Omega; \quad \sum \alpha_i = l.$$

2. By $L_p^{(l)}$ we will mean the set, the elements of which are the classes of elements in $W_p^{(l)}$ having all derivatives of order $l$ the same, i.e., $\phi_1$ and $\phi_2$ will be said to lie in the same class in $L_p^{(l)}$ if

[3] S. L. Sobolev, *On a theorem of functional analysis*, Mat. Sb. 4(1938), no. 3.

$$\frac{\partial^l \varphi_1}{\partial x_1^{\alpha_1} \dots \partial x_n^{\alpha_n}} = \frac{\partial^l \varphi_2}{\partial x_1^{\alpha_1} \dots \partial x_n^{\alpha_n}} \quad (\sum \alpha_i = l) \text{ almost everywhere in } \Omega.$$

The elements of $L_p^{(l)}$ will be denoted by the letter $\psi$. Functions $\phi$ of the same class $\psi$ will be said to be mutually equivalent.

The elements of $L_p^{(l)}$ may be added and multiplied by constants. Thus $L_p^{(l)}$ becomes a vector space.

For the multiplication of the element $\psi$ by a constant, it suffices to multiply by the constant all the functions $\phi$ lying in the class $\psi$. It is not difficult to see that in this way we will obtain elements of one and the same class.

The class $\psi_1+\psi_2$ is obtained if we add in pairs all the elements from the classes $\psi_1$ and $\psi_2$. It is not hard to see that thereby there will be obtained elements of one and the same class.

2. THE NORM IN $L_p^{(l)}$. By the norm of an element $\psi$ in $L_p^{(l)}$ we mean the number

$$\|\psi\|_{L_p^{(l)}} = \left\{ \int_\Omega \left[ \sum_{i_1 \dots i_l = 1}^{n} \left| \frac{\partial^l \varphi}{\partial x_{i_1} \dots \partial x_{i_l}} \right|^2 \right]^{\frac{p}{2}} dv \right\}^{\frac{1}{p}} =$$

$$= \left\{ \left[ \int_\Omega \sum_{\sum \alpha_i = l} \frac{l!}{\alpha_1! \dots \alpha_n!} \left( \frac{\partial^l \varphi}{\partial x_1^{\alpha_1} \dots \partial x_n^{\alpha_n}} \right)^2 \right]^{\frac{p}{2}} dv \right\}^{\frac{1}{p}}. \tag{7.1}$$

In some cases, where it causes no confusion, we shall write $\|\phi\|_{L_p^{(l)}}$ for $\phi \in W_p^{(l)}$, meaning by this the norm of the class $\psi$ to which $\phi$ belongs.

We note some of the simplest properties of the norm:

I. $\|a\phi\| = |a| \cdot \|\phi\|$, if $a$ is a constant.

II. $\|\phi_1+\phi_2\| \leqq \|\phi_1\| + \|\phi_2\|$ (the triangle inequality).

III. If $\|\phi\| = 0$, then $\phi$ is equivalent to zero, i.e., it is given by a polynomial of order at most $l-1$.

IV. The norm is invariant under every orthogonal transformation of the space of $x_1, x_2, \cdots, x_n$.

The first property is completely obvious. We prove the validity of the triangle inequality, for which we employ the first representation for the norm. Then using the fact that (by the Minkowski inequality for $p=2$):

$$\sqrt{\sum \left( \frac{\partial^l \varphi_1}{\partial x_{i_1} \dots \partial x_{i_l}} + \frac{\partial^l \varphi_2}{\partial x_{i_1} \dots \partial x_{i_l}} \right)^2} \leqslant \sqrt{\sum \left( \frac{\partial^l \varphi_1}{\partial x_{i_1} \dots \partial x_{i_l}} \right)^2} + \\ + \sqrt{\sum \left( \frac{\partial^l \varphi_2}{\partial x_{i_1} \dots \partial x_{i_l}} \right)^2}$$

we will have:

$$\|\varphi_1+\varphi_2\|_{L_p^{(l)}}\leqslant\Big\{\int_\Omega\Big[\sqrt{\sum\Big(\frac{\partial^l\varphi_1}{\partial x_{i_1}\ldots\partial x_{i_l}}\Big)^2}+$$

$$+\sqrt{\sum\Big(\frac{\partial^l\varphi_2}{\partial x_{i_1}\ldots\partial x_{i_l}}\Big)^2}\Big]^{\frac{p}{2}}dv\Big\}^{\frac{1}{p}}.$$

Applying the Minkowski inequality once more, we obtain

$$\|\varphi_1+\varphi_2\|_{L_p^{(l)}}\leqslant\Big\{\int_\Omega\Big[\sum\Big(\frac{\partial^l\varphi_1}{\partial x_{i_1}\ldots\partial x_{i_l}}\Big)^2\Big]^{\frac{p}{2}}dv\Big\}^{\frac{1}{p}}+$$

$$+\Big\{\int_\Omega\Big[\sum\Big(\frac{\partial^l\varphi_2}{\partial x_{i_1}\ldots\partial x_{i_l}}\Big)^2\Big]^{\frac{p}{2}}dv\Big\}^{\frac{1}{p}}=\|\varphi_1\|_{L_p^{(l)}}+\|\varphi_2\|_{L_p^{(l)}}.$$

The third property of the norm follows from the fact that if all the generalized derivatives of $l$th order of some function $\phi$ are equal to zero, then all the generalized derivatives of order $l$ of the averaged function $\phi_n$ are equal to zero. This means that the averaged function is a polynomial of degree at most $l-1$. However, the limit of a sequence of polynomials of degree at most $l-1$ can only be a polynomial of degree at most $l-1$, and it follows that $\phi$ is a polynomial of degree at most $l-1$.

The invariance of the norm under orthogonal transformations of coordinates follows easily from the first representation of the norm, since the expression

$$\sum_{i_1,i_2,\ldots,i_l=1}^{n}\Big(\frac{\partial^l\varphi}{\partial x_{i_1}\partial x_{i_2}\ldots\partial x_{i_l}}\Big)^2$$

is one of the invariants of the tensor

$$\frac{\partial^l\varphi}{\partial x_{i_1}\partial x_{i_2}\ldots\partial x_{i_l}}=\nabla_{i_1i_2\ldots i_l}\varphi.$$

We note also two inequalities:

$$\|\varphi\|_{L_p^{(l)}}\leqslant K_1\max_{\alpha_1,\ldots\alpha_n}\Big\|\frac{\partial^l\varphi}{\partial x_1^{\alpha_1}\ldots\partial x_n^{\alpha_n}}\Big\|_{L_p}\tag{7.2}$$

$$\Big\|\frac{\partial^l\varphi}{\partial x_1^{\alpha_1}\ldots\partial x_n^{\alpha_n}}\Big\|_{L_p}\leqslant\|\varphi\|_{L_p^{(l)}}.\tag{7.3}$$

In fact, the inequality (7.3) is obvious. The inequality (7.2) follows from the fact that

$$\sum_{i_1,\ldots,i_l=1}^{n}\Big(\frac{\partial^l\varphi}{\partial x_{i_1}\ldots\partial x_{i_l}}\Big)^2\leqslant\Big[\sum_{i_1,\ldots,i_l=1}^{n}\Big|\frac{\partial^l\varphi}{\partial x_{i_1}\ldots\partial x_{i_l}}\Big|\Big]^2.$$

Applying the Minkowski inequality to the first part of the last inequality, we obtain:

$$\|\varphi\|_{L_p^{(l)}} \leqslant \sum_{i_1, \ldots i_l = 1}^{n} \left\| \frac{\partial^l \varphi}{\partial x_{i_1} \ldots \partial x_{i_l}} \right\|_{L_p} \leqslant K_1 \max_{i_1, \ldots, i_l} \left\| \frac{\partial^l \varphi}{\partial x_{i_1} \ldots \partial x_{i_l}} \right\|_{L_p},$$

where $K_1$ is the number of terms in the last sum. Thus (7.2) is proved. After the introduction of the norm, the manifold $L_p^{(l)}$ becomes a functional space. Later we shall introduce a norm in $W_p^{(l)}$. Therefore we may thenceforward consider $W_p^{(l)}$ as a functional space.

3. DECOMPOSITIONS OF $W_p^{(l)}$ AND ITS NORMING. We consider also the space $S_l$ of all polynomials of degree at most $l-1$. This space can be considered, obviously, as a subspace of the space $W_p^{(l)}$. The space $L_p^{(l)}$ appears, speaking algebraically, as the factor-space of $W_p^{(l)}$ by $S_l$.

By a projection operator in the space $W_p^{(l)}$ is meant an operator whose square coincides with itself

$$\Pi^2 \varphi = \Pi \Pi \varphi = \Pi \varphi.$$

If some projection operator $\prod_1$ carries the space $W_p^{(l)}$ on the whole space $S_l$, then it will be the identity operator on $S_l$.

With the aid of each such a projection operator it is easy to construct a decomposition of the space $W_p^{(l)}$. We put:

$$\Pi_1^* \varphi = \varphi - \Pi \varphi,$$

i.e.,

$$\Pi_1^* = E - \Pi,$$

where $E$ is the identity operator.

The operator $\prod_1^*$ will be in its turn a projection. In fact

$$\Pi_1^* \Pi_1^* = (E - \Pi_1)(E - \Pi_1) =$$

$$= E - 2\Pi_1 + \Pi_1^2 = E - \Pi_1 = \Pi_1^*,$$

which was to be proved.

An arbitrary element $\phi$ of $W_p^{(l)}$ may be represented in the form

$$\varphi = \Pi_1 \varphi + \Pi_1^* \varphi.$$

The elements of the form $\phi^* = \prod_1^* \phi$ constitute a subspace of the vector space $W_p^{(l)}$, since the sum of two such elements $\prod_1^* \phi_1 + \prod_1^* \phi_2$ may be put in the form $\prod_1^*(\phi_1 + \phi_2)$ and thus again lies in this space.

It is not hard to see that the space of elements of the form $\prod_1^* \phi$ is isomorphic to the space $L_p^{(l)}$, because each class in $L_p^{(l)}$ will correspond to only one element of the form $\prod_1^* \phi$. This follows from the fact that $\prod_1^* \phi = 0$

if $\phi \in S_l$. The sum of elements of the form $\prod_1^* \phi$ corresponds to the sum of classes in $L_p^{(l)}$ and conversely. Analogously, multiplication by a constant of corresponding elements leads again to corresponding elements.

The space $S_l$ can be normed, as can every finite dimensional space. It is convenient to define a norm on it in the following manner. Suppose that $P$ is a polynomial of degree less than $l$ having the form:

$$P = \sum_{k=0}^{l-1} \sum_{\sum \alpha_s = k} a_{\alpha_1 \alpha_2 \ldots \alpha_n} x_1^{\alpha_1} \ldots x_n^{\alpha_n}.$$

Then we put:

$$\|P\|_{S_l}^p = \sum_{k=0}^{l-1} \Big\{ \sum_{\sum \alpha_s = k} a_{\alpha_1 \ldots \alpha_n}^2 \frac{k!}{\alpha_1! \ldots \alpha_n!} \Big\}^{\frac{p}{2}}. \tag{7.4}$$

The definition of the norm in this way will be invariant under all rotations of axes of coordinates, while in distinction from the norm on $L_p^{(l)}$ it will not be invariant under the translation of the origin. Indeed, the quantity $\sum k!/(\alpha_1! \cdots \alpha_n!)\ a^2_{\alpha_1 \cdots \alpha_n}$ is one of the invariants of the tensor $a_{\alpha_1 \ldots \alpha_n}$ and therefore this quantity is preserved under orthogonal transformations.

We may verify that again for such a norm all the three properties hold:

(a) $\|P_1 + P_2\| \leqq \|P_1\| + \|P_2\|$,

(b) $\|aP\| = |a| \cdot \|P\|$,

(c) If $\|P\| = 0$, then $P = 0$.

The validity of the conditions (b) and (c) is obvious. We establish also the triangle inequality.

Suppose

$$P_1 = \sum_{k=0}^{l-1} \sum_{\sum \alpha_s = k} a^{(1)}_{\alpha_1 \ldots \alpha_n} x_1^{\alpha_1} \ldots x_n^{\alpha_n};$$

$$P_2 = \sum_{k=0}^{l-1} \sum_{\sum \alpha_s = k} a^{(2)}_{\alpha_1 \ldots \alpha_n} x_1^{\alpha_1} \ldots x_n^{\alpha_n}.$$

We have

$$\|P_1 + P_2\|^p = \sum_{k=0}^{l-1} \Big\{ \sum_{\sum \alpha = k} \frac{k!}{\alpha_1! \ldots \alpha_n!} [a^{(1)}_{\alpha_1 \ldots \alpha_n} + a^{(2)}_{\alpha_1 \ldots \alpha_n}]^2 \Big\}^{\frac{p}{2}} \leqslant$$

$$\leqslant \sum_{k=0}^{l-1} \Big\{ \Big[ \sum_{\sum \alpha = k}^{*} a^{(1)2}_{\alpha_1 \ldots \alpha_n} \Big]^{\frac{1}{2}} + \Big[ \sum_{\sum \alpha = k}^{*} a^{(2)2}_{\alpha_1 \ldots \alpha_n} \Big]^{\frac{1}{2}} \Big\}^p;$$

$$\|P_1 + P_2\| \leqslant \Big\{ \Big\{ \sum_{k=0}^{l-1} \Big\{ \Big[ \sum_{\sum \alpha = k}^{*} a^{(1)2}_{\alpha_1 \ldots \alpha_n} \Big]^{\frac{1}{2}} + \Big[ \sum_{\sum \alpha = k}^{*} a^{(2)2}_{\alpha_1 \ldots \alpha_n} \Big]^{\frac{1}{2}} \Big\}^p \Big\} \Big\}^{\frac{1}{p}} \leqslant$$

$$\leqslant \Big\{\sum_{k=0}^{l-1}\Big[\sum_{\sum\alpha=k}^{*} a^{(1)2}_{\alpha_1\ldots\alpha_n}\Big]^{\frac{p}{2}}\Big\}^{\frac{1}{p}} + \Big\{\sum_{k=0}^{l-1}\Big[\sum_{\sum\alpha=k}^{*} a^{(2)2}_{\alpha_1\ldots\alpha_n}\Big]^{\frac{p}{2}}\Big\}^{\frac{1}{p}},$$

which was to be proved (here $\sum^*$ denotes summations with the weights $k!/\alpha_1!\cdots\alpha_n!$).

The establishment of a norm on $S_l$ enables us to carry through also the norming of the space $W_p^{(l)}$. This norming may be carried out if we are given any projection operator whatever.

Naturally, we put:

$$\|\varphi\|^p_{W_p^{(l)}} = \|\textstyle\prod_1\varphi\|^p_{S_l} + \|\textstyle\prod_1^*\varphi\|^p_{L_p^{(l)}} = \|\textstyle\prod_1\varphi\|^p_{S_l} + \|\varphi\|^p_{L_p^{(l)}}. \qquad (7.5)$$

This method of introducing a norm depends upon the given projection operator. Later we consider the question of what relations will hold between norms constructed with the aid of different projection operators. In the meantime, it is necessary for us to verify that the three basic properties of the norm are satisfied for our definition.

In fact, it is obvious that:

$$\|a\varphi\|_{W_p^{(l)}} = |a|\cdot\|\varphi\|_{W_p^{(l)}}.$$

Furthermore, if $\|\phi\|_{W_p^{(l)}}=0$, then $\phi=0$.

The triangle inequality follows in an obvious way from the Minkowski inequality.

4. SPECIAL DECOMPOSITIONS OF $W_p^{(l)}$. It is convenient for our purposes to use one special form for the operator $\prod_1$. Let us concern ourselves with this form. To begin with, we impose some restrictions upon the domain in space on which we consider our functions.

Let $\Omega$ be a star domain with respect to each point of the sphere $C$ of radius $H$ lying within $\Omega$. For convenience, we assume to begin with that the center of this sphere lies at the origin of coordinates.

Let $\vec{P}$ and $\vec{Q}$ be two arbitrary points of $\Omega$. We set $r=|\vec{P}-\vec{Q}|$ and let $\vec{l}=(\vec{Q}-\vec{P})/r$ be the unit vector having the direction from $\vec{P}$ to $\vec{Q}$. Each function of two variable points $\mu(\vec{Q},\vec{P})$ may be represented as a function of $\vec{P}$, $\vec{l}$, and $r$, setting $\vec{Q}=\vec{P}+r\vec{l}$, and

$$\mu(\vec{Q},\vec{P}) = \mu(\vec{P}+r\vec{l},\vec{P}) = \bar{\mu}(r,\vec{l},\vec{P}),$$

where the bar over $\mu$ indicates that $\vec{Q}$ is replaced by $\vec{P}$, $r$, $\vec{l}$. Conversely every function $\bar{v}(r,\vec{l},\vec{P})$ may be considered as a function of $\vec{Q}$ and $\vec{P}$. We consider the function

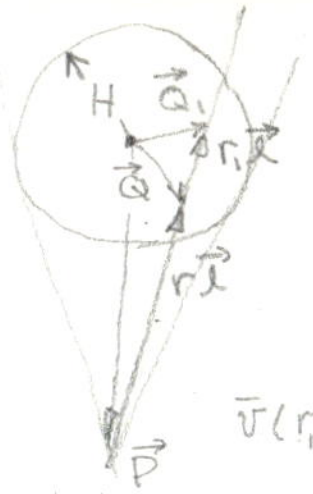

$$v(\vec{Q}) = \begin{cases} e^{\frac{R^2}{R^2 - H^2}} & \text{for } R < H; \\ 0 & \text{for } R \geqslant H, \end{cases}$$

where $R$ is the distance of the point $Q$ from the origin of coordinates.

$v(\vec{Q})$ is continuous with its derivatives of all orders and differs from zero only in the sphere $C$.

We form a new function of the two points $\vec{P}$ and $\vec{Q}$, setting

$$\bar{\chi}(r, \vec{l}, \vec{P}) = -\int_r^{\infty} \bar{v}(r_1, \vec{l}, \vec{P})\, r_1^{n-1}\, dr_1. \tag{7.6}$$

The integral, obviously, reduces to an integral over a finite interval since $\bar{v}$ differs from zero only on a bounded domain.

For fixed $\vec{P}$, the function $\bar{\chi}(r, \vec{l}, \vec{P})$ differs from zero only for those $r$ and $\vec{l}$ for which $\vec{Q} = \vec{P} + r\vec{l}$ lies in the interior of the domain bounded by the cone with vertex at $\vec{P}$ whose generators terminate on the sphere $C$ (see Figure 5). Indeed, on all rays which do not intersect the sphere $C$, the function $\bar{v}$ is identically zero, and on those rays which do intersect $C$, for points $\vec{Q}$ lying beyond the sphere $C$, the integrand in (7.6) is also null for $r_1 > r$. It is obvious that $\bar{\chi}(r, \vec{l}, \vec{P})$ is continuously differentiable. We introduce also a function:

$$\bar{\psi}(r, \vec{l}, \vec{P}) = \frac{1}{(l-1)!}\, r^{l-1}\, \bar{\chi}(r, \vec{l}, \vec{P}).$$

For any function $\phi(\vec{Q})$ continuously differentiable up to order $l$ on the domain $\Omega$, we may construct a corresponding function $\bar{\Phi}$ by the formula

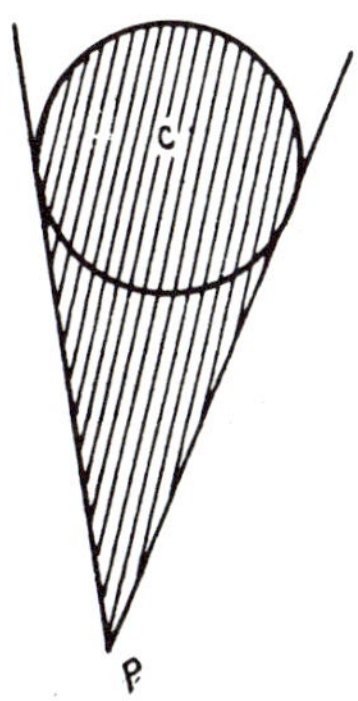

Figure 5

$$\bar{\Phi} = \bar{\varphi}\, \frac{\partial^{l-1}\bar{\psi}}{\partial r^{l-1}} - \frac{\partial \bar{\varphi}}{\partial r}\, \frac{\partial^{l-2}\bar{\psi}}{\partial r^{l-2}} + \cdots + (-1)^{l-1} \frac{\partial^{l-1}\bar{\varphi}}{\partial r^{l-1}}\, \bar{\psi}.$$

Obviously, we have

$$\frac{\partial \bar{\Phi}}{\partial r} = \bar{\varphi}\, \frac{\partial^{l}\bar{\psi}}{\partial r^{l}} + (-1)^{l-1} \frac{\partial^{l}\bar{\varphi}}{\partial r^{l}}\, \bar{\psi}. \tag{7.7}$$

In addition:

$$\bar{\psi}\Big|_{r=0} = \frac{\partial \bar{\psi}}{\partial r}\Big|_{r=0} = \cdots = \frac{\partial^{l-2}\bar{\psi}}{\partial r^{l-2}}\Big|_{r=0} = 0.$$

Calculating $\partial^{l-1}\bar{\psi}/\partial r^{l-1}$, we obtain:

$$\left.\frac{\partial^{l-1}\overline{\psi}}{\partial r^{l-1}}\right|_{r=0}=\overline{\chi}(0,\vec{l},\vec{P})=-\int_0^\infty \overline{v}(r_1,\vec{l},\vec{P})\,r_1^{n-1}\,dr_1,$$

from which it follows that

$$\overline{\Phi}(0,\vec{l},\vec{P})=-\varphi(\vec{P})\int_0^\infty \overline{v}(r_1,\vec{l},\vec{P})\,r_1^{n-1}\,dr_1;\quad \overline{\Phi}(\infty,\vec{l},\vec{P})=0.$$

Integrating (7.7) in $r$ from 0 to $\infty$, we find:

$$\varphi(\vec{P})\int_0^\infty \overline{v}(r_1,\vec{l},\vec{P})\,r_1^{n-1}\,dr_1=\int_0^\infty\left[\overline{\varphi}\,\frac{\partial^l\overline{\psi}}{\partial r^l}+(-1)^{l-1}\frac{\partial^l\overline{\varphi}}{\partial r^l}\,\overline{\psi}\right]dr. \qquad (7.8)$$

Multiplying (7.8) by the element of solid angle $d\omega_{\vec{l}}$ and integrating over the unit sphere, we obtain:

$$\varphi(\vec{P})\int_{\omega_{\vec{l}}} d\omega_{\vec{l}}\int_0^\infty \overline{v}(r_1,\vec{l},\vec{P})\,r_1^{n-1}\,dr_1=$$

$$=\int_{\omega_{\vec{l}}} d\omega_{\vec{l}}\int_0^\infty\left[\overline{\varphi}\,\frac{\partial^l\overline{\psi}}{\partial r^l}+(-1)^{l-1}\frac{\partial^l\overline{\varphi}}{\partial r^l}\,\overline{\psi}\right]dr.$$

Taking into consideration that $r_1^{n-1}dr_1d\omega_{\vec{l}}=dv_{\vec{Q}}$, where $dv_{\vec{Q}}$ is the volume element at the point $\vec{Q}$, we find:

$$\int_{\omega_{\vec{l}}} d\omega_{\vec{l}}\int_0^\infty \overline{v}(r_1,\vec{l},\vec{P})\,r_1^{n-1}\,dr_1=\int_{R\leqslant H}\overline{v}(\vec{Q})\,dv_{\vec{Q}}=\varkappa\neq 0,$$

i.e., the size of this integral does not depend upon the position of the point $\vec{P}$. In this fashion, we obtain

$$\varphi(\vec{P})=\frac{1}{\varkappa}\int_{\omega_{\vec{l}}} d\omega_{\vec{l}}\int_0^\infty\left[\overline{\varphi}\,\frac{\partial^l\overline{\psi}}{\partial r^l}+(-1)^{l-1}\frac{\partial^l\overline{\varphi}}{\partial r}\,\overline{\psi}\right]dr$$

or, introducing $dv_{\vec{Q}}$ into the integral on the right, we obtain:

$$\varphi(\vec{P})=\frac{1}{\varkappa}\int_\Omega \overline{\varphi}\,\frac{\partial^l\overline{\psi}}{\partial r^l}\,\frac{1}{r^{n-1}}\,dv_{\vec{Q}}+\frac{(-1)^{l-1}}{\varkappa}\int_\Omega \overline{\psi}\,\frac{\partial^l\overline{\varphi}}{\partial r^l}\,\frac{1}{r^{n-1}}\,dv_{\vec{Q}}. \qquad (7.9)$$

We now show that the first integral in (7.9) is a polynomial of degree $l-1$ in the coordinates $x_1,x_2,\cdots,x_n$ of the point $\vec{P}$.

In fact, from the definition of $\overline{\psi}$, it follows that

$$\frac{\partial^l \bar{\psi}}{\partial r^l} = -\sum_{k=1}^{l} B_k \frac{\partial^{l-k} r^{l-1}}{\partial r^{l-k}} \frac{\partial^k}{\partial r^k} \int_r^\infty \bar{v}(r_1, \vec{l}, \vec{P}) r_1^{n-1} dr_1 =$$

$$= \sum_{k=1}^{l} C_k r^{k-1} \frac{\partial^{k-1}}{\partial r^{k-1}} (r^{n-1} \bar{v}(r, \vec{l}, \vec{P})) =$$

$$= \sum_{k=1}^{l} \sum_{s=1}^{l} D_{k,s} r^{k-1} r^{n-k+s} \frac{\partial^s \bar{v}(r, \vec{l}, \vec{P})}{\partial r^l} =$$

$$= \sum_{k=1}^{l} \sum_{s=0}^{k-1} D_{k,s} r^{n-1+s} \frac{\partial^s \bar{v}(r, \vec{l}, \vec{P})}{\partial r^s}, \tag{7.10}$$

where $B_k$, $C_k$, and $D_{k,s}$ are constants which are binomial coefficients or combinations of binomial coefficients.

But $\bar{v}(r, \vec{l}, \vec{P}) = v(\vec{Q})$, with the substitution $\vec{Q} = \vec{P} + r\vec{l}$, $\vec{Q}(y_1, y_2, \cdots, y_n)$. Therefore

$$\frac{\partial \bar{v}}{\partial r} = \sum_{j=1}^{n} \frac{\partial v}{\partial y_j} l_j = \sum_{j=1}^{n} \frac{\partial v}{\partial y_j} \frac{y_j - x_j}{r},$$

where $l_j = (y_j - x_j)/r$ are the components of $\vec{l}$, and we find analogously:

$$\frac{\partial^s \bar{v}}{\partial r^s} = \sum_{i_1, i_2, \ldots, i_s = 1}^{n} \frac{\partial^s v}{\partial y_{i_1} \partial y_{i_2} \ldots \partial y_{i_s}} l_{i_1} l_{i_2} \ldots l_{i_s} =$$

$$= \frac{1}{r^s} \sum_{\Sigma \alpha_i = s} C_{\alpha_1 \ldots \alpha_n} \frac{\partial^s v}{\partial y_1^{\alpha_1} \ldots \partial y_n^{\alpha_n}} (y_1 - x_1)^{\alpha_1} \ldots (y_n - x_n)^{\alpha_n}.$$

Substituting in (7.10), we find:

$$\frac{\partial^l \bar{\psi}}{\partial r^l} = r^{n-1} \sum_{\Sigma \alpha_i \leq l-1} x_1^{\alpha_1} x_2^{\alpha_2} \ldots x_n^{\alpha_n} \cdot \zeta_{\alpha_1 \ldots \alpha_n}(y_1, y_2, \ldots, y_n),$$

where $\zeta_{\alpha_1 \cdots \alpha_n}(y_1, \cdots, y_n)$ are bounded and continuous functions of $(y_1, \cdots, y_n)$:

$$\zeta_{\alpha_1 \ldots \alpha_n}^{(y_1, \ldots, y_n)} = \sum_{\substack{\alpha_i \leq \beta_i \\ \Sigma \beta_i l - 1}} C_{\alpha_1, \ldots, \alpha_n}^{\beta_1, \ldots, \beta_n} \frac{\partial^{\beta_1 + \ldots + \beta_n} v}{\partial y_1^{\beta_1} \ldots \partial y_n^{\beta_n}} y_1^{\beta_1 - \alpha_1} \ldots y_n^{\beta_n - \alpha_n}.$$

Therefore the first integral in (7.9) takes the form:

$$\frac{1}{\varkappa}\int_{\Omega}\bar{\varphi}\,\frac{\partial^l\bar{\psi}}{\partial r^l}\,\frac{1}{r^{n-1}}\,dv_{\vec{Q}}=\sum_{\Sigma\alpha_i\leqslant l-1}x_1^{\alpha_1}\ldots x_n^{\alpha_n}\frac{1}{\varkappa}\int_{\Omega}\varphi(\vec{Q})\,\zeta_{\alpha_1\ldots\alpha_n}(\vec{Q})\,dv_{\vec{Q}},$$

i.e., it represents a polynomial of degree $l-1$ in $x_1,x_2,\cdots,x_n$. We note that $\zeta_{\alpha_1,\cdots,\alpha_n}(\vec{Q})\equiv 0$ outside of the sphere $C$, since this property holds for $v(\vec{Q})$ and all of its derivatives.

The operator

$$\prod\nolimits_1\varphi=\frac{1}{\varkappa}\int_{\Omega}\bar{\varphi}\,\frac{\partial^l\bar{\psi}}{\partial r^l}\,\frac{dv_{\vec{Q}}}{r^{n-1}}$$

has the property that it carries an arbitrary polynomial of degree not higher than $l-1$ into itself. This follows easily from the fact that if one substitutes such a polynomial in (7.9), the second term vanishes. Consequently $\prod_1\phi$ is a projection operator and formula (7.9) gives the decomposition of interest to us.

We turn now to the investigation of the second integral in (7.9), i.e., to the study of the operator $\prod_1\phi$. For this purpose, we note that

$$\frac{\partial^l\bar{\varphi}}{\partial r^l}=\sum_{\Sigma\alpha_i=l}C_{\alpha_1,\ldots,\alpha_n}\frac{(y_1-x_1)^{\alpha_1}\ldots(y_n-x_n)^{\alpha_n}}{r^l}\,\frac{\partial^l\varphi}{\partial y_1^{\alpha_1}\ldots\partial y_n^{\alpha_n}}.$$

This last formula is obtained like the one for $\partial^s\bar{v}/\partial r^s$.

Furthermore, it follows from this that

$$\frac{1}{r^{n-1}}\,\frac{\partial^l\bar{\varphi}}{\partial r^l}\bar{\psi}=\frac{1}{r^{n-1+l}}\sum_{\Sigma\alpha_i=l}C_{\alpha_1,\ldots,\alpha_n}(y_1-x_1)^{\alpha_1}\ldots(y_n-x_n)^{\alpha_n}\bar{\psi}\times$$

$$\times\frac{\partial^l\varphi}{\partial y_1^{\alpha_1}\ldots\partial y_n^{\alpha_n}}=\sum_{\Sigma\alpha_i=l}\mu_{\alpha_1,\ldots,\alpha_n}(\vec{Q},\vec{P})\,\frac{\partial^l\varphi}{\partial y_1^{\alpha_1}\ldots\partial y_n^{\alpha_n}}.$$

We have

$$\mu_{\alpha_1,\ldots,\alpha_n}(\vec{Q},\vec{P})=C_{\alpha_1,\ldots,\alpha_n}\frac{(y_1-x_1)^{\alpha_1}\ldots(y_n-x_n)^{\alpha_n}}{r^{n-1+l}}\bar{\psi}=$$

$$=C_{\alpha_1\ldots\alpha_n}\frac{l_1^{\alpha_1}l_2^{\alpha_2}\ldots l_n^{\alpha_n}}{r^{n-1}}\bar{\psi}=$$

$$=\frac{1}{r^{n-l}}[(C^1_{\alpha_1,\ldots,\alpha_n}l_1^{\alpha_1}l_2^{\alpha_2}\ldots l_n^{\alpha_n}\bar{\chi}(r,\vec{l},\vec{P})]=$$

$$=\frac{1}{r^{n-l}}\,\bar{w}_{\alpha_1,\ldots,\alpha_n}(r,\vec{l},\vec{P}),$$

where $\bar{w}_{\alpha_1,\dots,\alpha_n}(r, \vec{l}, \vec{P})$ is a bounded and infinitely differentiable function of its arguments. Considered as a function of $(\vec{Q},\vec{P})$, it appears as a bounded function of its argument. Then the second integral in (7.9) takes the form:

$$\frac{(-1)^{l-1}}{\varkappa}\int_{\Omega}\frac{1}{r^{n-l}}\sum_{\Sigma\alpha_i=l} w_{\alpha_1,\dots,\alpha_n}(\vec{Q},\vec{P})\frac{\partial^l\varphi}{\partial x_1^{\alpha_1}\partial x_2^{\alpha_2}\dots\partial x_n^{\alpha_n}}\,dv_{\vec{Q}}. \quad (7.11)$$

Introducing the coefficients $1/\kappa$ and $(-1)^{l-1}/\kappa$ into the functions $\zeta$ and $w$, we may rewrite (7.9) in the form:

$$\varphi(\vec{P}) = \sum_{\Sigma\alpha_i\leq l-1} x_1^{\alpha_1}x_2^{\alpha_2}\dots x_n^{\alpha_n}\int_C \zeta_{\alpha_1\dots\alpha_n}(\vec{Q})\,\varphi(\vec{Q})\,dv_{\vec{Q}} +$$

$$+\int_{\Omega}\frac{1}{r^{n-l}}\sum_{\Sigma\alpha_i=l} w_{\alpha_1\dots\alpha_n}(\vec{Q},\vec{P})\frac{\partial^l\varphi}{\partial y_1^{\alpha_1}\dots\partial y_n^{\alpha_n}}\,dv_{\vec{Q}}. \quad (7.12)$$

The formula (7.12) was established by us for functions having continuous derivatives to the $l$th order. It is not hard to see that it remains correct for an arbitrary function from $W_p^{(l)}$. Indeed, let $\phi\in W_p^{(l)}$ and let $\phi_h$ be its averaged function. For $\phi_h$, (7.12) is correct, i.e., we have

$$\varphi_h(\vec{P}) = \sum_{\Sigma\alpha_i\leq l-1} x_1^{\alpha_1}\dots x_n^{\alpha_n}\int_C \zeta_{\alpha_1,\dots,\alpha_n}(\vec{Q})\,\varphi_h(\vec{Q})\,dv_{\vec{Q}} +$$

$$+\int_{\Omega}\frac{1}{r^{n-l}}\sum_{\Sigma\alpha_i=l} w_{\alpha_1,\dots,\alpha_n}\frac{\partial^l\varphi_h}{\partial y_1^{\alpha_1}\dots\partial y_n^{\alpha_n}}\,dv_{\vec{Q}} = S^{(h)}+\varphi^{*(h)}. \quad (7.13)$$

We set

$$S = \sum_{\Sigma\alpha_i\leq l-1} x_1^{\alpha_1}\dots x_n^{\alpha_n}\int_C \zeta_{\alpha_1,\dots,\alpha_n}(\vec{Q})\,\varphi(\vec{Q})\,dv_{\vec{Q}};$$

$$\varphi^*(\vec{P}) = \int_{\Omega}\frac{1}{r^{n-l}}\sum_{\Sigma\alpha_i=l} w_{\alpha_1\dots\alpha_n}(\vec{Q},\vec{P})\frac{\partial^l\varphi}{\partial y_1^{\alpha_1}\dots\partial y_n^{\alpha_n}}\,dv_{\vec{Q}}.$$

It is obvious that $\int_C \zeta_{\alpha_1,\dots,\alpha_n}\phi_h dv \to \int_C \zeta_{\alpha_1,\dots,\alpha_n}\phi\, dv$, and consequently, $S^{(h)}\to S$ uniformly. In addition, by the basic property of the derivatives of the averaged functions, we have

$$\left\| \frac{\partial^l \varphi_h}{\partial y_1^{\alpha_1} \dots \partial y_n^{\alpha_n}} - \frac{\partial^l \varphi}{\partial y_1^{\alpha_1} \dots \partial y_n^{\alpha_n}} \right\|_{L_p} \to 0 \quad \text{for} \quad h \to 0.$$

To the integral $\phi^*$ we apply the theorem on integrals of potential type. For this, $\lambda = n - l$.

Consequently, if $\lambda = n - l < n(1-(1/p))$, i.e., if $lp > n$, then on the basis of Theorem 1 and estimate (6.2) §6, we conclude that $\phi^*$ is continuous and besides that $\phi^{*(h)} \to \phi^*$ uniformly. In this case, $\phi$ may be considered as a continuous function and then the equality (7.12) holds for all the points of $\Omega$.

Using Theorem 2 and estimating (6.4) §6, we conclude: if $\lambda = n - l \geqq n(1-(1/p))$, i.e., if $lp < n$ and if $s > n - (n-\lambda)p$, i.e., if $s > n - lp$, then $\phi^* \in L_{q^*}(q^* < sp/(n-lp))$ on every $s$ dimensional manifold. In addition, $\phi^{*(h)} \Rightarrow \phi^*$ in $L_{q^*}$ on every $s$-dimensional manifold.

In this case $\phi \in L_{q^*}$ on the $s$-dimensional manifolds, and formula (7.12) holds almost everywhere.

An important property of normed functional spaces is completeness, i.e., the existence of a limit for every Cauchy sequence. The space $W_p^{(l)}$ with the norms which we have introduced is complete. We shall establish this somewhat later. Now we shall establish some important theorems.

## §8. **Imbedding Theorems.**

1. THE IMBEDDING OF $W_p^{(l)}$ IN $C$.

THEOREM 1. *If $\phi \in W_p^{(l)}$ and $n < lp$, then $\phi$ is a continuous function. Denoting the space of continuous functions as usual by $C$, and letting $\|\phi\|_C = \max_\Omega |\phi|$, we will have*

$$\|\varphi\|_C \leqslant M \|\varphi\|_{W_p^{(l)}}, \tag{8.1}$$

*where $M$ is a constant independent of the choice of the function $\phi$.*

Indeed, the continuity of $\phi$ was already proved in the case $n < lp$. Further, from (7.12) we conclude that:

$$|\varphi| \leqslant |S| + |\varphi^*|.$$

Setting $\max |x_1^{\alpha_1} \cdots x_n^{\alpha_n}| = A$ $(x_i \in \Omega,\ \sum \alpha_i \leqq l-1)$, applying the Minkowski inequality, and letting $N$ be the number of different monomials $(x_1^{\alpha_1} \cdots x_n^{\alpha_n})$, $\sum \alpha_i \leqq l-1$, we obtain:

$$|S| < A \sum_{\nu=0}^{l-1} \sum_{\sum \alpha_i = \nu} |a_{\alpha_1 \ldots \alpha_n}| \leqslant A \sum_{\nu=0}^{l-1} \sum_{\sum \alpha_i = \nu} \left( \frac{\nu!}{\alpha_1! \ldots \alpha_n!} \right)^{\frac{1}{2}} |a_{\alpha_1 \ldots \alpha_n}| \leqslant$$

$$\leqslant A \sqrt{N} \left\{ \sum_{\nu=0}^{l-1} \sum_{\sum \alpha_i = \nu} \frac{\nu!}{\alpha_1! \ldots \alpha_n!} |a_{\alpha_1 \ldots \alpha_n}|^2 \right\}^{\frac{1}{2}} < K' \|\varphi\|_{W_p^{(l)}}.$$

Applying the inequality (6.2), we find an estimate for $|\phi^*|$; we set $\max|w_{\alpha_1 \cdots \alpha_n}| = B$ $(\vec{P} \in \Omega, \vec{Q} \in \Omega, \sum \alpha_i = l)$:

$$|\varphi^*| < KB \left\| \sum_{\sum \alpha_i = l} \left| \frac{\partial^l \varphi}{\partial x_1^{\alpha_1} \ldots \partial x_n^{\alpha_n}} \right| \right\|_{L_p} \leqslant KB \sum_{\sum \alpha_i = l} \left\| \frac{\partial^l \varphi}{\partial x_1^{\alpha_1} \ldots \partial x_n^{\alpha_n}} \right\|_{L_p}$$

and using (7.3), we find:

$$|\varphi^*| \leqslant KBN_1 \|\varphi\|_{L_p^{(l)}} \leqslant KBN_1 \|\varphi\|_{W_p^{(l)}} < K'' \|\varphi\|_{W_p^{(l)}},$$

where $N_1$ is the number of distinct derivatives of order $l$.

Consequently $|\phi| \leqq S + |\phi^*| < (K' + K'') \|\phi\|_{W_p^{(l)}} = M \|\phi\|_{W_p^{(l)}}$, and the inequality (8.1) is established. The theorem is proved.

2. IMBEDDING OF $W_p^{(l)}$ IN $L_{q^*}$.

THEOREM 2. *If $\phi \in W_p^{(l)}$ and $n \geqq lp$, then $\phi \in L_{q^*}$ on any hyperplane of dimension $s$, if $s > n - lp$ and $q^* < q = sp/(n - lp)$.*

In addition, we have the inequality

$$\|\varphi\|_{L_{q^*}} \leqslant M \|\varphi\|_{W_p^{(l)}}. \tag{8.2}$$

Indeed, the fact that $\phi \in L_{q^*}$ was already proved in the proof of formula (7.12) for the case $\phi \in W_p^{(l)}$ for $n \geqq lp$. It remains to prove (8.2). We have:

$$\|\varphi\|_{L_{q^*}} \leqslant \|S\|_{L_{q^*}} + \|\varphi\|_{L_{q^*}}.$$

Setting $A = \max_{\sum \alpha_i \leqq l-1} \|x_1^{\alpha_1} \cdots x_n^{\alpha_n}\|_{L_{q^*}}$, just as in the proof of Theorem 1, we find

$$\|S\|_{L_{q^*}} \leqslant K \|\varphi\|_{W_p^{(l)}}.$$

Setting once more $\max|w_{\alpha_1 \cdots \alpha_n}| = B((x_i) \in \Omega, \sum \alpha_i = l)$ and using the estimate (6.4) for $\lambda = n - l$, we find:

$$\|\varphi^*\|_{L_{q^*}} \leqslant B \sum_{\sum \alpha_i = l} \left\| \int_{\Omega} \frac{1}{r^{n-l}} \left| \frac{\partial^l \varphi}{\partial x_1^{\alpha_1} \dots \partial x_n^{\alpha_n}} \right| dv \right\|_{L_{q^*}} <$$

$$< BK_1 \sum_{\sum \alpha_i = l} \left\| \frac{\partial^l \varphi}{\partial x_1^{\alpha_1} \dots \partial x_n^{\alpha_n}} \right\|_{L_p} \leqslant BK_1 N \|\varphi\|_{L_p^{(l)}} \leqslant K_2 \|\varphi\|_{W_p^{(l)}},$$

from which we find:
$\|\phi\|_{L_{q^*}} \leqq (K+K_2)\|\phi\|_{W_p^{(l)}} = M\|\phi\|_{W_p^{(l)}}$, i.e., (8.2) and the theorem is proved.

Theorems 1 and 2 indicate that the identity transformation carrying a function $\phi$ considered as an element of $W_p^{(l)}$ into the same function considered as an element of $C$ $(n<lp)$ or as an element of $L_{q^*}$ on an $s$-dimensional plane $(n \geqq lp,\ s>n-lp,\ q^*<(sp/(n-lp)))$, is a bounded operator. In a following paragraph it will be shown that it is a completely continuous operator, i.e., that it carries any bounded set into a compact set.

3. EXAMPLES. We have shown that if $\phi \in W_p^{(l)}$ and $n>lp$, then $\phi \in L_{q^*}$ on any plane of dimension $s$, where $s>n-lp$ and $q^*<q=sp/(n-lp)$, and then the inequality (8.2) holds. We present two examples showing that in Theorem 2 the number $q^*$ may not be replaced by $q+\epsilon$, no matter how small we choose $\epsilon>0$. Thus we will have shown that the given bound $q$ is precise and may not be increased.

EXAMPLE 1. Let $R<1$. We consider the hemisphere $\Omega$ of radius $R$ in $n$-dimensional space:

$$\sum_{i=1}^n x_i^2 = r^2 \leqslant R^2, \quad x_n \geqslant 0 \quad (n \geqslant 3).$$

Then $u=(r^{(n/2)-1}\ln r)^{-1} \in W_2^{(1)}$ on $\Omega$, since

$$\sum_{i=1}^n \left(\frac{\partial u}{\partial x_i}\right)^2 = \frac{1}{r^n}\left[\frac{\left(\frac{n}{2}-1\right)^2}{(\ln r)^2} + \frac{n-2}{(\ln r)^3} - \frac{1}{(\ln r)^4}\right]$$

and

$$\int_\Omega \sum \left|\frac{\partial u}{\partial x_i}\right|^2 dv = \frac{\sigma_n}{2}\int_0^R \left\{\frac{1}{r^n}\left[\frac{\left(\frac{n}{2}-1\right)^2}{(\ln r)^2} + \frac{n-2}{(\ln r)^3} + \frac{1}{(\ln r)^4}\right]\right\} r^{n-1}\,dr$$

converges.

Here $\sigma_n$ denotes the surface area of the hypersphere in $n$-space.

We consider the plane $x_1=0$ $(x_n \geqq 0)$; and set $\sum_{i=2}^n x_i^2=\rho^2$. By Theorem 2, $u \in L_{q^*}$ on the plane $x_1=0$, where

$$q^* < q = \frac{(n-1)2}{n-2\cdot 1} = \frac{2(n-1)}{n-2}.$$

We show that

$$\overbrace{\int \cdots \int}^{n-1}_{x_1=0} |u|^{q+\varepsilon}\, dv_{n-1}$$

will diverge if $\epsilon > 0$. Indeed

$$\overbrace{\int \cdots \int}^{n-1}_{x_1=0} |u|^{q+\varepsilon}\, dv = \frac{\sigma_{n-1}}{2} \int_0^R \frac{1}{\left[\rho^{\frac{n-2}{2}} |\ln \rho|\right]^{q+\varepsilon}} \rho^{n-2} d\rho =$$

$$= \frac{\sigma_{n-1}}{2} \int_0^R \frac{\rho^{-1-\frac{\varepsilon}{2}(n-2)}}{|\ln \rho|^{q+\varepsilon}} d\rho = \frac{\sigma_{n-1}}{2} \int_{-\ln R}^{\infty} \frac{e^{\xi (n-2)\frac{\varepsilon}{2}}}{\xi^{q+\varepsilon}} d\xi;$$

$$\left[n-2-(q+\varepsilon)\frac{n-2}{2} = -1 + \frac{\varepsilon}{2}(n-2)\right],$$

from which it follows that the integral diverges no matter how small $\epsilon > 0$. Thus $u$ is not summable to the power $q+\epsilon$ on the plane $x_1 = 0$ and thereby does not lie in $L_{q+\epsilon}$.

EXAMPLE 2. Let $\delta > 0$. We denote by $r_\delta$ the distance from the point $(0,0,\cdots,0,-\delta)$ to the point $(x_1,\cdots,x_n)$, i.e.,

$$r_\delta = \sqrt{(x_n+\delta)^2 + \sum_{i=1}^{n-1} x_i^2}.$$

Then on the hemisphere $\Omega$ the family of functions

$$u_\delta = \frac{1}{r_\delta^{\frac{n}{2}-1} \ln r_\delta} \qquad \left(0 < \delta < \frac{1-R}{2}\right)$$

is uniformly bounded in the norm in $W_2^{(1)}$.

Indeed, the integrals of $u_\delta$ and $\sum_{i=1}^n |\partial u_\delta / \partial x_i|^2$ on the hemisphere of radius $R+\delta < (1+R)/2 < 1$ and center at the point $(0,0,\cdots,-\delta)$ are uniformly bounded since the integrals

$$\int_0^{\frac{1+R}{2}} \frac{r^{\frac{n}{2}} dr}{|\ln r|} \quad \text{and} \quad \int_0^{\frac{1+R}{2}} \frac{dr}{r |\ln r|^2}$$

converge.

Therefore the integrals of $|u_\delta|$ and $\sum_{i=1}^n |\partial u_\delta / \partial x_i|^2$ on $\Omega$ are also uniformly bounded ($\Omega$ is a part of each hemisphere of radius $R+\delta$) and by virtue of (7.12) and (7.5) $\|u_\delta\|_{W_2^{(1)}}$ is uniformly bounded.

On the other hand, it is not difficult to see that

$$\overbrace{\int \cdots \int}^{n-1}_{x_1=0} |u_\delta|^{q+\epsilon}\, dv_{n-1} \to \infty \text{ for } \delta \to 0$$

and, consequently, $\|u_\delta\|_{L_{q+\epsilon}}$ on the hyperplane $x_1=0$ will not be bounded, i.e., for $q^*=q+\epsilon$ the inequality (8.2) will not hold.

**§9. General methods of norming $W_p^{(l)}$ and corollaries of the Imbedding Theorem.**

1. A THEOREM ON EQUIVALENT NORMS. Let $\prod_1$ and $\prod_2$ be two projection operators mapping $W_p^{(l)}$ onto $S_l$. These operators generate norms on $W_p^{(l)}$ which we designate by ${}^1\|\phi\|$ and ${}^2\|\phi\|$. We shall say that these norms are equivalent if one can give two positive numbers $m$ and $M$ such that

$$m \leqslant \frac{{}^2\|\varphi\|}{{}^1\|\varphi\|} \leqslant M \quad \text{for all } \varphi \in W_p^{(l)}. \tag{9.1}$$

We consider the operator $\Xi_{12}\phi=(\prod_1-\prod_2)\phi$. This operator possesses two properties.

(1) $\Xi_{12}S=0$, i.e., $\Xi_{12}$ carries into zero every polynomial in $S$. This follows from the fact that $\prod_1 S=\prod_2 S=S$. It follows from this that $\Xi_{12}$ puts in correspondence with one and the same polynomial of $S$ all the functions of one and the same class $\psi \in L_p^{(l)}$.

(2) $\Xi_{12}^2\phi=0$, i.e., the square of the operator $\Xi_{12}$ is the annihilation operator.

Indeed, $\Xi_{12}\phi=S$, and consequently

$$\Xi_{12}^2\varphi = \Xi_{12}S = 0.$$

THEOREM. *For the equivalence of the norms ${}^1\|\phi\|$ and ${}^2\|\phi\|$, generated by the operators $\prod_1$ and $\prod_2$, it is necessary and sufficient that there exists a number $M>0$ such that*

$$\|\Xi_{12}\varphi\|_s \leqslant M\|\varphi\|_{L_p^{(l)}} \quad \textit{for all } \varphi \in W_p^{(l)}. \tag{9.2}$$

PROOF OF NECESSITY. The proof proceeds by reductio ad absurdum. Suppose that the condition (9.2) does not hold. Then we may find a sequence of functions $\{\phi_k\}$, $\phi_k \in W_p^{(l)}$, for which

$$\|\Xi_{12}\varphi_k\|_S > k\|\varphi_k\|_{L_p^{(l)}} \qquad (k=1,\ 2,\ \ldots). \tag{9.3}$$

Normalizing $\phi_k$, we may assume that $\|\phi_k\|_{L_p^{(l)}}=1$.

We consider the functions $\psi_k=\prod_1^*\phi_k=\phi_k-\prod_1\phi_k$ and calculate both norms of $\psi_k$:

$${}^1\|\psi_k\| = 0+\|\varphi_k\|_{L_p^{(l)}} = 1, \text{ since } \textstyle\prod_1\psi_k = 0.$$

Since $\Pi_1\phi_k = S$, $\Pi_2\Pi_1\phi_k = \Pi_1\phi_k$. In addition, taking (9.3) into account, we obtain:

$$^2\|\psi_k\| = \|\Pi_2\psi_k\|_S + \|\varphi_k\|_{L_p^{(l)}} = \|\Pi_2\varphi_k - \Pi_2\Pi_1\varphi_k\|_S + 1 =$$
$$= \|\Pi_2\varphi_k - \Pi_1\varphi_k\|_S + 1 = \|\Xi_{12}\varphi_k\|_S + 1 > k + 1,$$

from which it follows that $^2\|\psi_k\| / {}^1\|\psi_k\| \to \infty$. This contradicts (9.1) and the norms $^1\|\phi\|$ and $^2\|\phi\|$ are not equivalent, contradicting the assumption of the equivalence of the norms. The contradiction shows that (9.3) is false, and therefore that (9.2) holds for all $\phi \in W_p^{(l)}$.

PROOF OF SUFFICIENCY. Suppose that (9.2) is satisfied. Then

$$^1\|\varphi\| = \|\Pi_1\varphi\|_S + \|\varphi\|_{L_p^{(l)}} \leqslant [\|\Pi_2\varphi\|_S + \|(\Pi_1 - \Pi_2)\varphi\|_S] +$$
$$+ \|\varphi\|_{L_p^{(l)}} \leqslant \|\Pi_2\varphi\|_S + M\|\varphi\|_{L_p^{(l)}} + \|\varphi\|_{L_p^{(l)}} =$$
$$= \|\Pi_2\varphi\|_S + (M+1)\|\varphi\|_{L_p^{(l)}} \leqslant$$
$$\leqslant (M+1)[\|\Pi_2\varphi\|_S + \|\varphi\|_{L_p^{(l)}}] = (M+1)\cdot {}^2\|\varphi\|.$$

Analogously, it may be shown that $^2\|\phi\| \leqq (M+1)\cdot {}^1\|\phi\|$. These two inequalities together are equivalent to (9.1) and the theorem is proved.

The condition (9.2) is conveniently written in some other forms. We have:

$$\Xi_{12}\varphi = \Pi_1\varphi - \Pi_2\varphi = \Pi_1\varphi - \Pi_1\Pi_2\varphi = \Pi_1(\varphi - \Pi_2\varphi) = \Pi_1\Pi_2^*\varphi$$

and analogously

$$\Xi_{12}\varphi = -\Pi_2\Pi_1^*\varphi.$$

Using these, we see easily that each one of the following three inequalities below is a necessary and sufficient condition for the equivalence of norms.

$$\|\Xi_{12}\varphi\|_S \leqslant M\|\varphi\|_{L_p^{(l)}}; \tag{9.2}$$

$$\|\Pi_1\Pi_2^*\varphi\|_S \leqslant M\|\varphi\|_{L_p^{(l)}}; \tag{9.4}$$

$$\|\Pi_2\Pi_1^*\varphi\|_S \leqslant \|\varphi\|_{L_p^{(l)}}. \tag{9.5}$$

We shall make use of this result below.

2. The general form of norms equivalent to a given one. We shall now concern ourselves with a more detailed study of the structure of projection operators yielding norms equivalent to a given one. Let $\prod_1$ be given.

The general form of a projection operator $\prod_2$ will be

$$\Pi_2\varphi = \sum_{\Sigma\alpha_i \leqslant l-1} x_1^{\alpha_1} \ldots x_n^{\alpha_n} l_{\alpha_1, \ldots, \alpha_n}\varphi,$$

where $l_{\alpha_1,\ldots,\alpha_n}$ represents an ordinary additive functional.

It is not difficult to see that these functionals have the property

$$l_{\alpha_1\alpha_2, \ldots, \alpha_n} x_1^{\beta_1} x_2^{\beta_2} \ldots x_n^{\beta_n} = \begin{cases} 1 & \text{if} \quad \sum(\alpha_i - \beta_i)^2 = 0 \\ 0 & \text{if} \quad \sum(\alpha_i - \beta_i)^2 > 0. \end{cases} \tag{9.6}$$

Theorem 1. *In order that the norms* $^1\|\phi\|$ *and* $^2\|\phi\|$ *should be equivalent, it is necessary and sufficient that all the functionals* $l_{\alpha_1\alpha_2\cdots\alpha_n}$ *should be bounded in the sense of the norm defined by the projection operator* $\prod_1$.

Proof. It is easy to see that the condition (9.5) for the equivalence of the norms is equivalent to the system of conditions

$$| l_{\alpha_1\alpha_2, \ldots, \alpha_n} \Pi_1^* \varphi | \leqslant M \|\varphi\|_{L_p^{(l)}}. \tag{9.7}$$

It remains for us to show that (9.7) is equivalent to the condition of the theorem.

Suppose that all the functionals $l_{\alpha_1\alpha_2\cdots\alpha_n}$ are bounded in the norm $^1\|\phi\|$. Then

$$| l_{\alpha_1\alpha_2 \ldots \alpha_n} \Pi^* \varphi | \leqslant M \, {}^1\| \Pi_1^* \varphi \|_{W_p^{(l)}} = M \, \|\varphi\|_{L_p^{(l)}},$$

i.e., (9.7) is verified.

Conversely, if (9.7) is verified, then taking (9.6) into account, we find:

$$| l_{\alpha_1 \ldots \alpha_n} \varphi | = | l_{\alpha_1 \ldots \alpha_n} \Pi_1 \varphi + l_{\alpha_1 \ldots \alpha_n} \Pi_1^* \varphi | \leqslant$$

$$\leqslant | l_{\alpha_1 \ldots \alpha_n} \Pi_1 \varphi | + | l_{\alpha_1 \ldots \alpha_n} \Pi_1^* \varphi | \leqslant \| \Pi_1 \varphi \|_S + M \|\varphi\|_{L_p^{(l)}} \leqslant$$

$$\leqslant M_1 \cdot {}^1\|\varphi\|_{W_p^{(l)}}, \quad (M_1 = \max(1, M)),$$

i.e., from (9.7) follows the boundedness of the functionals $l_{\alpha_1\cdots\alpha_n}$ and the theorem is completely proved.

Thus, an arbitrary system of linear (additive and bounded) functionals $l_{\alpha_1\cdots\alpha_n}$ on $W_p^{(l)}$ satisfying the equations (9.6) yields a projection operator $]\,[_2$ and consequently a new norm on $W_p^{(l)}$, and this norm is equivalent to the norm with respect to which the boundedness of the functionals was defined.

If we now consider an arbitrary system of linear functionals $h_{\alpha_1 \cdots \alpha_n}$, the number of which is precisely equal to the number of monomials of power not higher than $l-1$, and taken so that the determinant of the matrix

$$A = \| h_{\alpha_1 \ldots \alpha_n} x_1^{\beta_1} \ldots x_n^{\beta_n} \|,$$

is non-null, where each row of the matrix consists of the values of one and the same functional and each column of functionals acting on one and the same monomial, then from such a system we may always construct linear combinations $l_{\alpha_1 \cdots \alpha_n}$ which will satisfy the equations (9.6).

Each such system of linear functionals enables one to define a norm on $S$ by the formula

$$\| \varphi \|_S = \left| \sum (h_{\alpha_1, \ldots, \alpha_n} \varphi)^2 \right|^{\frac{1}{2}}. \tag{9.8}$$

Such a norm, as we saw above, will be equivalent to the initial norm on $S$.

The condition $|A| \neq 0$ may be reformulated in a form suitable for applications in two ways.

It is well-known that if the determinant of a square matrix is not equal to zero, then it is impossible to have a nontrivial linear combination of the rows of the matrix or of the columns of the matrix which is a null vector. Conversely, if the rows or the columns are linearly independent, the determinant must be different from zero. From this follow two remarks.

(a) The condition $|A| \neq 0$ is satisfied if the following holds: for any polynomial $P$ of degree not greater than $l-1$, we can find a functional such that $h_{\alpha_1 \cdots \alpha_n} P$ is not zero.

The value of any linear functional on a polynomial of $S$ corresponds to a fixed linear combination of the elements in a column of the matrix $\| A \|$ corresponding to various rows.

If the condition (a) is satisfied, then it is impossible to find a linear combination of the columns of the matrix $\| A \|$ consisting only of zeros. In other words, $|A| \neq 0$.

(b) The condition $|A| \neq 0$ is satisfied if the following is valid: for any linear combination of functionals

$$\rho = \sum A_{\alpha_1 \ldots \alpha_n} h_{\alpha_1 \ldots \alpha_n},$$

we may always find a monomial

$$x_1^{\beta_1} \ \ldots \ x_n^{\beta_n},$$

for which $\rho x_1^{\beta_1} \cdots x_n^{\beta_n} \neq 0$.

The values of various $\rho$ on the set of monomials is in one-to-one correspondence with the various linear combinations of the elements of the rows of the matrix $A$ lying in all possible columns. If condition (b) is satisfied, then it is impossible to find a nontrivial linear combination of the rows of the matrix $A$ consisting only of zeros. This means that $|A| \neq 0$.

3. NORMS EQUIVALENT TO THE SPECIAL NORM. Up to now we have considered the equivalence of norms arising from two arbitrary projection operators and the whole argument was suitable for an arbitrary domain $\Omega$. We turn now to domains star-like with respect to some sphere $C$ for which we constructed a special projection operator defined by formula (7.12).

For the norm given by this operator, the imbedding Theorems 1 and 2 are valid, i.e., inequality (8.1) for the case $n<lp$ and (8.2) in the case $n \geqq lp$.

Let $h_{\alpha_1,\cdots,\alpha_n}$ ($\sum \alpha_i \leqq l-1$) be a system of linear functionals on $C$ (if $n<lp$) or on $L_{q^*}$ (if $n \geqq lp$). Then, for example, in the case of $n<lp$, we find using (8.1):

$$|h_{\alpha_1, \ldots, \alpha_n} \varphi| < K_1 \|\varphi\|_C \leqslant K_2 \|\varphi\|_{W_p^{(l)}}, \qquad (K_2 = K_1 M),$$

i.e., $h_{\alpha_1\cdots\alpha_n}$ is bounded on $W_p^{(l)}$ in the norm of the projection operator (7.12). An analogous assertion is valid if $h_{\alpha_1\cdots\alpha_n}$ is linear on $L_{q^*}$ ($n \geqq lp$). From this, using the preceding theorem, we obtain the following theorem.

THEOREM 2. *Let* $h_{\alpha_1\cdots\alpha_n}$, ($\sum \alpha_i \leqq l-1$) *be linear functionals on* $C$ ($n<lp$) *or on* $L_{q^*}$ ($n \geqq lp$) *and suppose that for each polynomial from* $S$ *one of the* $h_{\alpha_1\cdots\alpha_n}$ *is different from zero. Then the norm defined by* (7.5) *and* (9.8) *is equivalent to the norm* (7.5) *obtained from the operator* (7.12), *and then* (8.1) *and* (8.2) *are valid if on the right hand side of the latter we understand the norm of* $\phi$ *to be given in its turn by the formulae* (7.5) *and* (9.8).

4. SPHERICAL PROJECTION OPERATORS. The argument in §§1, 2 is valid for arbitrary domains. Now we shall demand that the domain $\Omega$ is star-like with respect to some sphere $C$ of radius $H$. Formula (7.12) defines a projection operator $\prod$ of the form

$$\Pi\varphi = \sum_{\Sigma\alpha_i \leqslant l-1} x_1^{\alpha_1} \ldots x_n^{\alpha_n} \int_{R \leqslant H} \zeta^{(H)}_{\alpha_1 \ldots \alpha_n}(\vec{Q})\, \varphi(\vec{Q})\, dv_{\vec{Q}}, \tag{9.9}$$

assigning to each $\phi \in W_p^{(l)}$ a polynomial of $S$, where the coefficients of this polynomial depend only upon the values taken by $\phi$ on the sphere $R \leqq H$. On the set $\{S\}$, this operator is the identity operator, as follows from the integral identity (7.12). Therefore the functionals

$$h_{\alpha_1 \cdots \alpha_n} \varphi = \int\limits_{R \leqslant H} \zeta^{(H)}_{\alpha_1 \cdots \alpha_n} (\vec{Q}) \varphi (\vec{Q})\, dv_{\vec{Q}} \tag{9.10}$$

have the property that for any $S$, at least one of the functionals differs from zero. Obviously, this conclusion is true for functionals constructed with the aid of a sphere of arbitrary radius since we may include every such sphere in a domain star-like with respect to that sphere and obtain the integral equality (7.12).

Let $\vec{T}(y_1, \cdots, y_n)$ be such a point of the domain $\Omega$ that the sphere $C_1$ of radius $H_1$ with center at $\vec{T}$ lies within the domain $\Omega$. We construct the operator:

$$\Pi_{C_1} \varphi = \sum_{\Sigma \alpha_i \leqslant l-1} (x_1 - y_1)^{\alpha_1} \cdots (x_n - y_n)^{\alpha_n} \times$$

$$\times \int\limits_{R \geqq H_1} \zeta^{(H_1)}_{\alpha_1 \cdots \alpha_n} (\vec{Q}) \varphi (\vec{Q} + \vec{T})\, dv_{\vec{Q}} . \tag{9.11}$$

The operator (9.11) carries each $\phi \in W_p^{(l)}$ into some corresponding polynomial $S$. In addition, obviously $\prod_{C_1}\phi$ carries each polynomial $S$ into itself, since formula (9.11) goes over into (9.9) by a translation of the origin of coordinates and under this mapping $\{S\}$ is carried into itself. It follows that the operator $\prod_{C_1}\phi$ is a projection, since $\prod_{C_1}\prod_{C_1}\phi$. The operator (9.11) will be called a spherical projection operator.

THEOREM. *The norms constructed by means of arbitrary spherical projection operators are mutually equivalent.*

In order to prove this theorem, it suffices to establish the equivalence of the norm defined by means of arbitrary spherical projection operator with the norm $\|\phi\|^*_{W_p^{(l)}}$ defined by the formulae (7.5) and (9.8) with functionals $l_{\alpha_1, \alpha_2, \cdots, \alpha_n}$ linear on $C$ or on $L_{q^*}$, for $n \geqq lp$, $q^* \leqq q = np/(n - lp)$, $(s = n)$. We shall show that these norms are indeed equivalent.

In fact, $\zeta^{H_1}_{\alpha_1, \cdots, \alpha_n} (\vec{Q})$ is bounded and therefore the functional

$$\int\limits_{R \leqslant H_1} \zeta^{(H_1)}_{\alpha_1, \cdots, \alpha_n} (\vec{Q}) \varphi (\vec{Q} + \vec{T})\, dv_{\vec{Q}}$$

is bounded on $C$ $(n < lp)$ or on $L_{q^*}$ $(n \geqq lp,\ s = n,\ q^* \leqq q = np/(n - lp))$, and as a consequence, on the basis of Theorem 2, item 3 the norm generated by the operator $\prod_{C_1}\phi$ is equivalent to the norm $\|\phi\|^*$. Thus the norm generated by an arbitrary spherical projection operator is equivalent to the norm $\|\phi\|^*_{W_p^{(l)}}$ and all such norms being mutually equivalent, the theorem is proved.

5. NONSTAR-LIKE DOMAINS. We may now free ourselves of the restriction of star-likeness imposed upon the domain $\Omega$ in the proof of the Imbedding Theorem.

Let $\Omega$ be an arbitrary domain containing some sphere $C$ of radius $H$. Formula (9.11) defines a projection operator on $W_p^{(l)}$ and we define a norm by formula (7.5). It remains for us to show that under some restrictions upon $\Omega$, the Imbedding Theorems 1 and 2 are still valid as well as the inequalities (8.1) and (8.2). We remark that if for some domain the Imbedding Theorems hold, then for that domain the preceding theorem of item 4 is valid.

LEMMA. *Let $\Omega_1$ and $\Omega_2$ be two domains, for each of which the Imbedding Theorem holds. Suppose that $\Omega_1$ and $\Omega_2$ intersect in a nonempty domain $\Omega_{12}=\Omega_1\Omega_2$. Then for the domain $\Omega=\Omega_1+\Omega_2$ the Imbedding Theorems are valid and the norms defined by all spherical projection operators are equivalent.*

PROOF. Let $\phi\in W_p^{(l)}$ in $\Omega$. Then $\phi\in W_p^{(l)}$ in $\Omega_1$ and in $\Omega_2$ and therefore $\phi\in C$ (if $n<lp$) in $\Omega_1$ and $\Omega_2$ are therefore in $\Omega_1+\Omega_2$. One treats the case $n\geqq lp$ analogously. It remains to prove (8.1) and (8.2).

Assuming $n<lp$, we shall prove (8.1). Let $C_{12}$ be a sphere in $\Omega_{12}$ and $\prod_{C_{12}}\phi$ its spherical projection operator. Then

$$\|\varphi\|_C<K_i\left[\|\prod_{C_{12}}\varphi\|_S+{}^{(i)}\|\varphi\|_{L_p^{(l)}}\right] \text{ in } \Omega_i \qquad (i=1,2),$$

from which it follows that in $\Omega=\Omega_1+\Omega_2$ we have the inequality

$$\|\varphi\|_C\leqslant K_3\left[\|\prod_{C_{12}}\varphi\|_S+\|\varphi\|_{L_p^{(l)}}\right],$$

since ${}^{(i)}\|\phi\|_{L_p^{(l)}}\leqq\|\phi\|_{L_p^{(l)}}$ and where $K_3=\max\left[K_1,K_2\right]$.

As a consequence (8.1) is valid. Analogously for the case $n\geqq lp$, (8.2) may be proved. On the basis of the theorem in item 4 there follows the equivalence of all norms defined by spherical projection operators.

THEOREM. *Let $\Omega$ be represented as the sum of a finite number of domains $\Omega_1,\Omega_2,\cdots,\Omega_k$ each of which $\Omega_j$ is star-like with respect to some sphere $H_j$. Suppose that the sum is connected (by this we mean that each domain $\Omega_j$ intersects the sum of the preceding $\Omega_1+\Omega_2+\cdots+\Omega_{j-1}$ in some nonempty domain $\omega_j$). Then the Imbedding Theorem holds for the domain $\Omega$ and the norms of all the spherical projection operators are equivalent.*

PROOF. This theorem follows by the repeated application of the lemma to the domain $\Omega_1+\Omega_2$, $\Omega_1+\Omega_2+\Omega_3$, etc. In the following we shall consider only such domains, and will not speak of this on each special occasion.

We may without loss of generality consider only such norms on the space $W_p^{(l)}$ as are equivalent to the norms obtained from spherical projection operators. We will call such norms natural. Each time that we speak of a norm or of convergence in the space $W_p^{(l)}$ if no special provision is made, we shall have in mind an arbitrary natural norm and convergence with respect to an arbitrary natural norm.

6. EXAMPLES. We introduce two examples, illustrating applications of the above theorems.

EXAMPLE 1. Let $p=2$, $l=1$, $s=n\geqq 3$. Since $q=2n/(n-2)>2$, we may take $q^*=2$, i.e., $W_2^{(1)}\subset L_2$. Since $l=1$, for the definition of the norm it suffices to take one functional (polynomial of the zeroth degree, there exists only one polynomial of null degree). We choose

$$(h,\ \varphi)=\int_\Omega \varphi\, dv.$$

It is obvious that $(h,1)\neq 0$, and the functional $h$ is linear on $L_2$. Therefore on the basis of Theorem 2, item 2, we have

$$\left\{\int_\Omega |\varphi|^2 dv\right\}^{\frac{1}{2}} \leqslant M\left[\left|\int_\Omega \varphi\, dv\right|+\left\{\int_\Omega \sum_{i=1}^n\left(\frac{\partial\varphi}{\partial x_i}\right)^2 dv\right\}^{\frac{1}{2}}\right],$$

from which there follows

$$\int_\Omega |\varphi|^2 dv \leqslant M_1\left[\left\{\int_\Omega \varphi\, dv\right\}^2+\int_\Omega \sum_{i=1}^n\left(\frac{\partial\varphi}{\partial x_i}\right)^2 dv\right]. \tag{9.12}$$

The last inequality is well-known under the name of Poincaré's inequality.

EXAMPLE 2. Let $p=2$, $l=1$, $s\geqq(n-2)\cdot 1=n-2$; $s_1=n-1$, $s_2=n$.

Since $q_1=2(n-1)/(n-2)>2$, we have $W_2^{(1)}\subset L_2$ on manifolds of dimension $(n-1)$.

We put

$$(h,\ \varphi)=\overbrace{\int\cdots\int_S}^{n-1}\varphi\, dv_{n-1}.$$

We have $(h,1)\neq 0$, and the functional $h$ is bounded in $L_2$ on the manifold $S$.

On the basis of the same Theorem 2, item 2, the Imbedding Theorem holds for the norm

$$\left|\overbrace{\int\cdots\int_S}^{n-1}\varphi\, dv_{n-1}\right|+\|\varphi\|_{L_p^{(l)}},$$

and in particular, since $W_2^{(1)}\subset L_2$ on $S$ $(s_2=n)$

$$\left\{\int_\Omega |\varphi|^2 dv\right\}^{\frac{1}{2}} \leqslant M\left[\left|\overbrace{\int_S \cdots \int}^{n-1} \varphi\, dv_{n-1}\right| + \|\varphi\|_{L_p^{(l)}}\right],$$

from which follows

$$\int_\Omega |\varphi|^2 dv \leqslant M_1\left\{\left|\overbrace{\int_S \cdots \int}^{n-1} \varphi\, dv_{n-1}\right|^2 + \int_\Omega \sum_{i=1}^{n} \left(\frac{\partial \varphi}{\partial x_i}\right)^2 dv\right\}. \tag{9.13}$$

It is obvious that Theorem 2, item 2, can yield many inequalities for suitable choices of the functionals $h_{\alpha_1 \cdots \alpha_n}$.

§10. **Some consequences of the Imbedding Theorems.**

1. COMPLETENESS OF THE SPACE $W_p^{(l)}$. We consider a Cauchy sequence $\{\phi_k\}$, $\phi_k \in W_p^{(l)}$. Suppose that the norm in $W_p^{(l)}$ is defined by an arbitrary spherical projection operator $\prod$. We have:

$$\|\varphi_m - \varphi_k\|_{W_p^{(l)}} \to 0, \quad m, \ k \to \infty. \tag{10.1}$$

On the basis of the Imbedding Theorems, we conclude that

$$\|\varphi_m - \varphi_k\|_C \to 0 (n < lp); \quad \|\varphi_m - \varphi_k\|_{L_{q^*}} \to 0 \ \ (n \geqslant lp), \ \ m, \ k \to \infty,$$

and consequently, $\{\phi_k\}$ converges in $C$ (if $n < lp$) or in $L_{q^*}$ (if $n \geqq lp$). Let the limit of the functions be $\phi_0$. Setting $\prod \phi_k = S_k$, we obtain from (10.1):

$$\|S_m - S_k\|_S^p + \|\varphi_m - \varphi_k\|_{L_p^{(l)}}^p \to 0, \quad m, \ k \to \infty,$$

from which follows:

$$\|S_m - S_k\|_S \to 0, \qquad m, \ k \to \infty\ ; \tag{10.2}$$

$$\|\varphi_m - \varphi_k\|_{L_p^{(l)}} \to 0, \qquad m, \ k \to \infty. \tag{10.3}$$

From (10.2) it follows that the coefficients of $S_k$ converge to finite limits and therefore that $S_k$ converges uniformly to some polynomial $S_0$. Obviously, we have

$$\|S_k - S_0\|_S \to 0, \quad k \to \infty. \tag{10.4}$$

From (10.3) it follows that

$$\left\| \frac{\partial^l \varphi_m}{\partial x_1^{\alpha_1} \cdots \partial x_n^{\alpha_n}} - \frac{\partial^l \varphi_k}{\partial x_1^{\alpha_1} \cdots \partial x_n^{\alpha_n}} \right\|_{L_p} \to 0, \quad \sum \alpha_n = l, \ m, \ k \to \infty,$$

i.e., each of the derivatives $\partial^l \phi_k / \partial x_1^{\alpha_1} \cdots \partial x_n^{\alpha_n}$ converges in $L_p$ to some function $\omega_{\alpha_1 \cdots \alpha_n} \in L_p$. We show that

$$\omega_{\alpha_1 \ldots \alpha_n} = \frac{\partial^l \varphi_0}{\partial x_1^{\alpha_1} \partial x_2^{\alpha_2} \ldots \partial x_n^{\alpha_n}}. \tag{10.5}$$

Indeed, for each $\psi$ having continuous derivatives up to the $l$th order on the whole space and vanishing outside some domain $V_\psi \subset \Omega$, we have

$$\int_\Omega \varphi_k \frac{\partial^l \psi}{\partial x_1^{\alpha_1} \ldots \partial x_n^{\alpha_n}} dv = (-1)^l \int_\Omega \psi \frac{\partial^l \varphi_k}{\partial x_1^{\alpha_1} \ldots \partial x_n^{\alpha_n}} dv,$$

from which, taking the limit as $k \to \infty$, the correctness of (10.5) follows.

Thus, $\phi_0 \in W_p^{(l)}$ (since $\omega_{\alpha_1, \ldots, \alpha_n} \in L_p$). It is not difficult to verify that $\prod \phi_0 = S_0$. In fact, replacing $\phi$ in (9.11) by $\phi_k$ and passing to the limit, we obtain:

$$\prod \varphi_0 = \lim_{k \to \infty} \prod \varphi_k = \lim_{k \to \infty} S_k = S_0.$$

Since $\partial^l \phi_k / \partial x_1^{\alpha_1} \cdots \partial x_n^{\alpha_n} \Rightarrow \omega_{\alpha_1 \cdots \alpha_n}$ in $L_p$ and by virtue of (10.4), obviously, we have

$$\|\varphi_0 - \varphi_k\|_{W_p^{(l)}} \to 0, \quad k \to \infty.$$

The completeness of the space $W_p^{(l)}$ is proved.

2. THE IMBEDDING OF $W_p^{(l)}$ IN $W_{p_k}^{(k)}$. Up to this point in §§7,8,9 we have not discussed the derivatives of order less than $l$. In §5 it was observed that from the existence of the generalized derivatives of higher order, there does not follow in general the existence of derivatives of lower order. Now the following theorem will be proved.

THEOREM. *If $\phi \in W_p^{(l)}$, then $\phi$ has all generalized derivatives of order less than $l$. For these:*

(1) *If $lp \geq n, 0 \leq m < l - (n/p)$, then $\partial^m \phi / \partial x_1^{\alpha_1} \cdots \partial x_n^{\alpha_n}$ is continuous and*

$$\left| \frac{\partial^m \varphi}{\partial x_1^{\alpha_1} \ldots \partial x_n^{\alpha_n}} \right| < M \|\varphi\|_{W_p^{(l)}}. \tag{10.6}$$

(2) *If $m \geq 0$ and $m \geq l - (n/p)$, $s > n - (l - m)p$, then on every manifold of dimensions*

$$\frac{\partial^m \varphi}{\partial x_1^{\alpha_1} \ldots \partial x_n^{\alpha_n}} \in L_{q^*},$$

*where $q^* < q = sp/(n - (l - m)p)$, while*

$$\left\| \frac{\partial^m \varphi}{\partial x_1^{\alpha_1} \ldots \partial x_n^{\alpha_n}} \right\|_{L_{q^*}} \leqslant M \|\varphi\|_{W_p^{(l)}}. \tag{10.7}$$

REMARK. If $lp>n$, then $W_p^{(l)} \subset C^{l-[n/p]-1}$, i.e., $W_p^{(l)}$ is part of the space of functions having $l-[n/p]-1$ continuous derivatives. This follows from the first part of the theorem. Setting $s=n$ in the second part of the theorem and noting that in this case the permissibility of $q^*=q$ is established, we conclude that if $k \geqq 0$ and $k \geqq l-(n/p)$, then

$$W_p^{(l)} \subset W^{(k)}_{\frac{np}{n-(l-k)p}} \quad \text{for} \quad W_p^{(l)} \subset W^{(k)}_{\frac{1}{\frac{1}{p}-\frac{l-k}{n}}} \qquad (k<l).$$

PROOF. It suffices to prove the theorem for domains which are star-like with respect to some sphere.

Let $\phi(\vec{P})$ be continuous and have continuous derivatives up to order $l$. Then formula (7.12) holds:

$$\varphi(\vec{P}) = \sum_{\Sigma \alpha_i \leqslant l-1} x_1^{\alpha_1} \dots x_n^{\alpha_n} \int_{R \leqslant H} \zeta_{\alpha_1,\dots,\alpha_n}(\vec{Q})\, \varphi(\vec{Q})\, dv_{\vec{Q}} +$$

$$+ \int_{\Omega} \frac{1}{r^{n-l}} \sum_{\Sigma \alpha_i = l} w_{\alpha_1,\dots\alpha_n}(\vec{Q}, \vec{P}) \frac{\partial^l \varphi}{\partial x_1^{\alpha_1} \dots \partial x_n^{\alpha_n}}\, dv_{\vec{Q}} \qquad (7.12)$$

The theorem will be proved if we show that

$$\frac{\partial^m}{\partial x_1^{\beta_1} \dots \partial x_n^{\beta_n}} \int_{\Omega} \frac{1}{r^{n-l}} w_{\alpha_1,\dots,\alpha_n} \frac{\partial^l \varphi}{\partial x_1^{\alpha_1} \dots \partial x_n^{\alpha_n}}\, dv =$$

$$= \int_{\Omega} \frac{1}{r^{n-l+m}} w^{\beta_1,\dots,\beta_m}_{\alpha_1,\dots,\alpha_n} \frac{\partial^l \varphi}{\partial x_1^{\alpha_1} \dots \partial x_n^{\alpha_n}}\, dv, \qquad (10.8)$$

where $w^{\beta_1\cdots\beta_n}_{\alpha_1\cdots\alpha_n}$ are bounded functions.

Indeed, let us differentiate $m$ times both sides of (7.12) written for the averaged functions $\phi_h$. The limit of the first term of the right side for $h \to 0$ will be the polynomial

$$\frac{\partial^m S}{\partial x_1^{\beta_1} \dots \partial x_n^{\beta_n}},$$

the coefficients of which are simply expressed in terms of the coefficients of $S$ and as a result,

$$\left| \frac{\partial^m S}{\partial x_1^{\beta_1} \dots \partial x_n^{\beta_n}} \right| < M \| S \|_S.$$

The limit of the second term of the right side will be the sum of terms of the forms of the right side of (10.8). On the basis of the theorem of §6 on integrals of potential type, the assertion of the theorem follows.

For the proof of (10.8), it suffices to show that

$$\frac{\partial^m}{\partial x_1^{\beta_1} \ldots \partial x_n^{\beta_n}} \left( \frac{1}{r^{n-l}} w(\vec{Q}, \vec{P}) \right) = \frac{w^{(m)}(\vec{Q}, \vec{P})}{r^{n-l+m}}, \tag{10.9}$$

where for simplicity the subscripts are omitted. We have (§7, item 4):

$$\frac{1}{r^{n-l}} w(\vec{Q}, \vec{P}) = \frac{(x_1 - y)^{\alpha_1} \ldots (x_n - y_n)^{\alpha_n}}{r^n} \chi(\vec{Q}, \vec{P})$$

and shall show that each differentiation of $1/r^{n-l}\, w(\vec{Q},\vec{P})$ increases by one unit the negative exponent of $1/r$.

Indeed, for the first expression

$$\frac{(x_1 - y_1)^{\alpha_1} \ldots (x_n - y_n)^{\alpha_n}}{r^n}$$

this is obvious.

If we are given $\vec{Q}$ and $\vec{P}$, then $r = |\vec{Q} - \vec{P}|$; $\vec{l} = (\vec{Q} - \vec{P})/r$, $\overline{v}(r, \vec{l}, \vec{P}) = v(\vec{Q}_1)$, where $\vec{Q}_1 = \vec{P} + r_1 \vec{l} = \vec{P} + (r_1/r)(\vec{Q} - \vec{P})$. Therefore,

$$\chi(\vec{Q}, \vec{P}) = -\int_r^\infty v\left[\vec{P} + \frac{r_1}{r}(\vec{Q} - \vec{P})\right] r_1^{n-1}\, dr_1.$$

Differentiating with respect to $x_1$, we find:

$$\begin{aligned} \frac{\partial \chi(\vec{Q}, \vec{P})}{\partial x_1} &= v(\vec{Q})\, r^{n-1} \frac{x_1 - y_1}{r} - \int_r^\infty \sum_{j=1}^n \frac{\partial v(\vec{Q}_1)}{\partial y_j} \times \\ &\quad \times \left[\delta_{1j} + \frac{r_1}{r^2} \frac{(x_1 - y_1)(x_j - y_j)}{r} - \delta_{j1} \frac{r_1}{r}\right] r_1^{n-1}\, dr_1 = \\ &= \frac{1}{r}\left[\int_r^\infty \frac{\partial v(\vec{Q}_1)}{\partial y_1} r_1^n\, dr_1 - \sum_{j=1}^n \frac{(x_1 - y_1)(x_j - y_j)}{r^2} \times \right. \\ &\quad \left. \times \int_r^\infty \frac{\partial v(\vec{Q}_1)}{\partial y_j} r_1^n\, dr_1\right] + v(\vec{Q})\, r^{n-2}(x_1 - y_1) - \\ &\quad - \int_r^\infty \frac{\partial v(\vec{Q}_1)}{\partial y_1} r_1^{n-1}\, dr_1. \end{aligned}$$

Denoting by $\chi_1(\vec{Q},\vec{P})$ and $\chi_2(\vec{Q},\vec{P})$ and so on, integrals of the same type as $\chi(\vec{Q},\vec{P})$ [$v(\vec{Q}_1)$ is replaced by a suitably often differentiable function, $r_1^{n-1}$ is replaced by $r_1^n$, $r_1^{n+1}$ etc.] we obtain:

$$\frac{\partial\chi}{\partial x_1}=\frac{1}{r}\Big[-\chi_1(\vec{Q},\vec{P})+\sum_{j=1}^{n} l_1 l_j \chi_j(\vec{Q},\vec{P})\Big]+$$

$$+\chi_2'(\vec{Q},\vec{P})+v(\vec{Q})\,r^{n-1}l_1,$$

from which it follows easily that further differentiations with respect to $x_i$ will each not increase the polarity in $1/r$ by more than one unit. Thus:

$$\frac{\partial^m \chi(\vec{Q},\vec{P})}{\partial x_1^{\beta_1}\ \ldots\ \partial x_n^{\beta_n}}=\frac{1}{r^m}\,\omega(\vec{Q},\vec{P}),$$

where $\omega(\vec{Q},\vec{P})$ is a bounded function of its arguments. Thereby (10.9) is proved and consequently the theorem as well.

REMARK. All through §7, the theorems were formulated for hyperplanes of dimension $s$. These theorems are extendable to sufficiently smooth manifolds of dimension $s$. Namely, if the manifold of dimension $s$ lies in some domain for which there exists a topological mapping which is continuously differentiable with bounded derivatives and which carries the manifold under consideration into a hyperplane, then for that manifold all the theorems stated for plane domains are valid.

For the investigation of certain problems, it will be important to know the behaviour of functions on the boundaries of domains. We introduce a class of domains which for convenience will be called domains with simple boundary. We shall say that the domain $\Omega$ has a simple boundary in the case that the boundary can be decomposed into a finite number of manifolds $S_{n-1}^{(1)}, S_{n-2}^{(2)}, \cdots, S_0^{(1)}, \cdots, S_0^{(k)}$ of various dimensions and such that each manifold $S_{n-s}^{j}$ by means of a transformation of coordinates defined on part of the domain $\Omega$, and continuous with continuous derivatives up to $l$th order, and can be transformed into a hyperplane. For domains with simple boundary, we may assert that the Imbedding Theorem is valid also for the boundary manifolds.

3. INVARIANT NORMING OF $W_p^{(l)}$. For further discussion, it will be convenient for us to introduce a norm on $W_p^{(l)}$ in still another way.

Let $\phi\in W_p^{(l)}$. Then $\phi\in L_p^{(l)}$ and by virtue of the Imbedding Theorem

$$\varphi\in L_p.$$

We shall show that the norm $\|\phi\|_{W_p^{(l)}}^{(0)}$ given by the equality

$$\|\varphi\|_{W_p^{(l)}}^{(0)\,p}=\|\varphi\|_{L_p}^{p}+\|\varphi\|_{L_p^l}^{p},$$

is equivalent to an arbitrary norm constructed by means of a spherical projection operator, i.e., it turns out to be the natural norm.

The right hand side of the equality which gives the definition depends neither upon the choice of the origin of coordinates, nor upon the direction of the coordinate axes, and thereby turns out to be invariant under all possible changes of coordinates. From this we see that the natural norm may be defined in an invariant way.

We show the equivalence of $\|\phi\|^{(0)}_{W_p^{(l)}}$ to an arbitrary natural norm. Let $\|\phi\|^{(1)}_{W_p^{(l)}}$ be a norm defined by means of some arbitrary spherical projection operator. It is necessary to show that there exist constants $m$ and $M$ such that

$$m\|\varphi\|^{(0)}_{W_p^{(l)}} \leqslant \|\varphi\|^{(1)}_{W_p^{(l)}} \leqslant M\|\varphi\|^{(0)}_{W_p^{(l)}}.$$

We have:

$$\|\varphi\|^{(1)p}_{W_p^{(l)}} = \|\textstyle\prod_1^*\varphi\|^p_{L_p^{(l)}} + \|\prod_1\varphi\|^p_{S^{(l)}} = \|\varphi\|^p_{L_p^{(l)}} + \|\prod_1\varphi\|^{(1)p}_{S^{(l)}},$$

where $\prod_1$ is the given projection operator.

Further:

$$\|\varphi\|_{L_p^{(l)}} \leqslant \|\varphi\|^{(0)}_{W_p^{(l)}}.$$

We shall prove also that

$$\|\textstyle\prod_1\varphi\|^{(1)}_{S^{(l)}} \leqslant K\|\varphi\|_{L_p},$$

where $K$ is a constant depending upon the shape of the domain $\Omega$.

Let the coefficients of the polynomial $\prod_1\phi$ be $a_{\alpha_1\alpha_2\cdots\alpha_n}$; then obviously

$$\|\textstyle\prod_1\varphi\|^{(1)}_{S^{(l)}} \leqslant K_1 \max|a_{\alpha_1\alpha_2\ldots\alpha_n}|. \tag{10.10}$$

On the other hand, each of the coefficients $a_{\alpha_1\alpha_2\cdots\alpha_n}$ may be represented as an integral

$$a_{\alpha_1\alpha_2\ldots\alpha_n} = \int\cdots\int \zeta_{\alpha_1\alpha_2\ldots\alpha_n}(x_1x_2,\ldots,x_n)\,\varphi(x_1,x_2,\ldots,x_n)\times$$
$$\times\, dx_1\ldots dx_n,$$

where $\zeta(x_1,\cdots,x_n)$ is a bounded continuous function of its arguments (cf. (7.12)). Applying Hölder's inequality, we obtain

$$a_{\alpha_1\alpha_2\ldots,\alpha_n} \leqslant K_2\|\varphi\|_{L_p}. \tag{10.11}$$

By (10.10) and (10.11) we have

$$\|\textstyle\prod_1\varphi\|^{(1)p}_{S^{(l)}} \leqslant K\|\varphi\|^p_{L_p}.$$

From this, finally:

$$\|\varphi\|^{(1)p}_{W_p^{(l)}} \leqslant M[\|\varphi\|^p_{L_p^{(l)}} + \|\varphi\|^p_{L_p}] \leqslant M\|\varphi\|^{(0)p}_{W_p^{(l)}}.$$

It remains to obtain the opposite estimate. Since we always have $W_p^{(l)} \subset L_p$, by the Imbedding Theorem we have:

$$\|\varphi\|_{L_p} \leqslant K\|\varphi\|^{(1)}_{W^{(l)}_{(p)}}.$$

In addition,

$$\|\varphi\|_{L_p^{(l)}} = |\Pi_1^* \varphi\|_{L_p^{(l)}} \leqslant \|\varphi\|^{(1)}_{W_p^{(l)}},$$

from which obviously follows:

$$\|\varphi\|^{(0)}_{W_p^{(l)}} \leqslant K_1\|\varphi\|^{(1)}_{W_p^{(l)}},$$

which was to be proved.

§11. **The complete continuity of the imbedding operator (Kondrašev's Theorem).**

1. FORMULATION OF THE PROBLEM. In this section it will be shown that every bounded set in $W_p^{(l)}$ (bounded in norm) turns out to be bounded and equicontinuous in $C$, if $n < lp$, or in $L_{q^*}$ in $n > lp$. Obviously, if the domain $\Omega$ is the union of a finite number of domains for each of which this assertion is correct, then it will be correct also for the domain $\Omega$. For this reason, in the proof of the theorem we restrict ourselves to the case of a domain $\Omega$ which is star-like with respect to a sphere $C$ of radius $H$. As a preparation we shall establish a lemma on integrals of a special form. We introduce some notations. Let $\vec{P}$ and $\vec{P} + \Delta\vec{P}$ be two arbitrary points, $\vec{Q}$ the point corresponding to the variable of integration, $r = |\vec{P} - \vec{Q}|$, $r_1 = |\vec{P} + \Delta\vec{P} - \vec{Q}|$. Let $f(\vec{P}) \in L_p$ on $\Omega$. We shall consider the function $f(\vec{P})$ to be equal to zero outside of $\Omega$ and extend it thereby to the whole space. Let $0 < \lambda < n$. We consider the integral

$$U(\vec{P}, \Delta\vec{P}) = \int_{r \leqslant R} \frac{(r + r_1)^{\lambda - 1}}{r^\lambda r_1^\lambda} f(\vec{Q})\, dv_{\vec{Q}}, \tag{11.1}$$

where $R$ is an arbitrary number larger than the diameter of the domain $\Omega$. We represent it in the form

$$U(\vec{P}, \Delta\vec{P}) = \int\limits_{\substack{r \leqslant \frac{|\Delta\vec{P}|}{2} \\ r \leqslant R}} + \int\limits_{\substack{r_1 \leqslant \frac{|\Delta\vec{P}|}{2} \\ r \leqslant R}} + \int\limits_{\substack{r \leqslant R \\ r > \frac{|\Delta\vec{P}|}{2} \\ r_1 > \frac{|\Delta\vec{P}|}{2}}} \frac{(r_1 + r)^{\lambda - 1}}{r^\lambda r_1^\lambda} f(\vec{Q})\, dv_{\vec{Q}}.$$

For fixed $\Delta\vec{P}$, the last integral is bounded, the first integral is bounded by $1/r_1^\lambda$ and the second by $1/r^\lambda$. Each of these integrals yields a function in $C$ if $n<(n-\lambda)p$, or a function from $L_{q^*}$ if $n \geqq (n-\lambda)p$, since the variable domain of integration may be replaced by a fixed domain by means of introducing multipliers which are equal to zero or unity on the corresponding domains. Therefore $U(\vec{P}, \Delta\vec{P})$ will lie in $C$ or $L_{q^*}$ in the corresponding cases.

2. A LEMMA ON THE COMPACTNESS OF THE SPECIAL INTEGRALS IN $C$. *If* $n<(n-\lambda)p$ [*i.e., if* $\lambda < n/p'$], *then*

$$|\Delta\vec{P}|\,|U(\vec{P}, \Delta\vec{P})| \leqslant C\|f\|\,|\Delta\vec{P}|^\beta, \tag{11.2}$$

*where* $\beta$ *is a constant with* $0<\beta\leqq 1$.

PROOF. Since $\lambda < n/p', \lambda p' < n$, and as a result $1/r^\lambda \in L_{p'}$ on a bounded domain. We have

$$|U(\vec{P}, \Delta\vec{P})| \leqslant \int\limits_{r \leqslant R} \frac{(r + r_1)^{\lambda - 1}}{r^\lambda r_1^\lambda} |f(\vec{Q})|\, dv_{\vec{Q}}.$$

Applying the Hölder inequality with the exponents $p$ and $p'$, we find:

$$|U(\vec{P}, \Delta\vec{P})| \leqslant \|f\| \left\{ \int\limits_{r \leqslant R} \frac{(r + r_1)^{\lambda p' - p'}}{r^{\lambda p'} r_1^{\lambda p'}}\, dv \right\}^{\frac{1}{p'}} = \|f\|\, J_1^{\frac{1}{p'}}. \tag{11.3}$$

We investigate the integral

$$J_1 = \int\limits_{r \leqslant R} \frac{(r + r_1)^{\lambda p' - p'}}{r^{\lambda p'} r_1^{\lambda p'}}\, dv_{\vec{Q}}.$$

For this purpose, we pass to new variables, setting $x_i = |\Delta\vec{P}|\xi_i$, $y_j = |\Delta\vec{P}|\eta_i$, where the coordinates of the point $\vec{P}$ are $x_i$ and the coordinates of $\vec{Q}$ are $y_i$. Under this change the point $\vec{P}$ goes over into the point $\vec{P}_1$, the point $\vec{P} + \Delta\vec{P}$ into the point $\vec{P}_2$, such that $|\vec{P}_1 - \vec{P}_2| = 1$. The point of integration $\vec{Q}$ goes over into the point $\vec{Q}_1$.

We will have $|\vec{Q}_1 - \vec{P}_1| = \rho = r/|\Delta\vec{P}|$, $|\vec{Q}_1 - \vec{P}_2| = \rho_1 = r_1/|\Delta\vec{P}|$, $dv_{\vec{Q}} = |\Delta\vec{P}|^n dv_{\vec{Q}_1}$. We obtain:

$$J_1 = \int\limits_{\rho \leqslant \frac{R}{|\Delta \vec{P}|}} \frac{(\rho+\rho_1)^{\lambda p'-p'} |\Delta \vec{P}|^{\lambda p'-p'}}{\rho^{\lambda p'} \rho_1^{\lambda p'} |\Delta \vec{P}|^{2\lambda p'}} |\Delta \vec{P}|^n dv_{\vec{Q}_1} =$$

$$= |\Delta \vec{P}|^{n-\lambda p'-p'} \int\limits_{\rho \leqslant \frac{R}{|\Delta \vec{P}|}} \frac{(\rho+\rho_1)^{\lambda p'-p'}}{\rho^{\lambda p'} \rho_1^{\lambda p'}} dv_{\vec{Q}_1} = |\Delta \vec{P}|^{n-\lambda p'-p'} J_2. \quad (11.4)$$

Dividing the integral $J_2$ into two parts, we obtain:

$$J_2 = \int\limits_{\rho \leqslant \frac{R}{|\Delta \vec{P}|}} \frac{(\rho+\rho_1)^{\lambda p'-p'}}{\rho^{\lambda p'} \rho_1^{\lambda p'}} dv_{\vec{Q}_1} =$$

$$= \int\limits_{\rho \leqslant 2} \frac{(\rho+\rho_1)^{\lambda p'-p'}}{\rho^{\lambda p'} \rho_1^{\lambda p'}} dv_{\vec{Q}_1} + \int\limits_{2 \leqslant \rho \leqslant \frac{R}{|\Delta \vec{P}|}} \frac{(\rho+\rho_1)^{\lambda p'-p'}}{\rho^{\lambda p'} \rho_1^{\lambda p'}} dv_{\vec{Q}_1}.$$

The first term does not depend on $|\Delta \vec{P}|$, so that this is a convergent integral.

Estimating the second term, we remark that outside the sphere $\rho \geqq 2$, $\rho_1/\rho$ is included within fixed bounds, for since $\rho - 1 \leqq \rho_1 \leqq \rho + 1$, from which follows: $1 - 1/\rho \leqq \rho_1/\rho \leqq 1 + 1/\rho$, i.e., $1/2 \leqq \rho_1/\rho \leqq 3/2$. Therefore

$$\int\limits_{2 \leqslant \rho \leqslant \frac{R}{|\Delta \vec{P}|}} \frac{(\rho+\rho_1)^{\lambda p'-p'}}{\rho^{\lambda p'} \rho_1^{\lambda p'}} dv_{\vec{Q}_1} \leqslant K \int\limits_2^{\frac{R}{|\Delta \vec{P}|}} \rho^{n-p'-\lambda p'-1} d\rho,$$

where $K$ is a constant.

Thus

$$J_2 \leqslant K_1 + K_2 \left( \frac{R}{|\Delta \vec{P}|} \right)^{n-p'-\lambda p'} \quad (11.5)$$

Taking account of (11.4) and (11.5), we obtain:

$$J_1 \leqslant C_2 + C_1 |\Delta \vec{P}|^{n-p'-\lambda p'}.$$

From formula (11.3) there follows

$$|U(\vec{P}, \Delta \vec{P})| \leqslant \|f\| [C_2 + C_1 |\Delta \vec{P}|^{n-p'-\lambda p'}]^{\frac{1}{p'}},$$

from which we find:

$$|\Delta \vec{P}| |U(\vec{P}, \Delta \vec{P})| \leqslant \|f\| [C_2 |\Delta \vec{P}|^{p'} + C_1 |\Delta \vec{P}|^{n-\lambda p'}]^{\frac{1}{p'}}.$$

If we set $\min(1, (n-\lambda p')/p') = \beta$, we obtain

$$|\Delta\vec{P}|\,|U(\vec{P},\ \Delta\vec{P})| \leqslant C\|f\|\,|\Delta\vec{P}|^{\beta}, \quad \text{where} \quad 0<\beta\leqslant 1,$$

and the lemma is proved.

REMARK. If $n-\lambda p'-p'=0$, then by a modification of the argument we obtain $\beta<1$.

3. A LEMMA ON THE COMPACTNESS OF INTEGRALS IN $L_{q^*}$. From Theorem 2, §6, we have that if $n \geqq (n-\lambda)p$ (i.e., $\lambda \geqq n/p'$) and if in addition $s<n-(n-\lambda)p$, then $U(\vec{P}, \Delta\vec{P}) \in L_{q^*}$ as a function of the point $\vec{P}$ for fixed $\Delta\vec{P}$ on an arbitrary hyperplane of dimension $s$ for which

$$q^* < q = \frac{sp}{n-(n-\lambda)\,p}.$$

LEMMA. *Under the conditions of Theorem* 2, §6 *we have the inequality*

$$|\Delta\vec{P}| \cdot \|U(\vec{P},\ \Delta\vec{P})\|_{L_{q^*}} \leqslant C\|f\|\,|\Delta\vec{P}|^{\beta}, \tag{11.6}$$

*where* $\beta=\min(1,2\epsilon)$, *if* $2\epsilon=(n/p')+(s/q^*)-\lambda\neq 1$, *and* $\beta=1-\eta$ *where* $\eta$ *is arbitrary with* $0<\eta<1$ *if* $(n/p')+(s/q^*)-\lambda=1$.

We shall prove the lemma in the case $2\epsilon\neq 1$, leaving the proof for the special case $2\epsilon=1$ to the reader.

PROOF. We have as before in the proof of Theorem 2, §6:

$$\lambda > \frac{n}{p'};\quad \frac{s}{q} = \frac{n}{p} - n + \lambda = \lambda - \frac{n}{p'},\quad \text{i.e.,}\quad \lambda = \frac{n}{p'} + \frac{s}{q}.$$

We put $\lambda=(n/p')+(s/q^*)-2\epsilon$, where $2\epsilon=(s/q^*)-(s/q)>0$. We choose, as before, $q^*>p$ $(p<q^*<q)$.

First of all, we establish an auxiliary inequality for the function $|U(\vec{P}, \Delta\vec{P})|^{q^*}$. Namely, we show that:

$$|U(\vec{P},\ \Delta\vec{P})|^{q^*} \leqslant \|f\|^{q^*-p}\left[A_1 + A_2\,|\Delta\vec{P}|^{\left(\varepsilon-\frac{1}{2}\right)p'}\right]^{\frac{q^*}{p'}} \times$$

$$\times \int_{r\leqslant R} |f|^p \frac{(r+r_1)^{s-\left(\varepsilon+\frac{1}{2}\right)q^*}}{r^{s-\varepsilon q^*} r_1^{s-\varepsilon q^*}}\, dv_{\vec{Q}}. \tag{11.7}$$

In fact, we obviously have

$$|U(\vec{P}, \Delta\vec{P})| \leqslant \int\limits_{r \leqslant R} \left[|f|^{p\left(\frac{1}{p}-\frac{1}{q^*}\right)}\right]\left[|f|^{\frac{p}{q^*}} \frac{(r+r_1)^{\frac{s}{q^*}-\varepsilon-\frac{1}{2}}}{r^{\frac{s}{q^*}-\varepsilon} r_1^{\frac{s}{q^*}-\varepsilon}}\right] \times$$

$$\times \left[\frac{(r+r_1)^{\frac{n}{p'}-\varepsilon-\frac{1}{2}}}{r^{\frac{n}{p'}-\varepsilon} r^{\frac{n}{p'}-\varepsilon}}\right] dv_{\vec{Q}}.$$

We apply Hölder's inequality to the last integral with the exponents

$$\frac{1}{\frac{1}{p}-\frac{1}{q^*}} = \frac{pq^*}{q^*-p}, \; q^*, \; p'.$$

As it is easy to see

$$\frac{q^*-p}{pq^*}+\frac{1}{q^*}+\frac{1}{p'} = \frac{1}{p}-\frac{1}{q^*}+\frac{1}{q^*}+\frac{1}{p'} = 1.$$

We obtain

$$|U(\vec{P}, \Delta\vec{P})| \leqslant \left[\int\limits_{r \leqslant R} |f|^p \, dv_{\vec{Q}}\right]^{\frac{1}{p}-\frac{1}{q^*}} \times$$

$$\times \left[\int\limits_{r \leqslant R} |f|^p \frac{(r+r_1)^{s-\left(\varepsilon+\frac{1}{2}\right)q^*}}{r^{s-\varepsilon q^*} r_1^{s-\varepsilon q^*}} \, dv_{\vec{Q}}\right]^{\frac{1}{q^*}} \times$$

$$\times \left[\int\limits_{r \leqslant R} \frac{(r+r_1)^{n-\left(\varepsilon+\frac{1}{2}\right)p'}}{r^{n-\varepsilon p'} r_1^{n-\varepsilon p'}} \, dv_{\vec{Q}}\right]^{\frac{1}{p'}} =$$

$$= \|f\|^{1-\frac{p}{q^*}} \omega(\Delta\vec{P}) \left[\int\limits_{r \leqslant R} |f|^p \frac{(r+r_1)^{s-\left(\varepsilon+\frac{1}{2}\right)q^*}}{r^{s-\varepsilon q^*} r_1^{s-\varepsilon q^*}} \, dv_{\vec{Q}}\right]^{\frac{1}{q^*}}, \tag{11.8}$$

where

$$\omega(\Delta\vec{P}) = \left[\int\limits_{r \leqslant R} \frac{(r+r_1)^{n-\left(\varepsilon+\frac{1}{2}\right)p'}}{r^{n-\varepsilon p'} r_1^{n-\varepsilon p'}} \, dv_{\vec{Q}}\right]^{\frac{1}{p'}}$$

Estimating $\omega(\Delta\vec{P})$ by the same means as we estimated $J_1$ in the proof of Lemma 1, we obtain:

$$|\omega(\Delta\vec{P})|^{p'} \leqslant |\Delta\vec{P}|^{\left(\varepsilon-\frac{1}{2}\right)p'}\left(\int\limits_{\rho \leqslant \frac{R}{|\Delta\vec{P}|}} \frac{(\rho+\rho_1)^{n-\left(\varepsilon+\frac{1}{2}\right)p'}}{\rho^{n-\varepsilon p'}\rho_1^{n-\varepsilon p'}}\, dv_{\vec{Q}_1}\right) \leqslant$$

$$\leqslant |\Delta\vec{P}|^{\left(\varepsilon-\frac{1}{2}\right)p'}\left[K_1 + K_2 \int\limits_2^{\frac{R}{|\Delta\vec{P}|}} \rho^{\left(\varepsilon-\frac{1}{2}\right)p'-1}\, d\rho\right].$$

From (11.8) and the estimate which we have obtained for $|\omega(\Delta\vec{P})|$, (11.7) follows. We remark that in the estimate for $|\omega(\Delta\vec{P})|^{p'}$ for $\epsilon<\frac{1}{2}$ the first term is the important one, while for $\epsilon>\frac{1}{2}$ it is the second term.

Using (11.7), we pass to the estimates of the integrals of $|U(\vec{P},\Delta\vec{P})|^{q^*}$ on an arbitrary manifold of dimension $s$.

For simplicity, we restrict the argument to the case of a hyperplane. The general case may be reduced to this by a transformation of coordinates.

Integrating the inequality (11.7) in the variable $\vec{P}$ and interchanging the order of integration, we obtain:

$$\overbrace{\int\cdots\int}^{s} |U(\vec{P},\Delta\vec{P})|^{q^*} dv_s \leqslant \|f\|^{q^*-p}\left[A_1 + A_2|\Delta\vec{P}|^{\left(\varepsilon-\frac{1}{2}\right)p'}\right]^{\frac{q^*}{p'}} \times$$

$$\times \overbrace{\int\cdots\int}^{s} \left\{\int\limits_{r\leqslant R} |f|^p \frac{(r+r_1)^{s-\left(\varepsilon+\frac{1}{2}\right)q^*}}{r^{s-\varepsilon q^*} r_1^{s-\varepsilon q^*}}\, dv_{\vec{Q}}\right\} dv_s =$$

$$= \|f\|^{q^*-p}\left[A_1 + A_2|\Delta\vec{P}|^{\left(\varepsilon-\frac{1}{2}\right)p'}\right]^{\frac{q^*}{p'}} \times$$

$$\times \int\limits_{r\leqslant R} |f|^p \left\{\overbrace{\int\cdots\int}^{s} \frac{(r+r_1)^{s-\left(\varepsilon+\frac{1}{2}\right)q^*}}{r^{s-\varepsilon q^*} r_1^{s-\varepsilon q^*}}\, dv_s\right\} dv_{\vec{Q}}. \qquad (11.9)$$

In the interior integral in the variables of the point $\vec{P}$, two singular points $\vec{Q}$ and $\vec{Q}-\Delta\vec{P}$ will be possible. Each of these points for various values of $\vec{Q}$ may fall either on or outside the hyperplane over which $\vec{P}$ runs. We estimate the interior integral by a device which does not depend upon the positions of the points $\vec{Q}$ and $\vec{Q}-\Delta\vec{P}$.

First of all, applying our previously described change of coordinates, we obtain:

$$\overbrace{\int \cdots \int}^{s} \frac{(r+r_1)^{s-\left(\varepsilon+\frac{1}{2}\right)q^*}}{r^{s-\varepsilon q^*} r_1^{s-\varepsilon q^*}} dv_s = |\Delta\vec{P}|^{\left(\varepsilon-\frac{1}{2}\right)q^*} \times$$

$$\times \underset{\rho \leqslant \frac{R}{|\Delta\vec{P}|}}{\overbrace{\int \cdots \int}^{s}} \frac{(\rho+\rho_1)^{s-\left(\varepsilon+\frac{1}{2}\right)q^*}}{\rho^{s-\varepsilon q^*} \rho_1^{s-\varepsilon q^*}} dv_s = |\Delta\vec{P}|^{\left(\varepsilon-\frac{1}{2}\right)q^*} J_1. \qquad (11.10)$$

We decompose once more the domain of integration into the portion lying within the sphere $\rho \leqq 2$ and the portion lying outside that sphere

$$J_1 = J_2 + J_3 = \underset{\rho \leqslant 2}{\overbrace{\int \cdots \int}^{s}} + \underset{\rho > 2}{\overbrace{\int \cdots \int}^{s}} \frac{(\rho+\rho_1)^{s-\left(\varepsilon+\frac{1}{2}\right)q^*}}{\rho^{s-\varepsilon q^*} \rho_1^{s-\varepsilon q^*}} dv_s. \qquad (11.11)$$

We shall show now that the integral $J_2$ is bounded for arbitrary positions of the points $\vec{Q}$ and $\vec{Q}-\Delta\vec{P}$. We introduce on the plane a system of polar coordinates, taking for its center the projection of the point $\vec{Q}$ on the plane. Suppose that this projection is $\vec{Q}_0$. The distance in the plane $P_s$ to the point $\vec{Q}$ we denote by $\mathfrak{R}$. Then $\rho = (\mathfrak{R}^2 + h^2)^{1/2}$, where $h$ is the distance of the point $\vec{Q}$ from the plane.

Suppose, further, that the projection of $\vec{Q}-\Delta\vec{P}$ on the plane $P_s$ is $\vec{Q}_1$, the distance on the plane $P_s$ from the point $\vec{Q}_1$ is denoted by $\mathfrak{R}_1$, and the distance of $\vec{Q}-\Delta\vec{P}$ from the plane by $h_1$. Then

$$\rho_1 = \sqrt{\mathfrak{R}_1^2 + h_1^2}.$$

For $h > \frac{1}{4}$, $h_1 > \frac{1}{4}$, the integral $J_2$ will be bounded by virtue of the fact that $\rho$ and $\rho_1$ will both be bounded from above, as well as from below (by the value $\frac{1}{4}$). For $h < \frac{1}{4}, h_1 > \frac{1}{4}$, to estimate the integral we remark that $\rho + \rho_1 \leqq 5$, $\rho_1 > \frac{1}{4}$ , $\rho_1 > \mathfrak{R}$. Thereby:

$$J_2 \leqslant K \int\limits_{\mathfrak{R} < 2} \frac{\mathfrak{R}^{s-1}}{\mathfrak{R}^{s-\varepsilon q^*}} d\mathfrak{R} = K_1,$$

where $K_1$ is some constant.

Similarly we estimate $J_2$ in the case in which $h > \frac{1}{4}$, $h_1 < \frac{1}{4}$. Finally in the last case, in which $h < \frac{1}{4}$, $h_1 < \frac{1}{4}$, the distance between the point $\vec{Q}_0$ and $\vec{Q}_1$ is greater than $\frac{1}{2}$ and, consequently,

$$J_2 \leqslant K \overbrace{\int \cdots \int}^{s}_{\mathfrak{R} < 2} \frac{1}{\mathfrak{R}^{s-\varepsilon q^*} \mathfrak{R}_1^{s-\varepsilon q^*}} dv_s,$$

i.e., $J_2$ will be a convergent integral.

We now estimate the integral $J_3$. We remark as before that the ratio $\rho_1/\rho$ is included within fixed bounds, and, using polar coordinates in the plane $P_s$, we shall have:

$$J_3 \leqslant K \int\limits_{2 \leqslant \rho \leqslant \frac{R}{|\Delta\vec{P}|}} \rho^{-s+\left(\varepsilon-\frac{1}{2}\right)q^*} \mathfrak{R}^{s-1} \, d\mathfrak{R} \leqslant$$

$$\leqslant K \int_a^{\frac{R}{|\Delta\vec{p}|}} \mathfrak{R}^{s-1} \left[\sqrt{\mathfrak{R}^2 + h^2}\right]^{-s+\left(\varepsilon-\frac{1}{2}\right)q^*} d\mathfrak{R},$$

where $a = 0$ if $h > 2$ and $a = (4-h^2)^{1/2}$ for $h \leqq 2$. For $h > 2$ this integral is a decreasing function of $h$. Therefore it suffices to verify the estimate for $h \leqq 2$. For these values the integrand is bounded on the segment $a \leqq \mathfrak{R} \leqq 2$.

Furthermore, if $\mathfrak{R} \geqq 2$, we obtain by means of some estimates:

$$J_3 \leqslant C_1 \int_2^{\frac{R}{|\Delta\vec{P}|}} \mathfrak{R}^{\left(\varepsilon-\frac{1}{2}\right)q^*-1} d\mathfrak{R} = C_3 \left(\frac{R}{|\Delta\vec{P}|}\right)^{\left(\varepsilon-\frac{1}{2}\right)q^*} + C_4 =$$

$$= C_4 + C_5 |\Delta\vec{P}|^{-\left(\varepsilon-\frac{1}{2}\right)q^*}$$

From this, using (11.11) and the boundedness of $J_2$, we find:

$$J_1 \leqslant C_6 + C_5 |\Delta\vec{P}|^{-\left(\varepsilon-\frac{1}{2}\right)q^*}.$$

Thereby, on the basis of (11.10), we obtain:

$$J \leqslant C_5 + C_6 |\Delta\vec{P}|^{\left(\varepsilon-\frac{1}{2}\right)q^*}$$

and if we apply (8.9), we obtain:

$$\overbrace{\int \cdots \int}^{s} |U(\Delta\vec{P})|^{q^*} dv_s \leqslant \|f\|^{q^*} \left[A_1 + A_2 |\Delta\vec{P}|^{\left(\varepsilon-\frac{1}{2}\right)p'}\right]^{\frac{q^*}{p'}} \times$$
$$\times [C_5 + C_6 |\Delta\vec{P}|^{\left(\varepsilon-\frac{1}{2}\right)q^*}].$$

From this we conclude easily that:

$$\left[\overbrace{\int \cdots \int}^{s} |U(\vec{P}, \Delta\vec{P})|^{q^*} dv_s\right]^{\frac{1}{q^*}} \leqslant \|f\| \cdot \begin{cases} B_1, & \text{if } \varepsilon > \frac{1}{2}; \\ B_2 |\Delta\vec{P}|^{2\varepsilon-1}, & \text{if } \varepsilon < \frac{1}{2}, \end{cases}$$

from which follows:

$$|\Delta\vec{P}| \left[\overbrace{\int \cdots \int}^{s} |U(\vec{P}, \Delta\vec{P})|^{q^*} dv_s\right]^{\frac{1}{q^*}} \leqslant B\|f\| |\Delta\vec{P}|^{\beta},$$

where $\beta = \min(1, 2\epsilon)$ and the lemma is proved.

4. COMPLETE CONTINUITY OF THE IMBEDDING OPERATOR IN $C$.

THEOREM 1. *If $n < lp$, then the operator of imbedding $W_p^{(l)}$ in $C$ is completely continuous, i.e., for any set $\{\phi\} \subset W_p^{(l)}$ with bounded norms $\|\phi\|_{W_p^{(l)}} \leqq N$, are compact in $C$.*

From the Imbedding Theorems it follows that $\{\phi\}$ is bounded in $C$. It suffices to prove the equicontinuity in $C$ of the set $\{\phi\}$. We have $\phi = S + \phi^*$, the decomposition of $\phi$ by the formula (7.12).

Since $\|S\|_S + \|\phi\|_{L_p^{(l)}} \leqq N$ for all $\phi \in \{\phi\}$, $\|S\|_S \leqq N$ and, as a result the coefficients of the polynomials $S$ are uniformly bounded. From this follows easily the equicontinuity of the polynomials $S$. It remains to show that from $\|\phi\|_{L_p^{(l)}} \leqq N$ follows the equicontinuity of the $\phi^*$. We have:

$$\varphi^*(\vec{P}) = \int_{\Omega} \frac{1}{r^{n-l}} \sum_{\Sigma\alpha_i = l} w_{\alpha_1 \cdots \alpha_n} \frac{\partial^l \varphi}{\partial x_1^{\alpha_1} \cdots \partial x_n^{\alpha_n}} dv_{\vec{Q}}. \tag{11.12}$$

Set

$$r_1 = |\vec{Q} - (\vec{P} + \Delta\vec{P})|.$$

Then

$$\frac{w_{\alpha_1,\ldots,\alpha_n}(\vec{Q},\vec{P}+\Delta\vec{P})}{r_1^{n-l}}-\frac{w_{\alpha_1,\ldots,\alpha_n}(\vec{Q},\vec{P})}{r^{n-l}}=\frac{\overline{w}_{\alpha_1,\ldots,\alpha_n}(r_1,\vec{l}_1,\vec{P}+\Delta\vec{P})}{r_1^{n-l}}-$$

$$-\frac{\overline{w}_{\alpha_1,\ldots,\alpha_n}(r,\vec{l},\vec{P})}{r^{n-l}}=\overline{w}_{\alpha_1,\ldots,\alpha_n}(r,\vec{l},\vec{P})\left(\frac{1}{r_1^{n-l}}-\frac{1}{r^{n-l}}\right)+$$

$$+\frac{1}{r_1^{n-l}}\Big\{\Big[\overline{w}_{\alpha_1,\ldots,\alpha_n}(r_1,\vec{l}_1,\vec{P}+\Delta\vec{P})-\overline{w}_{\alpha_1\ldots\alpha_n}(r,\vec{l}_1,\vec{P}+\Delta\vec{P})\Big]+$$

$$+\Big[\overline{w}_{\alpha_1,\ldots,\alpha_n}(r,\vec{l}_1,\vec{P}+\Delta\vec{P})-\overline{w}_{\alpha_1\ldots\alpha_n}(r,\vec{l},\vec{P}+\Delta\vec{P})\Big]+$$

$$+\Big[\overline{w}_{\alpha_1,\ldots,\alpha_n}(r,\vec{l},\vec{P}+\Delta\vec{P})-\overline{w}_{\alpha_1,\ldots,\alpha_n}(r,\vec{l},\vec{P})\Big]\Big\}. \qquad (11.13)$$

Since $\overline{w}_{\alpha_1\cdots\alpha_n}(r,\vec{l},\vec{P})$ and its derivatives with respect to $r$, $\vec{l}$, and $\vec{P}$ are bounded, while in addition $|r_1-r|\leqq|\Delta\vec{P}|$, $(r+r_1/r)\geqq 1$, and $1/(r+r_1)>A>0$ (the domain being bounded), it follows that

$$\frac{1}{r_1^{n-l}}\Big[\overline{w}_{\alpha_1,\ldots,\alpha_n}(r_1,\vec{l}_1,\vec{P}+\Delta\vec{P})-\overline{w}_{\alpha_1,\ldots,\alpha_n}(r,\vec{l}_1,\vec{P}+\Delta\vec{P})\Big]\Big|\leqslant$$

$$\leqslant\frac{A_1|\Delta\vec{P}|}{r_1^{n-l}}\leqslant B\,\frac{(r+r_1)^{n-l-1}}{r^{n-l}\,r_1^{n-l}}.$$

Analogously we find:

$$\Big|\frac{1}{r_1^{n-l}}\Big[\overline{w}_{\alpha_1,\ldots,\alpha_n}(r,\vec{l},\vec{P}+\Delta\vec{P})-\overline{w}_{\alpha_1,\ldots,\alpha_n}(r,l,\vec{P})\Big]\Big|\leqslant$$

$$\leqslant\frac{A_1|\Delta\vec{P}|}{r_1^{n-l}}\leqslant B\,\frac{(r+r_1)^{n-l-1}}{r^{n-l}\,r_1^{n-l}}.$$

Furthermore, since $\vec{l}=(\vec{Q}-\vec{P})/r$, $\vec{l}_1=(\vec{Q}-(\vec{P}+\Delta\vec{P}))/r_1$,

$$\vec{l}_1-\vec{l}=\frac{r\vec{Q}-r\vec{P}-r\Delta\vec{P}-r_1\vec{Q}+r_1\vec{P}}{rr_1}=\frac{-(r_1-r)(\vec{Q}-\vec{P}-\Delta\vec{P})-r_1\Delta\vec{P}}{rr_1},$$

from which it follows that

$$|\vec{l}_1-\vec{l}|\leqslant\frac{r_1|\Delta\vec{P}|}{rr_1}+r_1\,\frac{|\Delta\vec{P}|}{r_1 r}=2\,\frac{|\Delta\vec{P}|}{r}.$$

Therefore

$$\Big|\frac{1}{r_1^{n-l}}\Big[\overline{w}_{\alpha_1,\ldots,\alpha_n}(r,\vec{l}_1,\vec{P}+\Delta\vec{P})-\overline{w}_{\alpha_1\ldots\alpha_n}(r,\vec{l},\vec{P}+\Delta\vec{P})\Big]\Big|\leqslant$$

$$\leqslant\frac{A_1|\vec{l}_1-\vec{l}|}{r_1^{n-l}}\leqslant 2A_1|\Delta\vec{P}|\,\frac{1}{rr_1^{n-l}}\leqslant 2A_1|\Delta\vec{P}|\,\frac{(r+r_1)^{n-l-1}}{r^{n-l}\,r_1^{n-l}},$$

and, finally,

$$\left| \overline{w}_{\alpha_1, \ldots, \alpha_n}(r, \vec{l}, \vec{P}) \left( \frac{1}{r_1^{n-l}} - \frac{1}{r^{n-l}} \right) \right| \leqslant \frac{A_1}{r^{n-l} r_1^{n-l}} | r^{n-l} - r_1^{n-l} | =$$

$$= \frac{A_1}{r^{n-l} r_1^{n-l}} | r_1 - r | (r^{n-l-1} + r^{n-l-2} r_1 + \ldots + r r_1^{n-l-2} + r_1^{n-l-1}) \leqslant$$

$$\leqslant \frac{A | \Delta \vec{P} |}{r^{n-l} r_1^{n-l}} (r + r_1)^{n-l-1};$$

$$n > l, \quad n - l - 1 \geqslant 0.$$

Substituting these estimates in (11.13), we obtain:

$$\left| \frac{w_{\alpha_1, \ldots, \alpha_n}(\vec{Q}, \vec{P} + \Delta \vec{P})}{r_1^{n-l}} - \frac{w_{\alpha_1, \ldots, \alpha_n}(\vec{Q}, \vec{P})}{r^{n-l}} \right| \leqslant C \frac{(r + r_1)^{n-l-1}}{r^{n-l} r_1^{n-l}},$$

and from (11.12) we find

$$| \varphi^* (\vec{P} + \Delta \vec{P}) - \varphi^* (\vec{P}) | \leqslant C | \Delta \vec{P} | \int_{\Omega} \frac{(r + r_1)^{n-l-1}}{r^{n-l} r_1^{n-l}} \times$$

$$\times \sum_{\Sigma \alpha_i = l} \left| \frac{\partial^l \varphi}{\partial x_1^{\alpha_1} \ldots \partial x_n^{\alpha_n}} \right| dv_{\vec{Q}}. \tag{11.14}$$

In the derivation of the inequality (11.14) we never used the assumption that $lp > n$. We apply Lemma 1 to (11.14), putting $\lambda = n - l$. We obtain: $n - \lambda = l$, $(n - \lambda)p = lp$, and by the hypothesis of the theorem, $lp > n$; consequently the inequality $n < (n - \lambda)p$ is satisfied. Then

$$| \varphi^* (\vec{P} + \Delta \vec{P}) - \varphi^* (\vec{P}) | < K \| \varphi \|_{L_p^{(l)}} | \Delta \vec{P} |^{\beta}; \; \beta = \min \left( 1, l - \frac{n}{p} \right), \tag{11.15}$$

from which follows the equicontinuity of $\phi^*$ for a set $\{\phi\}$ bounded in $W_p^{(l)}$, since this estimate of the oscillation does not depend upon the particular member $\phi$ of the family of functions $\{\phi\}$.

Theorem 1 is proved.

5. COMPLETE CONTINUITY OF THE OPERATOR OF IMBEDDING IN $L_{q^*}$.

THEOREM 2. *If $n > lp$, $s > n - lp$, $q^* < sp/(n - lp)$, then the operator of imbedding the space $W_p^{(l)}$ in $L_{q^*}$ for an arbitrary hyperplane of dimension $s$ is completely continuous, i.e., each set $\{\phi\} \subset W_p^{(l)}$ with bounded $\| \phi \|_{W_p^{(l)}}$ turns out to be strongly compact in $L_{q^*}$.*

The proof of this theorem is essentially a repetition of the proof of Theorem 1. It all reduces to the proof of the equicontinuity in the large of the functions $\phi^*$ in $L_{q^*}$ if $\| \phi \|_{L_p^{(l)}} \leqq N$. We need to show that

$$\|\varphi^* (\vec{P}^{(s)} + \Delta\vec{P}) - \varphi^* (\vec{P}^{(s)})\|_{L_{q^*}} < \varepsilon, \tag{11.16}$$

if only $|\Delta\vec{P}| < \delta(\epsilon)$, where $\delta \to 0$ when $\epsilon \to 0$. For this purpose, we apply Lemma 2 to the right side of (11.14), putting $\lambda = n - l$; then the inequality $n \geqq (n-\lambda)p$ is satisfied since by the hypothesis of the theorem $n \geqq lp$. We obtain, using the inequality (11.6):

$$\|\varphi^* (\vec{P}^{(s)} + \Delta\vec{P}) - \varphi^* (\vec{P}^{(s)})\|_{L_{q^*}} \leqslant K \|\varphi\|_{L_p^{(l)}} |\Delta\vec{P}|^{\beta};$$

$$\beta = \min\left[1; \frac{s}{q^*} - \left(\frac{n}{p} - l\right)\right],$$

from which there follows (11.6). Theorem 2 is proved.

REMARK. Suppose that on the domain $\Omega$ there is given a summable function $\phi(x_1, \cdots, x_n)$. Let $E$ be a smooth manifold of dimension $s$. We shall say that $\phi(x_1, \cdots, x_n)$ is continuous in the sense of $L_{q^*,s}$ if

$$\int_E \left| \varphi(\vec{P} + \Delta\vec{P}) - \varphi(\vec{P}) \right|^{q^*} dE \to 0 \text{ for } |\Delta\vec{P}| \to 0,$$

for any manifold $E$ for which only the translation of $E$ by the vector $\Delta\vec{P}$ is contained in $\Omega$. From the theorems proved above it follows that every $\phi \in W_p^{(l)}$ is continuous in the sense of $L_{q^*,s}$ if $s > n - lp$ and $q^* < sp/(n-lp)$ (if $n - lp < 0$, then $\phi$ is continuous in the ordinary sense).

## CHAPTER II

# VARIATIONAL METHODS IN MATHEMATICAL PHYSICS

### §12. The Dirichlet problem.

1. INTRODUCTION. As is well-known, the equations of mathematical physics often appear as the Euler equations for certain variational problems.

In the calculus of variations, when looking for the extremum of some functional of the form

$$\overbrace{\int\int_{\Omega}\int\cdots\int}^{n} F\left(u, \frac{\partial u}{\partial x_j}, x_j\right) dx_1 \ldots dx_n +$$

$$+ \overbrace{\int_S\cdots\int}^{n-1} \Phi\left(u, \frac{\partial u}{\partial x_j}, x_j\right) dx_1 \ldots dx_n$$

in some class of functions, one finds the solution of some boundary value problem for the corresponding Euler equation.

On the one hand these variational problems for the extrema of functionals may often be solved by direct means. So it is natural to ask whether conversely in those cases where the basic equation is an Euler equation, one can reduce the given boundary value problem to a problem of the calculus of variations which may then be resolved by means of the direct method. This is precisely the idea of the variational methods in mathematical physics.

We begin the investigation of the variational methods of mathematical physics with the study of the simplest equation of elliptic type, namely the Laplace equation:

$$\Delta u = 0. \tag{12.1}$$

We consider for this equation the Dirichlet problem, i.e., the problem of finding the harmonic function taking on given values on the boundary.

Let $\Omega$ be a bounded domain of $n$-dimensional space bounded by a surface

$S$ which is a piecewise continuously differentiable or consists of a finite number of such surfaces. Let $S$ be a regular boundary in the sense defined above (cf. Chapter I, §10, item 2).

We consider on $\Omega$ a function $u(x_1,x_2,\cdots,x_n)$ which is summable and has square-summable generalized derivatives of first order. Let

$$D(u)=\int_{\Omega}\cdots\int\sum_{i=1}^{n}\left(\frac{\partial u}{\partial x_i}\right)^2 d\Omega<\infty. \tag{12.2}$$

This means that $u\in W_2^{(1)}$, and consequently by the basic Imbedding Theorems, $u\in L_2$ on every $(n-1)$-manifold since

$$2<q=\frac{(n-1)2}{n-1\cdot 2}=2+\frac{2}{n-2}.$$

Moreover, one may assert by virtue of the complete continuity of the imbedding operator that if a piece $S_1$ of some $(n-1)$-dimensional manifold is translated by a vector $\Delta\vec{P}$ so that it remains within $\Omega$, then

$$\overbrace{\int_{S_1}\cdots\int}^{n-1}|u(\vec{P}+\Delta\vec{P})-u(\vec{P})|^2\,dv_{\vec{P}}\to 0, \tag{12.3}$$

if

$$|\Delta\vec{P}|\to 0.$$

It is obvious therefore that not every function $\phi\in L_2$ given on the surface $S$ can be the limiting value of some function $v\in W_2^{(1)}$ given in the interior of the domain. Indeed there also follows from the Imbedding Theorem the summability of the limiting value $v$ on the surface to any power less than $2(n-1)/(n-2)$. Later we shall see (cf. item 5) that such summability and even continuity of the limiting values is still not sufficient in order that such a function may appear as the limiting value of a function in $W_2^{(1)}$.

Let us agree to call a function $\phi$ given on the boundary $S$ of the domain permissible if there exists a function $v$ in $W_2^{(1)}$ for which $\phi$ is the boundary value.

The Dirichlet problem consists in seeking a harmonic function from $W_2^{(1)}$ which assumes on the boundary the given permissible value $\phi$:

$$u|_S=\varphi. \tag{12.4}$$

Let us proceed to the solution of this problem.

For this purpose we first solve a variational problem and then show that the solution of the variational problem turns out to be a solution of the Dirichlet problem.

2. SOLUTION OF THE VARIATIONAL PROBLEM. We denote by $W_2^{(1)}(\phi)$ the set of functions $v$ in $W_2^{(1)}$ which assume the value $\phi$ on $S$. Since $\phi$ is a per-

missible function, $W_2^{(1)}(\phi)$ is not the empty set. For each $v \in W_2^{(1)}(\phi)$, we have

$$0 \leqslant D(v) < \infty,$$

and therefore there exists a greatest lower bound for the values of $D(v)$:

$$d = \inf_{v \in W_2^{(1)}(\phi)} D(v), \qquad d \geqslant 0. \tag{12.5}$$

From the set $W_2^{(1)}(\phi)$, we may choose a sequence $\{v_k\}$ for which

$$\lim_{k \to \infty} D(v_k) = d, \tag{12.6}$$

as follows from the definition of greatest lower bound. The sequence $\{v_k\}$ is called a minimizing sequence.

THEOREM 1. *The minimizing sequence $\{v_k\}$ converges in $W_2^{(1)}$; the limit function lies in $W_2^{(1)}(\phi)$ and yields a proper minimum for the functional $D(v)$ among all such functions.*

Let us prove this theorem. Indeed, we define the norm in $W_2^{(1)}$ by the formula

$$\|v\|_{W_2^{(1)}} = \left\{\left[\int \underset{S}{\cdots} \int v\, dS\right]^2 + D(v)\right\}^{\frac{1}{2}}, \tag{12.7}$$

obtained from formulas (7.5) and (9.8) for

$$(\rho, v) = \int \underset{S}{\cdots} \int v\, dS.$$

From the equation

$$\int \underset{S}{\cdots} \int (v_k - v_m)\, dS = 0$$

we obtain:

$$\|v_k - v_m\|_{W_2^{(1)}} = [D(v_k - v_m)]^{\frac{1}{2}}.$$

The convergence of $\{v_k\}$ in $W_2^{(1)}$ will be proved if we can show that

$$D(v_k - v_m) \to 0 \quad \text{for } k,\ m \to \infty.$$

Let $\epsilon > 0$ be given. We can find $N > 0$ such that $D(v_k) < d + \epsilon$ if $k > N$. Let

$k$ and $m>N$. Obviously $(v_k+v_m)/2\in W_2^{(1)}(\phi)$: and therefore

$$D\left(\frac{v_k+v_m}{2}\right)\geqslant d.$$

From the obvious equality

$$D\left(\frac{v_k+v_m}{2}\right)+D\left(\frac{v_k-v_m}{2}\right)=\frac{1}{2}D(v_k)+\frac{1}{2}D(v_m)$$

follows the inequality

$$d+D\left(\frac{v_k-v_m}{2}\right)<\frac{d+\varepsilon}{2}+\frac{d+\varepsilon}{2}=d+\varepsilon,$$

i.e.,

$$D\left(\frac{v_k-v_m}{2}\right)<\varepsilon, \quad \text{or} \quad D(v_k-v_m)<4\varepsilon,$$

and, consequently,

$$D(v_k-v_m)\to 0 \quad \text{for } k,\ m\to\infty.$$

From the completeness of the space $W_2^{(1)}$ it follows that $\{v_k\}$ converges to some function $v_0$ in $W_2^{(1)}$, i.e.,

$$\|v_0-v_k\|_{W_2^{(1)}}\to 0, \qquad k\to\infty.$$

We shall show that $D(v_0)=d$. For this purpose, we note that:

$$\begin{aligned}
|D(v_0)-D(v_k)|&=\left|\overbrace{\int\cdots\int_\Omega}^{n}\sum_{i=1}^{n}\left[\left(\frac{\partial v_0}{\partial x_i}\right)^2-\left(\frac{\partial v_k}{\partial x_i}\right)^2\right]d\Omega\right|=\\
&=\left|\overbrace{\int\cdots\int_\Omega}^{n}\sum_{i=1}^{n}\left[\left(\frac{\partial v_0}{\partial x_i}-\frac{\partial v_k}{\partial x_i}\right)\left(\frac{\partial v_0}{\partial x_i}+\frac{\partial v_k}{\partial x_i}\right)\right]d\Omega\right|\leqslant\\
&\leqslant\sum_{i=1}^{n}\left|\overbrace{\int\cdots\int_\Omega}^{n}\left(\frac{\partial v_0}{\partial x_i}-\frac{\partial v_k}{\partial x_i}\right)\left(\frac{\partial v_0}{\partial x_i}+\frac{\partial v_k}{\partial x_i}\right)d\Omega\right|\leqslant\\
&\leqslant\sum_{i=1}^{n}\left[\overbrace{\int\cdots\int_\Omega}^{n}\left(\frac{\partial v_0}{\partial x_i}-\frac{\partial v_k}{\partial x_i}\right)^2 d\Omega\right]^{\frac{1}{2}}\left[\overbrace{\int\cdots\int_\Omega}^{n}\left(\frac{\partial v_0}{\partial x_i}+\frac{\partial v_k}{\partial x_i}\right)^2 d\Omega\right]^{\frac{1}{2}}\leqslant\\
&\leqslant\sum_{i=1}^{n}\|v_0+v_k\|_{W_2^{(1)}}\left[\overbrace{\int\cdots\int_\Omega}^{n}\left(\frac{\partial v_0}{\partial x_i}-\frac{\partial v_k}{\partial x_i}\right)^2 d\Omega\right]^{\frac{1}{2}}\leqslant\\
&\leqslant\|v_0+v_k\|_{W_2^{(1)}}\cdot\|v_0-v_k\|_{W_2^{(1)}},
\end{aligned}$$

from which there follows

$$D(v_0)=\lim_{k\to\infty}D(v_k)=d.$$

It is not hard now to show that $v_0 \in W_2^{(1)}(\phi)$.

Indeed, $v_0 \in W_2^{(1)}$ and therefore, the function $v_0$ has a sense on every manifold of dimensional $(n-1)$ and on every such manifold lies in $L_2$.

The value of the function $v_0$ on the surface $S$ will be equal to $\phi$. Indeed,

$$\int_S \cdots \int (v_k - v_0)^2 \, dS \leqslant \| v_k - v_0 \|^2_{W_2^{(1)}}$$

and consequently

$$\int_S \cdots \int (v_k - v_0)^2 \, dS \to 0,$$

but $v_k|_S = \phi$, and thus,

$$\int_S \cdots \int (\varphi - v_0)^2 \, dS = 0.$$

The function $v_0$ takes on its boundary value $\phi$ by converging to it in the mean as was shown in the Imbedding Theorems.

Thus $v_0 \in W_2^{(1)}$ is such that

$$\left.\begin{array}{ll} (1) & v_0|_S = \varphi(\vec{P}), \\ (2) & D(v_0) = d. \end{array}\right\} \tag{12.8}$$

Therefore the variational problem is solved.

3. SOLUTION OF THE DIRICHLET PROBLEM.

THEOREM 2. *The function giving a minimum to $D(v)$ on $W_2^{(1)}(\phi)$ is the solution of the Dirichlet problem.*

We shall show that $v_0$ in the interior of $\Omega$ has continuous derivatives of arbitrary order and satisfies equation (12.1).

Let $\xi \subset W_2^{(1)}$, $\xi|_S = 0$ and otherwise arbitrary. Consider

$$D(v_0 + \lambda\xi) = D(v_0) + 2\lambda D(v_0, \xi) + \lambda^2 D(\xi), \tag{12.9}$$

where

$$D(v_0, \xi) = \overbrace{\int_\Omega \cdots \int}^{n} \sum_{i=1}^{n} \frac{\partial v_0}{\partial x_i} \frac{\partial \xi}{\partial x_i} \, d\Omega.$$

By virtue of the fact that $v_0 + \lambda\xi \in W_2^{(1)}(\phi)$, we have $D(v_0 + \lambda\xi) \geqq d = D(v_0)$ and therefore (12.9) has a minimum for $\lambda = 0$. By the theorem of Fermat, we have

$$D(v_0, \xi) = 0. \tag{12.10}$$

We shall choose $\xi(x_1, \cdots, x_n)$ of a special form. Let $\psi(\eta)$ be such that $\psi(\eta) = 1$

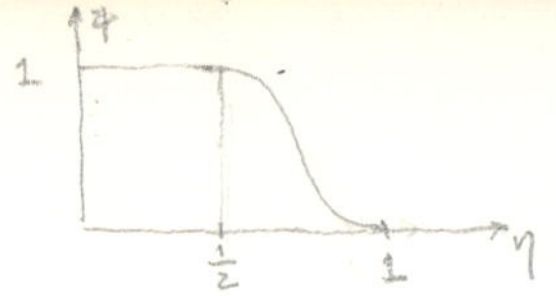



for $0 \leq \eta \leq \frac{1}{2}$, $\psi(\eta)=0$ for $\eta \geq 1$, $\psi(\eta)$ monotone on $\lfloor\frac{1}{2}, 1\rfloor$ and having continuous derivatives of arbitrary order for all $\eta \in [0, \infty]$. For example, put

$$\psi(\eta)=\frac{1}{2}\left[1+\operatorname{th}\frac{\eta-\frac{3}{4}}{\left(\eta-\frac{1}{2}\right)(\eta-1)}\right]; \quad \frac{1}{2}<\eta<1.$$

Consider some interior point $M_0$ of the domain $\Omega$; the distance from it to an arbitrary point we shall denote by $r$; suppose $M_0$ is at distance $\delta$ from $S$. We choose two numbers $h_1$ and $h_2$

$$0<h_1<h_2<\delta$$

and set

$$\xi=r^{2-n}\left[\psi\left(\frac{r}{h_1}\right)-\psi\left(\frac{r}{h_2}\right)\right].$$

It is obvious that $\xi|_S=0$ by virtue of the choice of $h_1$ and $h_2$. In addition $\xi=0$ for $r<h_1/2$ and consequently the function $\xi$ is continuous and has continuous derivatives of all orders, and thus $\xi \in W_2^{(1)}$. For such $\xi$, (12.10) holds. By virtue of the definition of the generalized derivative $\partial v_0/\partial x_i$, we have

$$\int_{\Omega}\cdots\int \frac{\partial v_0}{\partial x_i}\frac{\partial \xi}{\partial x_i}\,d\Omega=-\int_{\Omega}\cdots\int v_0\frac{\partial^2\xi}{\partial x_1^2}\,d\Omega.$$

Therefore the equality (12.10) gives

$$\int_{\Omega}\cdots\int v_0\Delta\xi\,d\Omega=0. \tag{12.11}$$

But

$$\Delta\xi=\Delta\left(r^{2-n}\psi\left(\frac{r}{h_1}\right)\right)-\Delta\left(r^{2-n}\psi\left(\frac{r}{h_2}\right)\right)=\frac{1}{h_1^n}\,\omega\left(\frac{r}{h_1}\right)-\frac{1}{h_2^n}\,\omega\left(\frac{r}{h_2}\right),$$

where

$$\omega\left(\frac{r}{h_i}\right)=h_i^n\Delta\left(r^{2-n}\psi\left(\frac{r}{h_i}\right)\right)=h_i^2\Delta\left[\left(\frac{r}{h_i}\right)^{2-n}\psi\left(\frac{r}{h_i}\right)\right],$$

and it is obvious that the right hand side is a function only of $r/h_i$

Using the fact that $\psi(r/h_i)=1$ for $r<h_i/2$ and $\Delta r^{2-n}=0$, we obtain:

$$\omega\left(\frac{r}{h_i}\right)=0; \quad r<\frac{h_i}{2}.$$

Thus $\Delta\xi$ is the difference of two arbitrarily often differentiable functions on the whole space and equation (12.11) gives

$$\frac{1}{h_1^n}\int_{\Omega}\cdots\int v_0\omega\left(\frac{r}{h_1}\right)d\Omega=\frac{1}{h_2^n}\int_{\Omega}\cdots\int v_0\omega\left(\frac{r}{h_2}\right)d\Omega. \tag{12.12}$$

Multiplying both sides of (12.12) by $1/(n-2)\sigma_n$ where $\sigma_n$ is the surface area of the unit sphere in $n$-dimensional space, we obtain:

$$\frac{1}{(n-2)\sigma_n h_1^n}\int_{\Omega}\cdots\int v_0\omega\left(\frac{r}{h_1}\right)d\Omega=$$

$$=\frac{1}{(n-2)\sigma_n h_2^n}\int_{\Omega}\cdots\int v_0\omega\left(\frac{r}{h_2}\right)d\Omega. \tag{12.13}$$

The function $((n-2)\sigma_n h^n)^{-1}\ \omega(r/h)$ may be considered as an averaging kernel (cf. Chapter I, §2, item 4) since its integral over the whole space is equal to 1. In fact,

$$\frac{1}{(n-2)\sigma_n h^n}\int_{\Omega}\cdots\int\omega\left(\frac{r}{h}\right)d\Omega=\frac{1}{(n-2)\sigma_n}\int_{\Omega}\cdots\int\Delta\left(r^{2-n}\psi\left(\frac{r}{h}\right)\right)d\Omega=$$

$$=\frac{1}{(n-2)\sigma_n}\overbrace{\int_{r=h}\cdots\int}^{n-1}\frac{\partial\left(r^{2-n}\psi\left(\frac{r}{h}\right)\right)}{\partial r}dS-$$

$$-\frac{1}{(n-2)\sigma_n}\overbrace{\int_{r=\frac{h}{2}}\cdots\int}^{n-1}\frac{\partial\left(r^{2-n}\psi\left(\frac{r}{h}\right)\right)}{\partial r}dS=\frac{1}{\sigma_n}\int_{r=\frac{h}{2}}\cdots\int r^{1-n}\,dS=$$

$$=\frac{1}{\sigma_n}\left(\frac{h}{2}\right)^{1-n}\left(\frac{h}{2}\right)^{n-1}\sigma_n=1.$$

Using this, the equation (12.13) may be rewritten in the form:

$$(v_0)_{h_1}=(v_0)_{h_2}. \tag{12.14}$$

We see that the averaged functions for $v_0$ do not change with a change in $h$ (if $h<\delta$) at points lying at a distance greater than $\delta$ from the boundary, and consequently the limit of the sequence $(v_0)_h$ coincides with the sequence, i.e., $(v_0)_h=v_0$. Since $(v_0)_h$ has continuous derivatives of all orders, the same is true for $v_0$.

Suppose now that $\xi$ is an arbitrary function continuous with its first derivatives in $\Omega$ and null outside some interior subdomain. Then obviously an integration by parts gives:

$$D(v_0,\xi)=-\overbrace{\int_{\Omega}\cdots\int}^{n}\xi\Delta v_0\,d\Omega=0,$$

from which by the arbitrariness of $\xi$, one concludes:

$$\Delta v_0=0,$$

i.e., $v_0$ is a solution of equation (12.1) and, as was shown earlier, assumes on $S$ the values $\phi(\vec{P})$ (in the sense of $L_{2,(n-1)}$). Thus $v_0$ is a solution of the Dirichlet problem.

4. UNIQUENESS OF THE SOLUTION OF THE DIRICHLET PROBLEM.

THEOREM. *The solution of the Dirichlet problem in the indicated formulation is unique.*

We establish as a preliminary one important lemma.

LEMMA. *Let $\xi \in W_2^{(1)}$ be continuous with its partial derivatives of first order within $\Omega$ and $\xi|_S = 0$ in the sense of $L_{2,n-1}$. We introduce the function*

$$\Psi_{2h} = \begin{cases} 1 & \text{within} \quad \Omega_{2h}, \\ 0 & \text{outside of} \quad \Omega_{2h}, \end{cases}$$

*where $\Omega_\delta$ is the set of all points whose distance from the boundary $S$ is not less than $\delta$. Form the average of $\psi_{2h}$ with respect to the kernel $(\kappa h^n)^{-1} \ \psi(r/h)$ where $\psi(\eta)$ is the function introduced earlier and $\kappa = \sigma_n \int_0^1 \eta^n \psi(\eta) d\eta$. This averaged function we denote by $\chi_h$:*

$\Omega_\delta = \{\vec{P} : d(\vec{P}, \partial\Omega) \geqslant \delta\}$

$$\chi_h = \frac{1}{\varkappa h^n} \int \cdots \int_{r \leqslant h} \Psi_{2h} \psi\left(\frac{r}{h}\right) d\Omega.$$

*Then for any function $v \in W_2^{(1)}$, we have the formula*

$$D(v, \xi) = \lim_{h \to 0} D(v, \xi\chi_h). \tag{12.15}$$

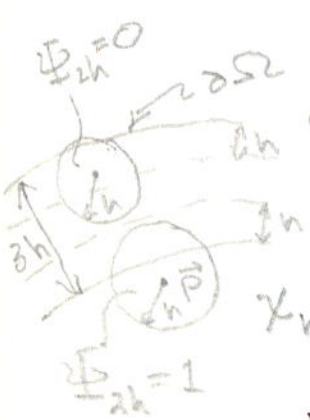

PROOF. Obviously $\chi_h$ has continuous derivatives of all orders and $\chi_h = 1$ on $\Omega_{3h}$, $\chi_h = 0$ outside of $\Omega_h$. From the equation:

$$\frac{\partial \chi_h}{\partial x_i} = \frac{1}{\varkappa h^n} \int \cdots \int_{r \leqslant h} \Psi_{2h} \psi\left(\frac{r}{h}\right) \frac{\partial r}{\partial x_i} \cdot \frac{1}{h} \, d\Omega$$

we conclude that

$$\left| \frac{\partial \chi_h}{\partial x_i} \right| < \frac{A}{h}. \tag{12.16}$$

The function $\xi \chi_h$ has continuous derivatives of first order in $\Omega$, and is equal to zero outside of $\Omega_h$ and equal to $\xi$ within $\Omega_{3h}$. We need to show that

$$D(v, \xi - \xi\chi_h) \to 0 \quad \text{for} \quad h \to 0, \tag{12.17}$$

for any $v \in W_2^{(1)}$. We estimate this expression. We have:

$$D(v, \xi - \xi\chi_h) = \overbrace{\int \cdots \int_\Omega}^{n} \sum_{i=1}^{n} \frac{\partial v}{\partial x_i} \frac{\partial (\xi - \xi\chi_h)}{\partial x_i} d\Omega =$$

$$= \int \cdots \int_\Omega \sum_{i=1}^{n} \frac{\partial v}{\partial x_i} \left[ \frac{\partial \xi}{\partial x_i} (1 - \chi_h) - \xi \frac{\partial \chi_h}{\partial x_i} \right] d\Omega =$$

$$=\int_{\Omega-\Omega_{3h}}\cdots\int\sum_{i=1}^{n}\frac{\partial v}{\partial x_i}\frac{\partial\xi}{\partial x_i}(1-\chi_h)\,d\Omega+\int_{\Omega_h-\Omega_{3h}}\cdots\int\xi\sum_{i=1}^{n}\frac{\partial v}{\partial x_i}\frac{\partial\chi_h}{\partial x_i}\,d\Omega.$$

As is easy to see, $|1-\chi_h|\leqq 1$, and therefore the first integral on the right side does not exceed

$$\left[\int_{\Omega-\Omega_{3h}}\cdots\int\sum_{i=1}^{n}\left(\frac{\partial v}{\partial x_i}\right)^2 d\Omega\right]^{\frac{1}{2}}\left[\int_{\Omega-\Omega_{3h}}\cdots\int\sum_{i=1}^{n}\left(\frac{\partial\xi}{\partial x_i}\right)^2 d\Omega\right]^{\frac{1}{2}},$$

and therefore converges to zero as $h\to 0$ by virtue of the convergence of the integrals $D(v)$ and $D(\xi)$.

Let us consider the second integral. We have by virtue of (12.16):

$$\left|\int_{\Omega_h-\Omega_{3h}}\cdots\int\xi\sum_{i=1}^{n}\frac{\partial v}{\partial x_i}\frac{\partial\chi_h}{\partial x_i}\,d\Omega\right|$$

$$<\frac{A}{h}\int_{\Omega_h-\Omega_{3h}}\cdots\int|\xi|\sum_{i=1}^{n}\left|\frac{\partial v}{\partial x_i}\right|d\Omega\leqslant$$

$$\leqslant\frac{A}{h}\int_{\Omega_h-\Omega_{3h}}\cdots\int|\xi|\sqrt{n\sum_{i=1}^{n}\left(\frac{\partial v}{\partial x_i}\right)^2}\,d\Omega\leqslant$$

$$\leqslant\frac{A_1}{h}\sqrt{\int_{\Omega_h-\Omega_{3h}}\cdots\int\xi^2 d\Omega\cdot\int_{\Omega_h-\Omega_{3h}}\cdots\int\sum_{i=1}^{n}\left(\frac{\partial v}{\partial x_i}\right)^2 d\Omega}.$$

Because of the convergence of $D(v)$:

$$\int_{\Omega_h-\Omega_{3h}}\cdots\int\sum_{i=1}^{n}\left(\frac{\partial v}{\partial x_i}\right)d\Omega\to 0\quad\text{for}\quad h\to 0.$$

It therefore suffices to show the boundedness for $h\to 0$ of the expression

$$\frac{1}{h^2}\int_{\Omega_h-\Omega_{3h}}\cdots\int\xi^2 d\Omega, \tag{12.18}$$

and the lemma will be proved.

Let $\bar{S}$ be a domain in the plane $y=$ const. and $\bar{\Omega}$ a cylindrical domain in the $n$-space of $(y_1,y_2,\cdots,y_n)$ given by the inequalities

$$0\leqslant y_n\leqslant h_0\quad(y_1,y_2,\ldots,y_{n-1})\in\bar{S}.$$

Let $\xi\in W_2^{(1)}$, $\xi|_S=0$ in the sense of $L_{2,n-1}$ and having continuous derivatives in $\bar{\Omega}$. Then we have, taking $\Delta y_n>0$,

$$|\xi(y_1, y_2 \ldots, y_{n-1}, y_n + \Delta y_n) - \xi(y_1 \ldots y_n)|^2 =$$

$$= \left| \int_{y_n}^{y_n+\Delta y_n} \frac{\partial \xi}{\partial y_n} dy_n \right|^2 \leqslant \Delta y_n \int_{y_n}^{y_n+\Delta y_n} \left(\frac{\partial \xi}{\partial y_n}\right)^2 dy_n.$$

Integrating over $\overline{S}$, we obtain:

$$\int \cdots \int_{\overline{S}} |\xi(y_1, y_2, \ldots y_{n-1}, y_n + \Delta y_n) - \xi(y_1 \ldots y_n)|^2 d\overline{S} \leqslant$$

$$\leqslant \Delta y_n \int \cdots \int_{\overline{S}} \int_{y_n}^{y_n+\Delta y_n} \left(\frac{\partial \xi}{\partial y_n}\right)^2 dy_n d\overline{S} \leqslant \Delta y_n \int \cdots \int_{\overline{\Omega}} \left|\frac{\partial \xi}{\partial y}\right|^2 dy\, d\overline{\Omega} \leqslant$$

$$\leqslant \Delta y_n \, \|\xi\|^2_{W_2^{(1)}} . \tag{12.19}$$

Taking $y_n \to 0$ and replacing $\Delta y_n$ by $h$, we obtain:

$$\int \cdots \int_{\overline{S}} |\xi(y_1, y_2, \ldots, y_{n-1}, h)|^2 d\overline{S} \leqslant h \|\xi\|^2_{W_2^{(1)}} .$$

Integrating in $h$ between the limits $Ah$ and $Bh$ $(A < B)$, we find:

$$\int_{Ah}^{Bh} \int \cdots \int_{\overline{S}} |\xi(y_1, y_2, \ldots, y_{n-1}, y_n)|^2 d\overline{S}\, dy_n \leqslant Kh^2$$

from which follows

$$\frac{1}{h^2} \int \cdots \int_{Ah < y_n < Bh} |\xi|^2 d\Omega \leqslant K. \tag{12.20}$$

We now prove the boundedness of (12.19). Let $h_0 > 0$ be a sufficiently small number. We decompose the domain $\Omega - \Omega_h$ into a finite number of intersecting domains $V_i$ $(i = 1, 2, \cdots, k)$ such that each piece is "based" upon some piece $S_i$ of the surface $S$. For each $V_i$, by virtue of our assumptions about the surface $S$, we can find a bicontinuous mapping of $V_i$ on a cylindrical set $\Omega_i$ with base $S_i$ such that the mapping and its inverse have piecewise continuous and bounded first derivatives. For this domain $V_i(\Omega_h - \Omega_{3h})$ (the intersection of $V_i$ with $(\Omega_h - \Omega_{3h})$) is mapped into some domain lying in the strip

$$A_i h \leqslant y_n \leqslant B_i h.$$

Applying (12.20), we conclude the boundedness of (12.19) for each $V_i$ and therefore for the whole domain $\Omega_h - \Omega_{3h}$. Thus for every $v \in W_2^{(1)}$ and $\xi \in W_2^{(1)}$; $\xi|_S = 0$, we have

$$D(v, \xi) = \lim_{h \to 0} D(v, \xi\chi_h)$$

and the lemma is proved.[1)] We now proceed to the proof of the uniqueness theorem.

Let us assume that in addition to $v_0$ there exists another function $u \in W_2^{(1)}(\phi)$ such that $\Delta u = 0$. For such a function $D(u) > d$. If $D(u) = d$, then $u$ could be interspersed infinitely often in a minimizing sequence converging to $v_0$, and then we should obtain the conclusion that $v_0 = u$ since the resulting minimizing sequence must converge and have the same limit as each of its subsequences.

Thus if we demand that $u$ be different from $v_0$, it is necessary that $D(u) > d$. We shall now show that this is not possible.

If $u \in W_2^{(1)}$ and in addition $\Delta u = 0$, and $\xi$ is the function from the lemma which we have just proved, then

$$D(u, \xi) = 0.$$

This equation is obtained by an obvious limiting process from $D(u, \xi\chi_h) = 0$ by virtue of the lemma.

Furthermore,

$$D(u + \lambda\xi) = D(u) + 2\lambda D(u, \xi) + \lambda^2 D(\xi) = D(u) + \lambda^2 D(\xi) \geqslant$$
$$\geqslant D(u) > d$$

by virtue of the condition $D(u) > d$. On the other hand, putting $\lambda = 1$ and $\xi = v_0 - u$ ($\xi|_S = 0$, we find:

$$D(v_0) > d,$$

which contradicts the fact shown earlier that $D(v_0) = d$.

The proof of the theorem is complete.

5. HADAMARD'S EXAMPLE. In conclusion, we present an example due to Hadamard showing that summability and even continuity of a function given on the surface is still insufficient to ensure that the function can be a boundary value for a function from $W_2^{(1)}$.

Let $\Omega$ be the circle $x^2 + y^2 \leq 1$ in the $(x,y)$ plane and $(\rho, \theta)$ polar coordinates in this plane.

Let

$$\varphi(\theta) = \sum_{n=1}^{\infty} \frac{\cos n^4 \theta}{n^2}.$$

---

[1)] Formula (12.20), and with it the whole lemma, may be proved not only for functions $\xi$ having continuous derivatives in the interior of the domain but for arbitrary functions from $W_2^{(1)}$ vanishing on the boundary. For this purpose it suffices to take a limiting process in (12.19) replacing $\xi$ by its averaged functions $\xi_h$.

Obviously $\phi(\theta)$ is a continuous function and the function

$$u_1(\rho,\theta)=\sum_{n=1}^{\infty}\frac{\cos n^4\theta}{n^2}\rho^{n^4}$$

is a harmonic function in the open circle $x^2+y^2<1$, continuous on the closed circle $x^2+y^2\leq 1$, and coinciding with $\phi(\theta)$ for $\rho=1$.

Furthermore, we have

$$\iint\limits_{\rho\leqslant\rho_0<1}\left[\left(\frac{\partial u_1}{\partial x}\right)^2+\left(\frac{\partial u_1}{\partial y}\right)^2\right]dx\,dy=\int_0^{\rho_0}\rho\left[\int_0^{2\pi}\left\{\left(\frac{\partial u_1}{\partial \rho}\right)^2+\frac{1}{\rho^2}\left(\frac{\partial u_1}{\partial \theta}\right)^2\right\}d\theta\right]d\rho=$$

$$=2\pi\int_0^{\rho_0}\sum_{n=1}^{\infty}n^4\rho^{2n^4-1}\,d\rho=\pi\sum_{n=1}^{\infty}\rho_0^{2n^4}\to\infty \text{ for } \rho_0\to 1,$$

from which it follows that $u_1\notin W_2^{(1)}$ in $\Omega$.

If $\phi(\theta)$ were a permissible function in the sense of item 1, then, solving the Dirichlet problem by the variational method, we would find a harmonic $u_2(x,y)\in W_2^{(1)}$ such that

$$\int_0^{2\pi}|u_2(\rho,\theta)-\varphi(\theta)|^2\,d\theta\to 0 \quad\text{for}\quad \rho\to 1,$$

and therefore

$$\int_0^{2\pi}|u_2(\rho,\theta)-\varphi(\theta)|\,d\theta\to 0.$$

The same condition holds for $u_1$ since $u_1(\rho,\theta)$ is uniformly continuous. Therefore

$$\int_0^{2\pi}|u_2(\rho,\theta)-u_1(\rho,\theta)|\,d\theta\to 0 \tag{12.21}$$

for $\rho\to 1$.

Let $\rho_0<\rho<1$. Then by the Poisson formula for harmonic functions on the circle, we have:

$$u_2(\rho_0,\theta_0)-u_1(\rho_0,\theta_0)=\frac{1}{2\pi}\int_0^{2\pi}\frac{\rho^2-\rho_0^2}{\rho_0^2-2\rho\rho_0\cos(\theta-\theta_0)+\rho_0^2}[u_2(\rho,\theta)-$$
$$-u_1(\rho,\theta)]\,d\theta. \tag{12.22}$$

For fixed $\rho_0$ and $\rho\to 1$, the function

$$\frac{\rho^2-\rho_0^2}{\rho_0^2-2\rho_0\rho\cos(\theta-\theta_0)+\rho^2}$$

remains bounded and therefore, on the basis of (12.21), the right side of (12.22) converges to zero for $\rho \to 1$. Since the left side of (12.22) does not depend on $\rho$, it must equal zero and by virtue of the arbitrariness of $\rho_0$ and $\theta_0$, we have: $u_1 \equiv u_2$. This is impossible since $u_1 \overline{\in} W_2^{(1)}$. As a result, $\phi$ cannot be a permissible function.

## §13. The Neumann problem.

1. FORMULATION OF THE PROBLEM. We have considered for the Laplace equation the simplest problem, the Dirichlet problem. Let us now pass to the study of another basic problem, the Neumann problem.

Let $\Omega$ be a domain of $n$-dimensional space bounded by a sufficiently smooth surface $S$.

Let $u \in W_2^{(1)}$. Consider the functional

$$H(u) = D(u) + 2(\rho, u), \tag{13.1}$$

where

$$D(u) = \int \underset{\Omega}{\cdots} \int \sum_{i=1}^{n} \left(\frac{\partial u}{\partial x_i}\right)^2 d\Omega$$

and $(\rho, u)$ is a linear functional on $W_2^{(1)}$. In all the following, we shall assume that

$$(\rho, 1) = 0. \tag{13.2}$$

THEOREM. *If* $(\rho, 1) = 0$, *then* $H(u)$ *is bounded from below.*

PROOF. If $u - v = \text{const.}$, then $(\rho, u) = (\rho, v)$, i.e., $(\rho, u)$ has a constant value on each class $\psi \in L_2^{(1)}$. By virtue of the linearity of the functional $\rho$, we have:

$$|(\rho, u)| \leqslant M \|u\|_{W_2^{(1)}}.$$

Since for every class $\psi \in L_2^{(1)}$, $(\rho, u)$ has a constant value, while for one of the functions $u_1$ of that class $\|u_1\|_{W_2^{(1)}} = \|u_1\|_{L_2^{(1)}}$, it follows that

$$|(\rho, u)| = |(\rho, u_1)| \leqslant M \|u_1\|_{L_2^{(1)}} = M \|u\|_{L_2^{(1)}} = M \sqrt{D(u)}.$$

Thus we have:

$$H(u) = D(u) + 2(\rho, u) \geqslant D(u) - 2|(\rho, u)| \geqslant D(u) - 2M\sqrt{D(u)} = \\ = (\sqrt{D(u)} - M)^2 - M^2 \geqslant -M^2,$$

i.e.,

$$H(u) \geqslant -M^2.$$

The theorem is proved.

As a result, there exists a greatest lower bound for $H(u)$ which we denote by $-d$:

$$\inf H(u) = -d. \tag{13.3}$$

2. SOLUTION OF THE VARIATIONAL PROBLEM.

THEOREM 2. *There exists* $u \in W_2^{(1)}$ *such that* $H(u) = -d$. *In addition, for an arbitrary* $\xi \in W_2^{(1)}$, *we have the equality*

$$D(\mu, \xi) + (\rho, \xi) = 0. \tag{13.4}$$

PROOF. Let $\{v_k\}$ be a minimizing sequence, i.e., $v_k \in W_2^{(1)}$ and $\lim_{k\to\infty} H(v_k) = -d$. Then from the obvious equations

$$D\left(\frac{v_k - v_m}{2}\right) = \frac{1}{2}[D(v_k) + 2(\rho, v_k)] + \frac{1}{2}[D(v_m) + 2(\rho, v_m)] -$$
$$-\left[D\left(\frac{v_k + v_m}{2}\right) + 2\left(\rho, \frac{v_k + v_m}{2}\right)\right] =$$
$$= \frac{1}{2}H(v_k) + \frac{1}{2}H(v_m) - H\left(\frac{v_k + v_m}{2}\right).$$

If we choose $k$ and $m$ sufficiently large so that $H(v_k) < -d + \epsilon$ and $H(v_m) < -d + \epsilon$, and take into account the fact that $-H((v_k + v_m)/2) < d$, we have:

$$D\left(\frac{v_k - v_m}{2}\right) < \frac{-d + \varepsilon}{2} + \frac{-d + \varepsilon}{2} + d = \varepsilon,$$

i.e.,

$$D(v_k - v_m) < 4\varepsilon.$$

Thus $\{v_k\}$ converges in $L_2^{(1)}$. By virtue of the fact that $H(u)$ has a constant value on each class $\psi \in L_2^{(1)}$ while for one of the functions $u_1$ of that class $\|u_1\|_{W_2^{(1)}} = \|u_1\|_{L_2^{(1)}}$, for a minimizing sequence we may always pick a sequence for which

$$\|v_k\|_{W_2^{(1)}} = \|v_k\|_{L_2^{(1)}}.$$

For such a choice $\{v_k\}$ converges in $W_2^{(1)}$. Let $u \in W_2^{(1)}$ be the limit function.

Proceeding as we did above in the consideration of the Dirichlet problem, we obtain:

$$|H(u) - H(v_k)| = |D(u) - D(v_k) + 2(\rho, u - v_k)| \leqslant$$
$$\leqslant |D(u) - D(v_k)| + 2|(\rho, u - v_k)| \leqslant (\|u\|_{L_2^{(1)}} +$$
$$+ \|v_k\|_{L_2^{(1)}}) \, | \, \|u - v_k\|_{L_2^{(1)}} | + 2M \|u - v_k\|_{W_2^{(1)}} \leqslant [\|u\|_{L_2^{(1)}} +$$
$$+ \|v_k\|_{L_2^{(1)}} + 2M] \cdot \|u - v_k\|,$$

or

$$|H(u)-H(v_k)|\xrightarrow[k\to\infty]{}0,$$

from which it follows that

$$H(u)=\lim_{k\to\infty}H(v_k)=-d.$$

Let $\xi\in W_2^{(1)}$. Then

$$H(u+\lambda\xi)=H(u)+2\lambda[D(u,\xi)+(\rho,\xi)]+\lambda^2D(\xi)$$

has a minimum for $\lambda=0$, and by Fermat's theorem we have

$$D(u,\xi)+(\rho,\xi)=0.$$

The theorem is proved.

Each function $u\in W_2^{(1)}$ is summable on $S$ to any power $q^*$ where

$$q^*<q=\frac{2(n-1)}{n-2}.$$

We denote by $L_{q^*}(S)$ the set of functions $u$ defined on $S$ and summable to the $q^*$th power. Let $(\rho_S, u)$ be a linear functional on $L_{q^*}(S)$. Then by the Imbedding Theorem $(\rho_S, u|_S)$ is linear on $W_2^{(1)}$. Indeed if $u\in W_2^{(1)}$, then

$$|(\rho_S, u|_S)|\leqslant K_1\|u\|_{L_{q^*}(S)}\leqslant K_1M\|u\|_{W_2^{(1)}}=K\|u\|_{W_2^{(1)}},$$

where $K_1$, $M$, and $K$ are various constants. The inequality so obtained establishes our assertion.

3. SOLUTION OF THE NEUMANN PROBLEM.

THEOREM 3. *If* $(\rho_S,1)=0$, *then there exists a function* $u\in W_2^{(1)}$ *such that:*
(1) *In* $\Omega$, $u$ *has continuous derivatives of all orders and satisfies the equation*

$$\Delta u=0. \tag{13.5}$$

(2) *Let* $\Omega_k$ *be an arbitrary increasing sequence of domains having sufficiently smooth boundaries, contained in* $\Omega$ *and converging to* $\Omega$. *Then*

$$\lim_{k\to\infty}\int\cdots\int_{S_k}\frac{\partial u}{\partial n}\xi\,dS_k=-(\rho_S,\xi), \tag{13.6}$$

*where* $n$ *is the exterior normal to* $S_k$ *and* $\xi\in W_2^{(1)}$ *is an arbitrary function. The problem of finding such a function* $u$ *we will call the Neumann problem.*

PROOF. On the basis of Theorem 2 there exists a function $u\in W_2^{(1)}$ such that

$$D(u,\xi)+(\rho_S,\xi)=0 \quad \text{for any} \quad \xi\in W_2^{(1)}. \tag{13.7}$$

If we choose $\xi\equiv 0$ on some strip around the boundary $S$, then $(\rho_S,\xi)=0$ and we obtain $D(u,\xi)=0$. From this it follows by a literal repetition of the argument of the preceding paragraph that $u$ is arbitrarily often continuously differentiable and satisfies the equation $\Delta u=0$. The first condition is established.

Let $\xi$ be an arbitrary function from $W_2^{(1)}$. We then have

$$D(u,\xi)=\lim_{k\to\infty}\int\cdots\int_{\Omega_k}\sum_{k=1}^{n}\frac{\partial u}{\partial x_i}\frac{\partial\xi}{\partial x_i}\,d\Omega_k=\lim_{k\to\infty}\Big[\int\cdots\int_{S_k}\xi\frac{\partial u}{\partial n}\,dS_k+$$

$$+\int\cdots\int_{\Omega_k}\xi\Delta u\,d\Omega\Big]=\lim_{k\to\infty}\int\cdots\int_{S_k}\xi\frac{\partial u}{\partial n}\,dS_k.$$

Taking (13.7) into consideration, we obtain (13.6).

THEOREM 4. *The solution of the Neumann problem is unique up to an arbitrary constant.*

Suppose that there exists two functions $u_1$ and $u_2\in W_2^{(1)}$ satisfying the equation $\Delta u=0$ and the same condition (13.6). Their difference $\psi=u_1-u_2$ is a harmonic function and satisfies the condition

$$\lim_{k\to\infty}\int\cdots\int_{S_k}\xi\frac{\partial\psi}{\partial n}\,dS_k=0.$$

Putting $\xi=\psi$, we find

$$\lim_{k\to\infty}\int\cdots\int_{S_k}\psi\frac{\partial\psi}{\partial n}\,dS_k=\lim_{k\to\infty}\int\cdots\int_{\Omega_k}|\operatorname{grad}\psi|^2\,d\Omega=D(\psi)=0,$$

from which it follows that $\psi=\text{const}$.

REMARK. If

$$(\rho_S,u)=\int\cdots\int_S\varphi u\,dS,$$

then the "boundary" condition (1) takes the form

$$\int\cdots\int_S\varphi\xi\,dS+\lim_{k\to\infty}\int\cdots\int_{S_k}\xi\frac{\partial u}{\partial n}\,dS_k=0.$$

The functional $(\rho_S,\xi_S)$ represents its generalized boundary condition for the function $u$ in which the quantity $\partial u/\partial n$ converges to its value in the weak sense. Such a formulation is completely natural in the given situation since limit values for $\partial u/\partial n$ if $u \in W_2^{(1)}$ may not exist or need not converge to their limit in the strong sense of the word.

## §14. **Polyharmonic equations.**

1. THE BEHAVIOUR OF FUNCTIONS FROM $W_2^{(m)}$ ON BOUNDARY MANIFOLDS OF VARIOUS DIMENSIONS. The consideration of variational methods may be carried over to boundary value problems for polyharmonic equations. We pass now to the study of the basic boundary value problems for such equations.

Suppose that the bounded domain $\Omega$ of $n$-dimensional space has a boundary $S$ consisting of a finite number of smooth pieces of varying dimensions. Suppose that $S$ is a regular boundary (cf. Chapter I, §10, item 2). The equation

$$\Delta^m u = \left( \sum_{i=1}^{n} \frac{\partial^2}{\partial x_i^2} \right)^m u = 0 \tag{14.1}$$

or

$$\sum_{\Sigma \alpha_i = m} \frac{m!}{\alpha_1! \, \alpha_2! \, \dots \, \alpha_n!} \frac{\partial^{2m} u}{\partial x_1^{2\alpha_1} \partial x_2^{2\alpha_2} \dots \partial x_n^{2\alpha_n}} = 0$$

is called a polyharmonic equation.

The equation (14.1) appears as the Euler equation for the integral

$$D(u) = \int_{\Omega} \cdots \int \sum_{\Sigma \alpha_i = m} \frac{m!}{\alpha_1! \, \alpha_2! \, \dots \, \alpha_n!} \left( \frac{\partial^m u}{\partial x_1^{\alpha_1} \partial x_2^{\alpha_2} \dots \partial x_n^{\alpha_n}} \right)^2 d\Omega. \tag{14.2}$$

Obviously

$$D(u) = \| u \|^2_{L_2^{(m)}}.$$

Let $u \in W_2^{(m)}$. If $n-2m<0$, it follows from the Imbedding Theorem that $u(x_1, x_2, \cdots, x_n)$ is continuous and has well-defined values on manifolds of arbitrary dimension. If $n-2m>0$ and $n-s>n-2m$, then $u(x_1, \cdots, x_n)$ is summable on manifolds of dimensions $(n-s)$ to any power

$$q^* < q = 2(n-s)/(n-2m).$$

Since $q>2$, we may take $q^*=2$. Therefore if $n-s>n-2m>0$, then $u \in L_{2,\, n-s}$ on manifolds of dimensions $(n-s)$, which we shall denote by $S_{n-s}$.

Thus, $u \in L_{2,\, n-s}$ on $S_{n-s}$ if

$$s < 2m. \tag{14.3}$$

Moreover, from the complete continuity of the imbedding operator it follows that $u(x_1, \cdots, x_n)$ is continuous in the sense of $L_{2,\, n-s}$ and therefore has limiting values in the sense of $L_{2,\,n-s}$ on portions of the boundary of $S$ having dimension $n-s$, where $s<2m$. As was shown earlier, $u(x_1, \cdots, x_n)$ has all derivatives of order less than $m$. We consider the derivatives of $k$th order, where $k<m$.

If $n<2m$, the derivatives of $k$th order belong to $L_2$ on manifolds $S_{n-s}$ of dimension $(n-s)$, where

$$n-s>n-2(m-k)$$

or

$$s<2(m-k). \tag{14.4}$$

From the complete continuity of the imbedding operator, it follows that the derivatives of $k$th order are continuous in the sense of $L_{2,\,n-s}$ where $s$ satisfies (14.4) and therefore having limiting values on portions of the boundary $S_{n-s}$ in the sense of $L_{2,\,n-s}$.

From the inequality (14.4) it follows that

$$k\leqslant m-\left[\frac{s}{2}\right]-1, \tag{14.5}$$

i.e., on manifolds of dimension $(n-s)$ there exist limiting values in the sense of $L_{2,\,n-s}$ for derivatives up to order $m-[s/2]-1$.

We form these results into a table.

| On manifolds of the boundary: | exist limiting values in the sense of: | for functions and all derivatives up to order (inclusive): |
|---|---|---|
| $S_{n-1}$ | $L_{2,\,n-1}$ | $m-1$ |
| $S_{n-2}$ | $L_{2,\,n-2}$ | $m-2$ |
| $S_{n-3}$ | $L_{2,\,n-3}$ | $m-2$ |
| $S_{n-4}$ | $L_{2,\,n-4}$ | $m-3$ |
| . . . . . . . . . . . . | . . . . . . . . . . . . | . . . . . . . . . . |
| $S_{n-2k}$ | $L_{2,\,n-2k}$ | $m-k-1$ |
| $S_{n-2k-1}$ | $L_{2,\,n-2k-1}$ | $m-k-1$ |
| . . . . . . . . . . . . | . . . . . . . . . . . . | . . . . . . . . . . |
| $S_{n-2m+2}$ | $L_{2,\,n-2m+2}$ | 0 |
| $S_{n-2m+1}$ | $L_{2,\,n-2m+1}$ | 0 |

This table is formed under the assumption that $n-2m+1>0$. If $n-2m+1\leqq 0$, then for some $k$ either $n-2k=0$ or $n-2k-1=0$.

Then we have $m-k-1<m-(n/2)$, and as a result, the functions and

all their derivatives up to order $m-k-1$ are continuous. The table then turns into one row: On $S_0$ there exist limiting values of the function and all derivatives up to order $m-k-1$, inclusive. On the manifolds $S_1, S_2, \cdots, S_{n-1}$, all the derivatives up to the $(m-k-1)$st order are also continuous.

Obviously from the continuity of the function there follows its continuity in the sense of $L_{2,n-s}$. Therefore in the following estimates ordinary continuity will not be mentioned and all arguments will be carried out under the assumption of continuity in the sense of $L_{2,n-s}$.

2. FORMULATION OF THE BASIC BOUNDARY VALUE PROBLEM. Suppose that the boundary of the domain consists of the manifolds $S_{n-1}, S_{n-2}, \cdots, S_{n-2m+1}$, if $n-2m+1>0$, or if the manifolds $S_{n-1}, S_{n-2}, \cdots, S_0$ if $n-2m+1\leqq 0$. Any one of these manifolds may be absent.

Suppose now that on each of the manifolds $S_{n-s}$ we are given functions

$$\varphi^{(n-s)}_{\alpha_1 \ldots \alpha_n}, \quad \text{where} \quad 0 \leqslant \sum \alpha_i \leqslant m - \left[\frac{s}{2}\right] - 1.$$

If there exists $v(x_1, \cdots, x_n) \in W_2^{(m)}$ such that

$$\left.\frac{\partial^{\alpha_1+\cdots+\alpha_n} v}{\partial x_1^{\alpha_1} \cdots \partial x_n^{\alpha_n}}\right|_{S_{n-s}} = \varphi^{(n-s)}_{\alpha_1 \ldots \alpha_n}, \tag{14.6}$$

then we shall say that the system of boundary values

$$\left\{ \varphi^{(n-s)}_{\alpha_1 \ldots \alpha_n} \right\}$$

is permissible.

For any permissible boundary values, it is obvious that if $n-2m+1>0$, then all the $\phi^{(n-s)}_{\alpha_1\cdots\alpha_n} \in L_{2,n-s}$, and if $n-2m+1\leqq 0$, then all the functions for which $0 \leqq \sum \alpha_i \leqq m-[n/2]-1$ are continuous and the remainder belong to $L_{2,n-s}$.

For the polyharmonic equations, in distinction to the Laplace equation, we may give boundary values not only on the boundary surfaces of dimension $n-1$ but also give permissible boundary values on surfaces of lower dimension, as a simple example will indicate.

We consider the equation $\Delta^2 u=0$ in a three-dimensional space. We have $n=3$, $m=2$, $n-2m+1=0$. In the role of boundary manifolds we necessarily have $S_2$, and we may have $S_1$ and $S_0$. Since $m-[n/2]-1=0$, all the functions $u$ will be continuous. On the manifold $S_2$, we must prescribe $u|_{S_2}$ continuous and $\partial u/\partial x_i|_{S_2} \in L_2$. On $S_1$ and $S_0$, we prescribe the function $u(x_1, x_2, x_3)$.

Let $\Omega$ be the sphere of unit radius with its center omitted:

$$0 < x_1^2 + x_2^2 + x_3^2 < 1.$$

In our example $S_1$ is missing. We consider the solution satisfying the conditions

$$u(0,0,0)=1 \quad (\text{given on } S_0),$$

$$u\Big|_{r=1}=\frac{\partial u}{\partial r}\Big|_{r=1}=0 \quad (\text{given on } S_2).$$

The data given on $S_2$ is obviously equivalent to the following:

$$u\Big|_{S_2}=\frac{\partial u}{\partial x_1}\Big|_{S_2}=\frac{\partial u}{\partial x_2}\Big|_{S_2}=\frac{\partial u}{\partial x_3}\Big|_{S_2}=0.$$

The function $u=(1-r)^2$ is a solution of the problem. Indeed, it satisfies the equation with the boundary conditions. At the point $r=0$, the derivatives do not exist. The second derivatives are summable in the square on $\Omega$, i.e., $u\in W_2^{(2)}$. As we shall show later, there are no other solutions in $W_2^{(2)}$. If we omit $S_0$, then the only solution is $u\equiv 0$.

3. SOLUTION OF THE VARIATIONAL PROBLEM. We proceed to the study of the basic boundary value problem for the polyharmonic equation from a general point of view.

THEOREM 1. *If the system* $\{\phi^{(n-s)}_{\alpha_1\cdots\alpha_n}\}$ *is permissible, then there exists an unique function* $u(x_1,\cdots,x_n)\in W_2^{(m)}$ *satisfying the conditions* (14.6) *and giving a minimum to the integral* $D$ *among all such functions.*

We denote by $W_2^{(m)}\{\phi^{(n-s)}_{\alpha_1\cdots\alpha_n}\}$ the set of functions $v\in W_2^{(m)}$ satisfying (14.6). Since the system $\phi^{(n-s)}_{\alpha_1\cdots\alpha_n}$ is permissible, the set $W_2^{(m)}\{\phi^{(n-s)}_{\alpha_1\cdots\alpha_n}\}$ is nonempty. For each function $v$ from this set we have $0\leq D(v)<\infty$. There must therefore exist a greatest lower bound for the values $D(v)$, which we denote by $d$:

$$d=\inf D(v),\quad v\in W_2^{(m)}\{\varphi^{(n-s)}_{\alpha_1\ldots\alpha_n}\}.$$

From the set $W_2^{(m)}\{\phi^{(n-s)}_{\alpha_1\cdots\alpha_n}\}$ we may choose a minimizing sequence $\{v_k\}$ such that

$$\lim_{k\to\infty} D(v_k)=d. \tag{14.7}$$

We shall show that $\{v_k\}$ converges in $W_2^{(m)}$.

In order that one may define any natural norm on the space $W_2^{(m)}$, as we have seen, it suffices to give a system of linear functionals

$$\rho_{\beta_1,\beta_2,\ldots,\beta_n},\quad \Big(\sum_{i=1}^n\beta_i\leqslant m-1\Big),$$

bounded in one of these natural norms and such that for an arbitrary linear condition

$$\rho = \sum_{\Sigma \beta_i \leqslant m-1} A_{\beta_1, \beta_2, \ldots, \beta_n} \rho_{\beta_1, \beta_2, \ldots, \beta_n}$$

one can find at least one monomial $x_1^{\alpha 1} \cdots x_n^{\alpha n}$ for which this combination does not vanish.

Consider the surface $S_{n-1}$. By assumption of this surface consists of a finite number of smooth pieces. Put

$$\rho_{\beta_1, \beta_2, \ldots, \beta_n} v = \overbrace{\int \cdots \int}^{n-1}_{S_{n-1}} \frac{\partial^{\beta_1 + \beta_2 + \ldots + \beta_n} v}{\partial x_1^{\beta_1} \ldots \partial x_n^{\beta_n}} \, d S_{n-1}, \tag{14.8}$$

where $v$ is an arbitrary element of $W_2^{(m)}$.

It is not difficult to see that all the functionals $\rho_{\beta_1, \beta_2, \cdots, \beta_n}$ are linear on $W_2^{(m)}$, if the norm is defined on the latter by means of the spherical projection operator. This follows immediately from the Imbedding Theorem.

Now let $\rho = \sum A_{\beta_1 \cdots \beta_n} \rho_{\beta_1, \cdots, \beta_n}$ be an arbitrary linear combination of the functionals $\rho_{\beta_1 \cdots \beta_n}$. Let $A_{\beta_1^{(0)}, \cdots, \beta_n^{(0)}}$ be one of its non-null coefficients for which the sum $\beta_1^{(0)} + \beta_2^{(0)} + \cdots + \beta_n^{(0)} = \sigma^{(0)}$ has its least value among all the sums

$$\beta_1 + \beta_2 + \ldots + \beta_n = \sigma.$$

Then

$$\rho \, x_1^{\beta_1^{(0)}} x_2^{\beta_2^{(0)}} \ldots x_n^{\beta_n^{(0)}} \neq 0.$$

In fact, all the $\rho_{\beta_1 \cdots \beta_n}$ containing derivatives of higher order, as well as those having the same sum $\sigma$ but for which for at least one coefficient $\beta_k > \beta_k^{(0)}$, will vanish on this monomial and we obtain

$$(\rho, x_1^{\beta_1^{(0)}} x_2^{\beta_2^{(0)}} \ldots x_n^{\beta_n^{(0)}}) = \int_{S_{n-1}} \cdots \int A_{\beta_1^{(0)}, \ldots, \beta_n^{(0)}} \, dS \cdot \beta_1^{(0)}! \, \beta_2^{(0)}! \ldots \beta_n^{(0)}! \neq 0,$$

as was to be shown.

Thus, we may define a norm on $W_2^{(m)}$ by the formula

$$\| v \|^2_{W_2^{(m)}} = \left\{ \sum_{\mu} (\rho_\mu v)^2 + D(v) \right\}, \tag{14.9}$$

where $\rho_\mu$ runs through the values $\rho_{\beta_1 \cdots \beta_n}$. Then for all functions $v_k$ and $v_l$ of the minimizing sequence, we obtain

$$(\rho_\mu, v_k) = (\rho_\mu, v_l)$$

and thereby

$$\|v_k - v_l\|_{W_2^{(m)}} = D(v_k - v_l).$$

We choose $k$ and $l$ so large that $D(v_k) < d + \epsilon$, $D(v_l) < d + \epsilon$, as is possible according to (14.7).

Obviously

$$\frac{v_k - v_l}{2} \in W_2^{(m)} \{\varphi_{\alpha_1, \ldots, \alpha_n}^{(n-s)}\}$$

and therefore

$$D\left(\frac{v_k + v_l}{2}\right) \geqslant d.$$

From the preceding equality

$$D\left(\frac{v_k - v_l}{2}\right) = \frac{1}{2} D(v_k) + \frac{1}{2} D(v_l) - D\left(\frac{v_k + v_l}{2}\right)$$

we find

$$D\left(\frac{v_k - v_l}{2}\right) < \frac{d + \varepsilon}{2} + \frac{d + \varepsilon}{2} - d = \varepsilon,$$

i.e.,

$$D(v_k - v_l) < 4\varepsilon,$$

from which by the arbitrariness of $\epsilon$ it follows that

$$\|v_k - v_l\|_{W_2^{(m)}} \to 0, \quad k, l \to \infty. \tag{14.10}$$

Since the space $W_2^{(m)}$ is complete, we conclude that there exists a limit function $u \in W_2^{(m)}$ for which $\|u - v_k\|_{W_2^{(m)}} \to 0$. Just as in the Dirichlet problem, we show that

$$(1)\ u \in W_2^{(m)} \{\varphi_{\alpha_1, \ldots, \alpha_n}^{(n-s)}\}; \tag{14.11}$$

$$(2)\ D(u) = d. \tag{14.12}$$

There do not exist two different functions satisfying (14.11) and (14.12). Indeed if $u_1$ and $u_2$ were two such functions, then the sequence $u_1, u_2, u_1, u_2, u_1, \cdots$ would be minimizing and therefore would converge, which would only be possible if $u_1 = u_2$.

The theorem is proved.

4. SOLUTION OF THE BASIC BOUNDARY VALUE PROBLEM.

THEOREM 2. *The function $u$ giving a minimum for $D(v)$ in $W_2^{(m)}\{\phi_{\alpha_1 \cdots \alpha_n}^{(n-s)}\}$ has continuous derivatives of all orders in the interior of $\Omega$ and satisfies equation* (14.1).

PROOF. Let $\xi \in W_2^{(m)}(0)$. Then

$$u+\lambda\xi \in W_2^{(m)}\{\varphi^{(n-s)}_{\alpha_1\ldots\alpha_n}\},\quad \lambda=\text{const.}$$

and hence

$$D(u+\lambda\xi)=D(u)+2\lambda D(u,\xi)+\lambda^2 D(\xi)\geqslant d$$

for all $\lambda$ and has a minimum equal to $d$ for $\lambda=0$. From this it follows that $d(D(u+\lambda\xi))/d\lambda|_{\lambda=0}=0$, which gives

$$D(u,\xi)=0. \tag{14.13}$$

We consider the elementary solution of the polyharmonic equation:

$$g(r)=\begin{cases} r^{2m-n}, & \text{if} \quad 2m-n \quad \text{is odd or negative} \\ r^{2m-n}\ln r, & \text{if} \quad 2m-n\geqslant 0 \quad \text{and is even.}\end{cases}$$

It is easy to verify that $\Delta^{m-1}g(r)=A/r^{n-2}$, where $A=\text{const.}\neq 0$, and as a result, $\Delta^m g(r)=0$.

Let $\delta>0$ be a sufficiently small number. We consider the domain $\Omega_\delta$ and form the function

$$\xi=g(r)\left[\psi\left(\frac{r}{h_1}\right)-\psi\left(\frac{r}{h_2}\right)\right],$$

where $0<h_1<h_2<\delta$ and $\psi$ is the averaging function considered earlier (cf. the Dirichlet problem, §12, item 3). From the properties of this function we conclude that $\xi=0$ for $r<h_1/2$, $r>h_2$, and that all the derivatives of $\xi$ are continuous. If the point from which $r$ is calculated lies within $\Omega_\delta$, then $\xi$ and all its derivatives vanish on the boundary of $\Omega$. Therefore (14.13) holds for $\xi$. Since $\xi$ has continuous derivatives of all orders and is zero outside of $\Omega_\delta$, by the definition of the generalized derivatives

$$\frac{\partial^m u}{\partial x_1^{\alpha_1}\ldots\partial x_n^{\alpha_n}}$$

we have:

$$\int\underset{\Omega}{\cdots}\int \frac{\partial^m u}{\partial x_1^{\alpha_1}\ldots\partial x_n^{\alpha_n}}\cdot\frac{\partial^m \xi}{\partial x_1^{\alpha_1}\ldots\partial x_n^{\alpha_n}}\,d\Omega=$$

$$=(-1)^m\int\underset{\Omega}{\cdots}\int u\,\frac{\partial^{2m}\xi}{\partial x_1^{2\alpha_1}\ldots\partial x_n^{2\alpha_n}}\,d\Omega.$$

Hence equation (14.13) gives

$$D(u,\xi)=(-1)^m\int\underset{\Omega}{\cdots}\int u\sum_{\Sigma\alpha_i=m}\frac{m!}{\alpha_1!\alpha_2!\ldots\alpha_n!}\frac{\partial^{2m}\xi}{\partial x_1^{2\alpha_1}\ldots\partial x_n^{2\alpha_n}}\,d\Omega=0$$

or

$$\int \underset{\Omega}{\cdots} \int u \Delta^m \xi \, d\Omega = 0. \tag{14.14}$$

But

$$\Delta^m \xi = \Delta^m \left[ g(r) \psi\left(\frac{r}{h_1}\right) \right] - \Delta^m \left[ g(r) \psi\left(\frac{r}{h_2}\right) \right]$$

and since $\psi(r/h_i) = 1$ for $r \leqq h_i/2$ and $\Delta^m g(r) = 0$,

$$\Delta^m \left[ g(r) \psi\left(\frac{r}{h_i}\right) \right] \equiv 0 \text{ в } r \leqslant \frac{h_i}{2} \ (i = 1, 2).$$

Hence, equation (14.14) may be put in the form

$$\int \underset{\Omega}{\cdots} \int u \Delta^m \left[ g(r) \psi\left(\frac{r}{h_1}\right) \right] d\Omega = \int \underset{\Omega}{\cdots} \int u \Delta^m \left[ g(r) \psi\left(\frac{r}{h_2}\right) \right] d\Omega, \tag{14.15}$$

where both sides of the equation have a sense.

Consider the function

$$\omega(r, h) = \frac{1}{(n-2) A \sigma_n} \Delta^m \left[ g(r) \psi\left(\frac{r}{h}\right) \right],$$

where $\sigma_n$ is the surface area of the unit sphere in $n$-dimensional space and $A$ is the constant from the equation $\Delta^{m-1} g(r) = A/r^{n-2}$. It is obvious that $\omega(r/h)$ has continuous derivatives of all order and is equal to zero for $r \leqq h/2$ or $r \geqq h$, and that the same properties hold for

$$\Delta^m \left\{ g(r) \psi\left(\frac{r}{h}\right) \right\}.$$

Furthermore

$$\int \underset{\Omega}{\cdots} \int \omega(r, h) \, d\Omega = \frac{1}{(n-2) \sigma_n A} \int \underset{\Omega}{\cdots} \int \Delta^m \left[ g(r) \psi\left(\frac{r}{h}\right) \right] d\Omega =$$

$$= \frac{1}{(n-2) A \sigma_n} \int \underset{\Omega}{\cdots} \int \Delta \left\{ \Delta^{m-1} \left[ g(r) \psi\left(\frac{r}{h}\right) \right] \right\} d\Omega =$$

$$= \frac{1}{(n-2) A \sigma_n} \int \underset{r=h}{\cdots} \int \frac{d}{dr} \left\{ \Delta^{m-1} \left[ g(r) \psi\left(\frac{r}{h}\right) \right] \right\} dS -$$

$$- \frac{1}{(n-2) A \sigma_n} \int \underset{r=\frac{h}{2}}{\cdots} \int \frac{d}{dr} \left\{ \Delta^{m-1} g(r) \psi\left(\frac{r}{h}\right) \right\} dS =$$

$$= \frac{h^{n-1} \sigma_n}{(n-2) \sigma_n A} \frac{d}{dr} \left\{ \Delta^{m-1} \left[ g(r) \psi\left(\frac{r}{h}\right) \right] \right\}_{r=h} -$$

$$- \frac{\left(\frac{h}{2}\right)^{n-1} \sigma_n}{(n-2) \sigma_n A} \frac{d}{dr} \left\{ \Delta^{m-1} \left[ g(r) \psi\left(\frac{r}{h}\right) \right] \right\}_{r=\frac{h}{2}}.$$

Since $\psi(r/h)$ and all its derivatives vanish for $r=h$, the first term must be zero. The second term differs from zero since

$$\psi\left(\frac{r}{h}\right)\Big|_{r=\frac{h}{2}}=1 \qquad \psi^{(k)}\left(\frac{r}{h}\right)\Big|_{r=\frac{h}{2}}=0,\ (k\geqslant 1).$$

As a result, we obtain

$$\int\underset{\Omega}{\cdots}\int \omega(r,\ h)\,d\Omega=-\frac{\left(\frac{h}{2}\right)^{n-1}}{(n-2)A}\frac{d}{dr}\left\{[\Delta^{m-1}g(r)]\,\psi\left(\frac{r}{h}\right)\right\}\Big|_{r=\frac{h}{2}}=$$

$$=-\frac{\left(\frac{h}{2}\right)^{n-1}}{(n-2)A}\frac{d}{dr}\left[\frac{A}{r^{n-2}}\psi\left(\frac{r}{h}\right)\right]\Big|_{r=\frac{h}{2}}=$$

$$=-\frac{\left(\frac{h}{2}\right)^{n-1}}{n-2}\left[\frac{-(n-2)}{r^{n-1}}\psi\left(\frac{r}{h}\right)+\frac{1}{hr^{n-2}}\psi'\left(\frac{r}{h}\right)\right]\Big|_{r=\frac{h}{2}}=1.$$

Thus

$$\int\underset{\Omega}{\cdots}\int \omega(r,\ h)\,d\Omega=1,$$

which shows that $\omega(r,h)$ is a regular averaging kernel (cf. Chapter I, §2, item 4). We multiply (14.15) by $1/(n-2)A\sigma_n$, rewriting it in the form

$$\int\underset{\Omega}{\cdots}\int u\omega(r,\ h_1)\,d\Omega=\int\underset{\Omega}{\cdots}\int u\omega(r,\ h_2)\,d\Omega.$$

This equation shows that the averaged function for $u$ does not change on $\Omega_\delta$ for a change in the averaging radius $h(h<\delta)$ and therefore on $\Omega_\delta$, $u$ is equal to its averaged functions. Since the averaged functions have continuous derivatives of all orders, the same must be true for $u$. In view of the arbitrariness of $\delta$, we conclude that $u$ has continuous derivatives of all orders at all interior points of $\Omega$.

Let $\xi$ have continuous derivatives up to $m$th order within $\Omega$ and vanish outside of some closed domain entirely contained within $\Omega$. For such a function $\xi$, (14.13) must hold since obviously

$$\xi\in W_2^{(m)}\{0\}.$$

Integrating by parts in (14.13), we find

$$\int\underset{\Omega}{\cdots}\int \xi\Delta^m u\,d\Omega=0,$$

from which, in view of the arbitrariness of $\xi$, there follows

$$\Delta^m u = 0.$$

We have shown that $u$ is a solution of equation (14.1) with the conditions (14.6). We shall call the problem just considered the basic boundary value problem for the polyharmonic equation.

## §15. Uniqueness of the solution of the basic boundary value problem for the polyharmonic equation.

1. FORMULATION OF THE PROBLEM.

THEOREM 3. *The solution of the basic boundary value problem is unique.*

If we assume that there exists within $W_2^{(m)}$ still another solution $w$ of equation (14.1) with the conditions (14.6), then we must have $D(w)>d$, since in the contrary case we may construct a minimizing sequence by interspersing with repetitions of $w$ a sequence converging to $u$. From the convergence of this sequence, we would then arrive at the equality: $u=w$. We shall show that for every solution $w \in W_2^{(m)}$ of the equation (14.1) and the conditions (14.6), it is impossible to have

$$D(w) > d. \tag{15.1}$$

Let $\xi \in W_2^{(m)}(0)$ and suppose that $\xi$ within $\Omega$ has continuous derivatives up to $m$th order.

If we can show that

$$D(w, \xi) = 0, \tag{15.2}$$

then, repeating word for word the corresponding argument for the proof of uniqueness in the Dirichlet problem, we would arrive at the result that

$$D(u) > d,$$

contradicting (14.12). Therefore for the proof of the impossibility of (15.1) and to establish uniqueness in the basic boundary value problem, it suffices to prove (15.2).

2. LEMMA. *It is convenient to proceed as we did in the proof of uniqueness in the Dirichlet problem: we introduce in $\Omega$ the function*

$$\Psi_{2h} = \begin{cases} 1 & \text{within} \quad \Omega_{2h}, \\ 0 & \text{outside of} \quad \Omega_{2h}, \end{cases}$$

*where $h>0$ is a small number.*

We form the average for $\Psi_{2h}$ with respect to the kernel $(\kappa h^n)^{-1}\psi(r/h)$. This averaged function we denote by $\chi_h$. It is obvious that $\chi_h$ has derivatives of arbitrary order, $\chi_h = 1$ in $\Omega_{3h}$ and vanishes outside of $\Omega_h$, and everywhere $|\chi_h| \leq 1$. In addition, it is not difficult to show that

$$\left| \frac{\partial^k \chi_h}{\partial x_1^{\alpha_1} \ldots \partial x_n^{\alpha_n}} \right| < \frac{A}{h^k}. \tag{15.3}$$

The function $\xi\chi_h$ has continuous derivatives up to $m$th order in $\Omega$, vanishes outside of $\Omega_h$, and coincides with $\xi$ on $\Omega_{3h}$. We shall prove the following basic lemma.

*For every $v \in W_2^{(m)}$, we can choose a sequence $\{h_r\}$ $(h_r \to 0)$ such that*

$$D(v, \xi) = \lim_{r \to \infty} D(v, \xi\chi_{h_r}). \tag{15.4}$$

*We have:*

$$D(v, \xi - \xi\chi_h) = \int_\Omega \cdots \int \sum \frac{m!}{\alpha_1! \ldots \alpha_n!} \times$$

$$\times \left( \frac{\partial^m v}{\partial x_1^{\alpha_1} \ldots \partial x_n^{\alpha_n}} \frac{\partial^m (\xi - \xi\chi_h)}{\partial x_1^{\alpha_1} \ldots \partial x_n^{\alpha_n}} \right) d\Omega =$$

$$= \int_{\Omega - \Omega_{3h}} \cdots \int \sum \frac{m!}{\alpha_1! \ldots \alpha_n!} \frac{\partial^m v}{\partial x_1^{\alpha_1} \ldots \partial x_n^{\alpha_n}} \frac{\partial^m (\xi - \xi\chi_h)}{\partial x_1^{\alpha_1} \ldots \partial x_n^{\alpha_n}} d\Omega =$$

$$= \int_{\Omega - \Omega_{3h}} \cdots \int \sum \frac{m!}{\alpha_1! \ldots \alpha_n!} \frac{\partial^m v}{\partial x_1^{\alpha_1} \ldots \partial x_n^{\alpha_n}} \frac{\partial^m \xi}{\partial x_1^{\alpha_1} \ldots \partial x_n^{\alpha_n}} d\Omega -$$

$$- \int_{\Omega - \Omega_{3h}} \cdots \int \sum \frac{m!}{\alpha_1! \ldots \alpha_n!} \frac{\partial^m v}{\partial x_1^{\alpha_1} \ldots \partial x_n^{\alpha_n}} \sum C^k_{\beta_1 \ldots \beta_n} \times$$

$$\times \frac{\partial^{m-k} \xi}{\partial x_1^{\beta_1} \ldots \partial x_n^{\beta_n}} \frac{\partial^k \chi_h}{\partial x_1^{\alpha_1 - \beta_1} \ldots \partial x_n^{\alpha_n - \beta_n}} d\Omega.$$

For $h \to 0$ the first integral tends to zero since $v$ and $\xi \in W_2^{(m)}$, while $\Omega - \Omega_{3h} \to 0$. For the proof of (15.4) it suffices to show that the second integral tends to zero.

For this purpose it suffices to prove that

$$J_k(h) = \left| \int_{\Omega_h - \Omega_{3h}} \cdots \int \frac{\partial^m v}{\partial x_1^{\alpha_1} \ldots \partial x_n^{\alpha_n}} \frac{\partial^{m-k} \xi}{\partial x_1^{\beta_1} \ldots \partial x_n^{\beta_n}} \times \right.$$

$$\left. \times \frac{\partial^k \chi_h}{\partial x_1^{\gamma_1} \ldots \partial x_n^{\gamma_n}} d\Omega \right| \to 0, \quad (h \to 0)$$

$$(k = 1, 2, \ldots, m)$$

(for $k = 0$ it is obvious that $J_0(h) \to 0$ since $v$, $\xi \in W_2^{(m)}$ and $|\chi_h| \leq 1$). By means of (15.3), we have

$$J_k(h) \leqslant \frac{A}{h^k} \int_{\Omega_h - \Omega_{3h}} \cdots \int \left| \frac{\partial^m v}{\partial x_1^{\alpha_1} \ldots \partial x_n^{\alpha_n}} \; \frac{\partial^{m-k} \xi}{\partial x_1^{\beta_1} \ldots \partial x_n^{\beta_n}} \right| d\Omega \leqslant$$

$$\leqslant \frac{A}{h^k} \sqrt{\int_{\Omega_h - \Omega_{3h}} \cdots \int \left| \frac{\partial^m v}{\partial x_1^{\alpha_1} \ldots \partial x_n^{\alpha_n}} \right|^2 d\Omega} \times$$

$$\times \sqrt{\int_{\Omega_h - \Omega_{3h}} \cdots \int \left| \frac{\partial^{m-k} \xi}{\partial x_1^{\beta_1} \ldots \partial x_n^{\beta_n}} \right|^2 d\Omega}. \tag{15.5}$$

Set $h'_\mu = 1/3^\mu$ $(\mu = 1, 2, \cdots)$.
Then

$$\Omega = \sum_{\mu=1}^{\infty} (\Omega_{h'_\mu} - \Omega_{3h'_\mu}) + \Omega_1,$$

and since $v \in W_2^{(m)}$, the series

$$\sum_{\mu=1}^{\infty} \int_{\Omega_{h'_\mu} - \Omega_{3h'_\mu}} \cdots \int \left| \frac{\partial^m v}{\partial x_1^{\alpha_1} \ldots \partial x_n^{\alpha_n}} \right|^2 d\Omega$$

converges. Hence we may find a infinite subsequence of the terms of this series less than the corresponding terms of the divergent series.

$$\sum_{\mu=2}^{\infty} \frac{1}{\mu \ln \mu}.$$

In other words, for the infinite sequence $\{\mu_r\}$ just obtained, and setting $h'_{\mu_r} = h_r$:

$$\left| \int_{\Omega_{h_r} - \Omega_{3h_r}} \cdots \int \frac{\partial^m v}{\partial x_1^{\alpha_1} \ldots \partial x_n^{\alpha_n}} \right|^2 d\Omega \leqslant \frac{2}{\mu_r \ln \mu_r} = \frac{\ln 3}{|\ln h_r| \cdot [\ln |\ln h_r| - \ln \ln 3]} \leqslant$$

$$\leqslant \frac{K}{|\ln h_r| \cdot \ln |\ln h_r|}. \tag{15.6}$$

If we can show that

$$\int_{\Omega_h - \Omega_{3h}} \cdots \int \left| \frac{\partial^{m-k} \xi}{\partial x_1^{\beta_1} \ldots \partial x_n^{\beta_n}} \right|^2 d\Omega \leqslant B h^{2k} \;\; \ln |h|, \tag{15.7}$$

$$(k = 1, \; 2, \; \ldots \; m)$$

then, obviously, we will have from (15.5) and (15.6)

$$J_k(h_r) \leqslant \frac{A}{h_r^k} \sqrt{\frac{K}{|\ln h_r| \cdot \ln |\ln h_r|}} \; \sqrt{B h_r^{2k} \; |\ln h_r|} = \frac{A \sqrt{BK}}{\sqrt{\ln |\ln h_r|}}$$

and it will have been shown that $J_k(h_r)\to 0$ for $r\to\infty$, from which (15.4) follows.

Thus for the proof of (15.4) and the basic lemma, it suffices to prove (15.7).

3. THE STRUCTURE OF THE DOMAINS $\Omega_h-\Omega_{3h}$. We shall concern ourselves with a more detailed study of the structure of the domain $\Omega_h-\Omega_{3h}$. We decompose the whole boundary of the domain $\Omega$ into a finite number of smooth pieces of various dimensions: $S^*_{n-s}$; to the collection of these pieces we adjoin all boundaries between two smooth pieces of the same dimension and all singular manifolds of the type of conical points or conical lines. The boundary between two pieces of dimension $l$ will be, generally speaking, of dimension $l-1$. For example, if the domain $\Omega$ is a cube, then the manifolds $S^*_2$ will be all the faces of the cube, $S^*_1$ all its edges, and the manifolds of $S^*_0$ all its vertices. If the domain is a right circular cone, then we have to consider two manifolds $S^*_2$, the lateral surface and the base surface, $S^*_1$ will be the base curve, and $S^*_0$ will be the vertex of the cone.

We construct for each of these manifolds the domain $(\Omega'_h-\Omega'_{3h})^*$ consisting of all points of the domain $\Omega$ whose distance from the manifold $S^*_{n-s}$ is less than $3h$ and not more than $h$. The domain $\Omega_h-\Omega_{3h}$ is covered by the sum of the domains $(\Omega'_h-\Omega'_{3h})^*$.

We extract from each of the domains $(\Omega'_h)^*_{n-s}$ a portion $(\Omega'_h)^{**}_{n-s}$ which, by the aid of a coordinate transformation with continuous bounded derivatives, may be transformed into a cylinder of radius $h$, the axis of which is a hyperplane of dimension $n-s$ and has as its image $S^*_{n-s}$, and we do this in such a way that the domains $(\Omega_h-\Omega_{3h})^{**}_{n-s}$ completely cover the whole domain $\Omega_h-\Omega_{3h}$. An intuitive figure (Figure 6) shows how this decomposition is to be carried out if the domain $\Omega$ is a nonconvex hexagon.

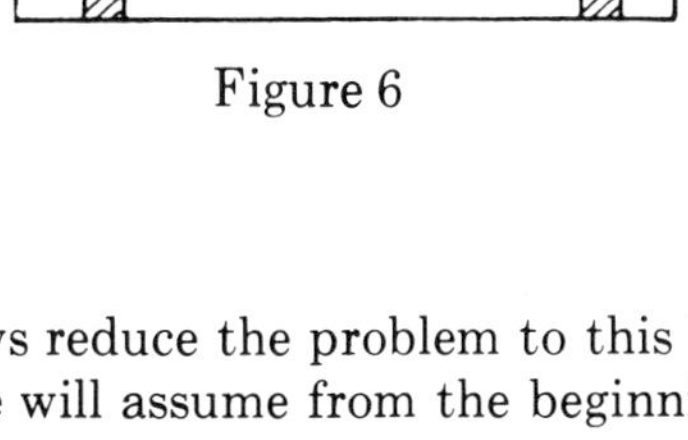

Figure 6

In Figure 6 the domain $(\Omega'_h-\Omega'_{3h})^{**}_{n-s}$ is shaded.

To prove the correctness of (15.7), it suffices to show its validity for each of the domains $(\Omega'_h-\Omega'_{3h})^{**}_{n-s}$.

To obtain the corresponding estimate, we may assume from the beginning that all the manifolds $S^{**}_{n-s}$ are points or hyperplanes of the corresponding dimension. Obviously, we may always reduce the problem to this by a change of coordinates.

We will assume from the beginning that the manifold $S_{n-s}$ is a domain in the hyperplane $x_1 = x_2 = \cdots = x_s = 0$. By virtue of the assumptions on the simplicity of the boundary $S$, the general case may be reduced to this. In the space $(x_1, x_2, \cdots, x_s, x_{s+1}, \cdots, x_n)$, we introduce cylindrical coordinates with "axis" $S_{n-s}$, i.e., put

$$
\begin{aligned}
x_1 &= \rho \cos \varphi_1; \\
x_2 &= \rho \sin \varphi_1 \cos \varphi_2; \\
x_3 &= \rho \sin \varphi_1 \sin \varphi_2 \cos \varphi_3; \\
&\cdots\cdots\cdots\cdots \\
x_{s-1} &= \rho \sin \varphi_1 \sin \varphi_2 \ \ldots\ \sin \varphi_{s-2} \cos \varphi_{s-1} \\
x_s &= \rho \sin \varphi_1 \sin \varphi_2 \ \ldots\ \sin \varphi_{s-2} \sin \varphi_{s-1} \\
x_{s+1} &= x_{s+1} \\
&\cdots\cdots \\
x_n &= x_n
\end{aligned}
$$

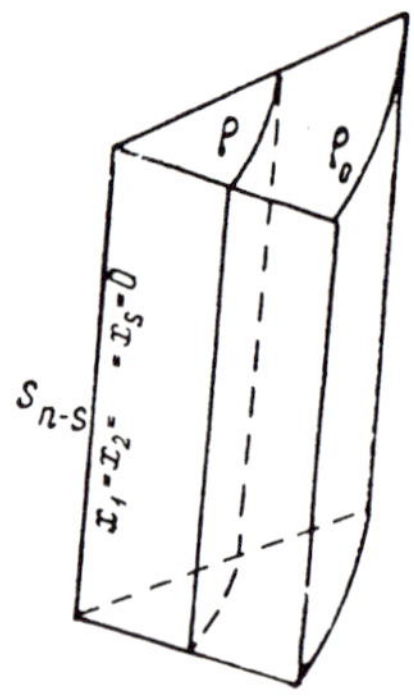


Figure 7

Then the manifold $x_1 = x_2 = \cdots = x_s = 0$ goes over into the manifold $\rho = 0$ (Figure 7).

The proof for the domains based on $S_{n-s}$ for the cases $s$ even and odd differs slightly. We shall carry through the proof in a unified form, indicating the differences where they are unavoidable.

Let $s = 2t$ or $s = 2t + 1$. Then it will be necessary for us to consider separately: Case I when $k = 1, 2, \cdots, t$, i.e., to estimate the integral (15.7) for derivatives, the values of which are not defined on $S_{n-s}$ (cf. §14, item 1), Case II when $K = t + 1$, and finally Case III: $k = t + 2, \cdots, m$, i.e., the case when the integrated derivatives are defined on $S_{n-s}$.

4. PROOF OF THE LEMMA FOR $k < [s/2]$. We consider first Case I: $k = 1, 2, \cdots, t$. Set

$$
Z(\rho, \varphi_i, x_j) = \frac{\partial^{m-k} \xi(\rho, \varphi_i, x_j)}{\partial x_1^{\beta_1} \ldots \partial x_n^{\beta_n}} \qquad \begin{pmatrix} k = 1, 2, \ldots, t \\ i = 1, 2, \ldots, s-1 \\ j = s+1, \ldots, n \end{pmatrix}
$$

Let $\rho_0 > 0$ be some number. We let $0 < \rho < \rho_0$. Since $\xi$ has continuous derivatives up to $m$th order, $Z$ will have continuous derivatives up to order $k$ everywhere except on the manifold $\rho = 0$.

Hence, applying Taylor's formula, we find

$$Z(\rho, \varphi_i, x_j) = Z(\rho_0, \varphi_i, x_j) + \frac{\rho-\rho_0}{1!}\frac{\partial Z}{\partial \rho}\bigg|_{\rho=\rho_0} + \dots +$$

$$+ \frac{(\rho-\rho_0)^{k-1}}{(k-1)!}\frac{\partial^{k-1} Z}{\partial \rho^{k-1}}\bigg|_{\rho=\rho_0} + \int_{\rho_0}^{\rho} \frac{(\rho-\rho_1)^{k-1}}{(k-1)!}\frac{\partial^k Z}{\partial \rho_1^k} d\rho_1. \tag{15.8}$$

Set

$$\| Z(\rho, \varphi_i, x_j) \|_{L_{2,n-s}} = \Big\{ \underset{\substack{\rho=\text{const.}\\ \varphi_i=\text{const.}}}{\int \dots \int} |Z|^2 dx_{s+1} \dots dx_n \Big\}^{\frac{1}{2}}.$$

Then equation (15.8) gives

$$\| Z(\rho, \varphi_i, x_j) \|^2_{L_{2,n-s}} \leqslant C \Big\{ \| Z(\rho_0, \varphi_i, x_j) \|^2_{L_{2,n-s}} +$$

$$+ \frac{(\rho-\rho_0)^2}{(1!)^2} \Big\| \frac{\partial Z}{\partial \rho_0} \Big\|^2_{L_{2,n-s}} + \dots + \frac{(\rho-\rho_0)^{2k-2}}{[(k-1)!]^2} \Big\| \frac{\partial^{k-1} Z}{\partial \rho_0^{k-1}} \Big\|^2_{L_{2,n-s}} +$$

$$+ \underset{\substack{\rho=\text{const.}\\ \varphi_i=\text{const.}}}{\int \dots \int} \Big[ \int_{\rho_0}^{\rho} \frac{(\rho-\rho_1)^{k-1}}{(k-1)!}\frac{\partial^k Z}{\partial \rho_1^k} d\rho_1 \Big]^2 dx_{s+1} \dots dx_n \Big\}. \tag{15.9}$$

Multiplying (15.9) by $\rho^{s-1} d\omega_{\phi_i}$ and integrating over the unit sphere $\omega_{\phi_i}$, and denoting by $S_\rho$ the $(n-1)$-dimensional manifold $\rho = \text{const.} \neq 0$, we obtain

$$\underset{S_\rho}{\int \dots \int} |Z|^2 dS_\rho \leqslant C \Big| \frac{\rho}{\rho_0} \Big|^{s-1} \Big\{ \underset{S_{\rho_0}}{\int \dots \int} |Z|^2 dS_{\rho_0} +$$

$$+ \frac{(\rho-\rho_0)^2}{(1!)^2} \underset{S_{\rho_0}}{\int \dots \int} \Big| \frac{\partial Z}{\partial \rho_0} \Big|^2 dS_{\rho_0} + \dots$$

$$+ \frac{(\rho-\rho_0)^{2k-2}}{[(k-1)!]^2} \underset{S_{\rho_0}}{\int \dots \int} \Big| \frac{\partial^{k-1} Z}{\partial \rho_0^{k-1}} \Big|^2 dS_{\rho_0} \Big\} +$$

$$+ C \underset{\omega_{\varphi_i}}{\int \dots \int} \rho^{s-1} d\omega_{\varphi_i} \underset{\substack{\rho=\text{const.}\\ \varphi_i=\text{const.}}}{\int \dots \int} \Big[ \int_{\rho_0}^{\rho} \frac{(\rho-\rho_1)^{k-1}}{(k-1)!}\frac{\partial^k Z}{\partial \rho_1^k} d\rho_1 \Big]^2 dx_{s+1} \dots dx_n. \tag{15.10}$$

Since $\xi \in W_2^{(m)}$, by the basic Imbedding Theorem all the derivatives of $\xi$ up to order $m-1$ belong to $L_{2,n-1}$ and all of them figuring in the right hand side of (15.10) may be estimated by

$$A \| \xi \|^2_{W_2^{(m)}}.$$

Consider the last term on the right side of (15.10). Noting that $\rho<\rho_0$, we obtain

$$\int_{\omega_{\varphi_i}}\cdots\int \rho^{s-1}\,d\omega_{\varphi_i}\int_{\substack{\rho=\text{const.}\\ \varphi_i=\text{const.}}}\cdots\int\left[\int_{\rho_0}^{\rho}\frac{(\rho-\rho_1)^{k-1}}{(k-1)!}\frac{\partial^k Z}{\partial\rho_1^k}\,d\rho_1\right]^2 dx_{s+1}\ldots dx_n\leqslant$$

$$\leqslant\rho^{s-1}\int_{\omega_{\varphi_i}}\cdots\int d\omega_{\varphi_i}\int_{\substack{\rho=\text{const.}\\ \varphi_i=\text{const.}}}\cdots\int\int_{\rho}^{\rho_0}\frac{(\rho-\rho_1)^{2k-2}}{[(k-1)!]^2\rho_1^{s-1}}\,d\rho_1\times$$

$$\times\int_{\rho}^{\rho_0}\left|\frac{\partial^k Z}{\partial\rho_1^k}\right|^2\rho_1^{s-1}\,d\rho_1\,dx_{s+1}\ldots dx_n=$$

$$=\left\{\rho^{s-1}\int_{\omega_{\varphi_i}}\cdots\int d\omega_{\varphi_i}\int_{\substack{\rho=\text{const.}\\ \varphi_i=\text{const.}}}\cdots\int\int_{\rho}^{\rho_0}\rho_1^{s-1}\left|\frac{\partial^k Z}{\partial\rho_1^k}\right|^2 d\rho_1\,dx_{s+1}\ldots dx_n\right\}\times$$

$$\times\int_{\rho}^{\rho_0}\frac{(\rho-\rho_1)^{2k-2}}{[(k-1)!]^2\rho_1^{s-1}}\,d\rho_1=\rho^{s-1}\overbrace{\int_{\rho^2\leqslant\sum x_j^2\leqslant\rho_0^2}\cdots\int}^{n}\left|\frac{\partial^k Z}{\partial\rho_1^k}\right|^2 d\Omega\times$$

$$\times\int_{\rho}^{\rho_0}\frac{(\rho-\rho_1)^{2k-2}}{[(k-1)!]^2\rho_1^{s-1}}\,d\rho_1\leqslant C\rho^{s-1}\|\xi\|^2_{W_2^{(m)}}\int_{\rho}^{\rho_0}\frac{(\rho-\rho_1)^{2k-2}}{\rho_1^{s-1}}\,d\rho_1. \qquad (15.11)$$

We estimate $\int_\rho^{\rho_0}((\rho-\rho_1)^{2k-2}/\rho_1^{s-1}d\rho_1$. For this purpose, set: $\rho_1=\rho x$, $\rho_0=\rho y$.

$$\int_{\rho}^{\rho_0}\frac{(\rho-\rho_1)^{2k-2}}{\rho_1^{s-1}}\,d\rho_1=\int_1^y\frac{\rho^{2k-2}(1-x)^{2k-2}}{\rho^{s-1}x^{s-1}}\rho\,dx=\rho^{2k-s-1}\int_1^y\frac{(1-x)^{2k-2}}{x^{s-1}}\,dx.$$

If $2k-2-s+1<-1$, then the integral converges as $y\to\infty$ (i.e., as $\rho\to 0$). If $2k-2-s+1=-1$, the integral grows like $\ln y$, i.e., like $|\ln\rho|$. Hence, if $2k\leqq s$, we have from (15.10) and (15.11)

$$\int_{S_\rho}\cdots\int|Z|^2\,dS_\rho\leqslant\rho^{s-1}\|\xi\|_{W_2^{(m)}}[C_1+C_2\rho^{2k-s}|\ln\rho|].$$

Integrating in $\rho$ from $Ah$ to $Bh$, we obtain:

$$\int_{\Omega_{Ah}-\Omega_{Bh}}\cdots\int|Z|^2 d\Omega\leqslant\|\xi\|_{W_2^{(m)}}[K_1h^s+K_2h^{2k}|\ln h|]\leqslant K_3h^{2k}|\ln h|.$$

Thus (15.7) is proved for $k=1,2,\cdots,t$ ($s=2t$ or $s=2t+1$).

5. PROOF OF THE LEMMA FOR $k=[s/2]+1$. We pass to the proof of (15.7) for Case II: $k=t+1$. In this case we already know that the expression

within the integral sign tends to zero on the boundary. However, the estimates which we obtained above in §11 turn out to be insufficient and we give some slightly strengthened forms.

Consider some plane $x_{s+1}=\text{const.},\cdots,x_n=\text{const.}$ and denote the distances from the points $x_1=x_2=\cdots=x_s=0$ and $(\Delta x_1,\Delta x_2,\cdots,\Delta x_s)$ $(\sum \Delta x_i^2=|\Delta P|^2)$ of this plane to an arbitrary point of the same plane by $r$ and $r_1$.

Set $(\partial^{m-t-1}\xi/\partial x_1^{\alpha_1}\cdots\partial x_n^{\alpha_n})=\phi(x_1,\cdots,x_s,\ x_{s+1},\cdots,x_n)$. Derivatives of order $t+1$ with respect to $x_1,\cdots,x_s$ of $\phi$ are derivatives of $m$th order of $\xi$. Hence, applying the formula for estimating

$$\varphi(\Delta x_1,\ \ldots,\ \Delta x_s,\ x_{s+1},\ \ldots,x_n)-\varphi(0,\ \ldots,\ 0,\ x_{s+1},\ \ldots,\ x_n)=$$
$$=\varphi(\Delta\vec{P},\ x_j)-\varphi(0,\ x_j),$$

we find that as in the proof of (11.14)

$$|\varphi(\Delta\vec{P},\ x_j)-\varphi(0,x_j)|\leqslant|\Delta\vec{P}|\Big[A\overbrace{\int\cdots\int_{\Omega_s}}^{s}\frac{(r+r_1)^{s-t-2}}{r^{s-t-1}r_1^{s-t-1}}\times$$
$$\times\sum\Big|\frac{\partial^m\xi}{\partial x_1^{\alpha_1}\ldots\partial x_n^{\alpha_n}}\Big|d\Omega_s+B\sqrt{\int\cdots\int_{\Omega_s}|\varphi|^2d\Omega_s}\Big]\leqslant|\Delta\vec{P}|\times$$
$$\times\Big[A\sqrt{\int\cdots\int_{\Omega_s}\Big\{\sum\Big|\frac{\partial^m\xi}{\partial x_1^{\alpha_1}\ldots\partial x_n^{\alpha_n}}\Big|^2\Big\}d\Omega_s}\times$$
$$\times\sqrt{\int\cdots\int_{\Omega_s}\frac{(r+r_1)^{2s-2t-4}}{r^{2s-2t-2}r_1^{2s-2t-2}}d\Omega_s}+B\sqrt{\int\cdots\int_{\Omega_s}|\varphi|^2d\Omega_s}\Big].$$

We consider the cases $s=2t$. Then using the representation which we encountered earlier in Chapter 1,§11, we obtain

$$\int\cdots\int_{\Omega_s}\frac{(r+r_1)^{2s-2t-4}}{r^{2(s-t-1)}r_1^{2(s-t-1)}}d\Omega_s=\int\cdots\int_{\Omega_s}\frac{(r+r_1)^{2t-4}}{r^{2(t-1)}r_1^{2(t-1)}}d\Omega_s\leqslant$$
$$\leqslant|\Delta P|\int\cdots\int_{\rho<\frac{R}{|\Delta P|}}\frac{(\rho+\rho_1)^{2t-4}}{\rho^{2t-2}\rho_1^{2t-2}}d\Omega_s\leqslant C_1'+C_2'\int_2^{\frac{R}{|\Delta P|}}\frac{d\rho}{\rho}=$$
$$=C_1+C_2|\ln|\Delta P||.$$

Thus we find from (15.11)

$$|\varphi(\Delta P,\ x_j)-\varphi(0,\ x_j)|^2\leqslant|\Delta P|^2\Big\{[K_1+K_2|\ln|\Delta P||]\times$$
$$\times\int\cdots\int_{\Omega_s}\sum\Big|\frac{\partial^m\xi}{\partial x_1^{\alpha_1}\ldots\partial x_n^{\alpha_n}}\Big|^2d\Omega_s+K_3\int\cdots\int_{\Omega_s}|\varphi|^2d\Omega_s\Big\}.$$

Integrating over the plane $S_{n-s}$ $(x_1 = x_2 = \cdots = x_s = 0)$, we see that

$$\int_{S_{n-s}} \cdots \int |\varphi(\Delta P, x_j) - \varphi(0, x_j)|^2 d\Omega_{n-s} \leqslant |\Delta P|^2 \Big\{ [K_1 + K_2 |\ln|\Delta P||] \times$$

$$\times \int_\Omega \cdots \int \sum \left| \frac{\partial^m \xi}{\partial x_1^{\alpha_1} \ldots \partial x_n^{\alpha_n}} \right| d\Omega + K_3 \int_\Omega \cdots \int |\varphi|^2 d\Omega \Big\} \leqslant$$

$$\leqslant |\vec{\Delta P}|^2 \{ [K_1 + K_2 |\ln|\Delta P||] \|\xi\|^2_{W_2^{(m)}} + K_4 \|\xi\|^2_{W_2^{(m)}} \} \leqslant$$

$$\leqslant K_5 \|\xi\|_{W_2^{(m)}} |\Delta P|^2 |\ln|\Delta P||. \tag{15.12}$$

This estimate is precisely that which would follow immediately from formula (11.15) provided that we obtained in it

$$|\Delta P|^{2-\epsilon} \text{ instead of } |\ln|\Delta P|| |\Delta P|^2.$$

In deducing (15.12), we used the fact that

$$\int_\Omega \cdots \int |\varphi|^2 d\Omega = \int_\Omega \cdots \int \left| \frac{\partial^{m-t-1} \xi}{\partial x_1^{\alpha_1} \ldots \partial x_n^{\alpha_n}} \right|^2 d\Omega \leqslant N \|\xi\|_{W_2^{(m)}}.$$

Using the fact that $\phi(0, x_j) = 0$ (in the sense of $L_{2,\, n-s}$), we give (15.12) the form

$$\int_{\substack{x_1 = \Delta x_1 \\ \cdots\cdots \\ x_s = \Delta x_s}} \cdots \int \left| \frac{\partial^{m-t-1} \xi}{\partial x_1^{\alpha_1} \ldots \partial x_n^{\alpha_n}} \right|^2 dx_{s+1} \ldots dx_n \leqslant$$

$$\leqslant K_5 \|\xi\|_{W_2^{(m)}} |\Delta P|^2 |\ln|\Delta P||. \tag{15.13}$$

Introducing cylindrical coordinates with axis $S_{n-s}$, replacing $|\Delta P|$ by $\rho$ in (15.13), multiplying by $\rho^{2t-1} d\omega_{\phi_i}$ and integrating over the unit sphere, we obtain:

$$\int_{S_\rho} \cdots \int \left| \frac{\partial^{m-t-1} \xi}{\partial x_1^{\alpha_1} \ldots \partial x_n^{\alpha_n}} \right|^2 dS_\rho \leqslant K_6 \|\xi\|_{W_2^{(m)}} \rho^{2t+1} |\ln \rho|. \tag{15.14}$$

Integrating in $\rho$ from $Ah$ to $Bh$, we find:

$$\int_{\Omega_{Ah} - \Omega_{Bh}} \cdots \int \left| \frac{\partial^{m-t-1} \xi}{\partial x_1^{\alpha_1} \ldots \partial x_n^{\alpha_n}} \right|^2 d\Omega \leqslant K h^{2t+2} |\ln h|. \tag{15.15}$$

Analogous arguments for the case $s = 2t+1$ give the estimate (15.14) with a right hand side of order $\rho^{2t+1}$ and (15.15) with a right hand side of order $h^{2t+2}$. Thus (15.7) is proved for $k = t+1$.

6. PROOF OF THE LEMMA FOR $[s/2] + 2 \leqq k \leqq m$. We pass to the proof of

(15.7) for Case III: $k=t+2, t+3,\cdots,m$. Set

$$Z(\rho, \varphi_i, x_j)=\frac{\partial^{m-k}\xi(\rho, \varphi_i, x_j)}{\partial x_1^{\alpha_1}\partial x_2^{\alpha_2}\ldots\partial x_n^{\alpha_n}}.$$

$Z(\rho, \phi_i, x_j)$ has continuous derivatives up to $k$th order, $k>t+1$, with respect to all arguments everywhere except on the manifold $\rho=0$. Applying Taylor's formula, we find

$$Z(\rho, \varphi_i, x_j)=Z(\rho_0, \varphi_i, x_j)+\frac{\rho-\rho_0}{1!}\left(\frac{\partial Z}{\partial \rho}\right)_{\rho=\rho_0}+\cdots+$$

$$+\frac{(\rho-\rho_0)^{k-t-2}}{(k-t-2)!}\left(\frac{\partial^{k-t-2}Z}{\partial\rho^{k-t-2}}\right)_{\rho=\rho_0}+\int_{\rho_0}^{\rho}\frac{(\rho-\rho_1)^{k-t-2}}{(k-t-2)!}\frac{\partial^{k-t-1}Z}{\partial\rho_1^{k-t-1}}d\rho_1. \quad (15.16)$$

From formula (15.16), we obtain:

$$\|Z(\rho, \varphi_i, x_j)\|_{L_{2,n-s}}\leqslant\|Z(\rho_0, \varphi_i, x_j)\|_{L_{2,n-s}}+\frac{|\rho-\rho_0|}{1!}\times$$

$$\times\left\|\left(\frac{\partial Z}{\partial\rho}\right)_{\rho=\rho_0}\right\|_{L_{2,n-s}}+\cdots+\frac{|\rho-\rho_0|^{k-t-2}}{(k-t-2)!}\left\|\left(\frac{\partial^{k-t-2}Z}{\partial\rho^{k-t-2}}\right)_{\rho=\rho_0}\right\|_{L_{2,n-s}}+$$

$$+\left\{\int_{S_{n-s}}\cdots\int\left[\int_{\rho_1}^{\rho}\frac{(\rho-\rho_1)^{k-t-2}}{(k-t-2)!}\frac{\partial^{k-t-1}Z}{\partial\rho_1^{k-t-1}}d\rho_1\right]^2 dx_{s+1}\ldots dx_n\right\}^{\frac{1}{2}}$$

By virtue of the boundary conditions for $\xi$ ($\xi$ and all its derivatives up to $(m-t-1)$st order are null in the sense of $L_{2,n-s}$ on the manifold $\rho=0$), in taking the limit $\rho_0\to 0$, we obtain

$$\|Z(\rho, \varphi_i, x_j)\|^2_{L_{2,n-s}}\leqslant$$

$$\leqslant\overbrace{\int_{S_{n-s}}\cdots\int}^{n-s}\left[\int_0^{\rho}\frac{(\rho-\rho_1)^{k-t-2}}{(k-t-2)!}\frac{\partial^{k-t-1}Z}{\partial\rho_1^{k-t-1}}d\rho_1\right]^2 dx_{s+1}\ldots dx_n\leqslant$$

$$\leqslant A\overbrace{\int_{S_{n-s}}\cdots\int}^{n-s}\left\{\int_0^{\rho}\left|\frac{\partial^{k-t-1}Z}{\partial\rho_1^{k-t-1}}\right|^2 d\rho_1\right\}dx_{s+1}\ldots dx_n\int_0^{\rho}(\rho-\rho_1)^{2k-2t-4}d\rho_1=$$

$$=A_1\rho^{2k-2t-3}\int_0^{\rho}\left[\overbrace{\int_{S_{n-s}}\cdots\int}^{n-s}\left|\frac{\partial^{k-t-1}Z}{\partial\rho_1^{k-t-1}}\right|^2 dx_{s+1}\ldots dx_n\right]d\rho_1. \quad (15.17)$$

We consider the case $s=2t$. Multiplying (15.17) by $\rho^{2t-1}d\omega_{\phi_i}$ and integrating over the unit sphere, we obtain

$$\overbrace{\int \cdots \int_{S_\rho}}^{n-1} |Z(\rho, \varphi_i, x_j)|^2 \, dS_\rho \leqslant$$

$$\leqslant A_2 \rho^{2k-4} \int_0^\rho \frac{1}{\rho_1^{2t-1}} \int \cdots \int_{S_{\rho_1}} \sum \left| \frac{\partial^{m-t-1}\xi}{\partial x_1^{\alpha_1} \ldots \partial x_n^{\alpha_n}} \right|^2 dS_{\rho_1} \, d\rho_1,$$

from which, taking (15.14) into consideration, we find:

$$\overbrace{\int \cdots \int_{S_\rho}}^{n-1} |Z(\rho, \varphi_i, x_j)|^2 \, dS_\rho \leqslant A_3 \|\xi\|^2_{W_2^{(m)}} \rho^{2k-4} \int_0^\rho \frac{\rho_1^{2t+1} |\ln \rho_1|}{\rho_1^{2t-1}} \, d\rho_1 =$$

$$= A_3 \|\xi\|^2_{W_2^{(m)}} \rho^{2k-4} \int_0^\rho \rho_1^2 |\ln \rho_1| \, d\rho_1 \leqslant A_4 \rho^{2k-1} |\ln \rho|.$$

Integrating in $\rho$ from $Ah$ to $Bh$, we find:

$$\int \cdots \int_{\Omega_{Ah} - \Omega_{Bh}} |Z|^2 \, d\Omega \leqslant A_5 h^{2k} |\ln h|.$$

In the case $s = 2t+1$, the right hand side is obtained of order $h^{2k}$. Thus (15.17) is proven for all $k = 1, 2, \cdots, m$.

From this the basic lemma immediately follows.[2)]

From the lemma an obvious method yields equation (15.2). Indeed for an arbitrary function $w$, satisfying the equation $\Delta^m w = 0$ and $\xi\chi_h$, we will have as before:

$$D(w, \xi\chi_h) = 0. \tag{15.2}$$

Passing to the limit, we establish (15.2) and with it the uniqueness theorem.

7. REMARKS ON THE FORMULATION OF THE BOUNDARY CONDITIONS. We shall formulate still another problem. We shall call a function $\xi$ a function from $W_2^{(m)}(0)$ if it is the limit in $W_2^{(m)}$ of functions vanishing on strips around the boundary. By $W_2^{(m)}(*)$ we denote the family of functions which are weak limits of functions vanishing on boundary strips.[3)] If

---

[2)] Here, as in the proof of the lemma of equation 12.4, we may dispense with the assumption of the continuous differentiability of up to order $m$. For this purpose one only needs to introduce a limit process on averaging functions in formulas (15.9), (15.12) and (15.17). The remainder of the proof proceeds without change.

[3)] Here, weak limits are taken in the following sense: a function $\xi \in W_2^{(m)}$ is the weak limit of the sequence $\{\xi_k\}$, $\xi_k \in W_2^{(m)}$, if for every $v \in W_2^{(m)}$, there holds $D(v, \xi) = \lim_{k \to \infty} D(v, \xi_k)$.

we are given some function $\psi$ belonging to $W_2^{(m)}$, then we shall say that $u$ coincides on the boundary with $\psi$ together with its derivatives in the sense of $W_2^{(m)}(0)$ or $W_2^{(m)}(*)$ if the difference $u-\psi$ lies in the corresponding family. From a theorem in the first chapter, it follows that correspondence on the boundary with a function $\psi$ in the sense of $W_2^{(m)}(0)$ implies a fortiori that $u$ belongs to $W_2^{(m)}(\psi^{(n-s)}_{\alpha_1\cdots\alpha_n})$, where $\psi^{(n-s)}_{\alpha_1\cdots\alpha_n}$ are the limiting values of the corresponding derivatives of the function $\psi$, which as we showed above always exist.

The lemma proved above for the uniqueness theorem now gives a partial converse to this last assertion. If the function $u\in W_2^{(m)}(\psi^{(n-s)}_{\alpha_1\cdots\alpha_n})$, then $u$ coincides with the function $\psi$ on the boundary together with its derivatives in the sense of $W_2^{(m)}(*)$. Obviously then $W_2^{(m)}(0)\subset W_2^{(m)}(*)$. We are not interested here in the question as to whether these two sets coincide or not.

In fact, from $(u-\psi)\in W_2^{(m)}(0)$ follows $u\in W_2^{(m)}(\psi^{(n-s)}_{\alpha_1\cdots\alpha_n})$, and from the latter follows $u-\psi\in W_2^{(m)}(*)$.

In a series of publications on variational methods, there has been considered the problem of finding functions giving an extremum to the functional $D(u)$ (for the case $m=1$) and coinciding on the boundary with a given function $\psi$ in the sense of $W_2^{(m)}(0)$. Such problems obviously have an unique solution, as one can easily see immediately. However, without our results it was not clear what are the real boundary conditions for a solution, or when $u-\psi\in W_2^{(m)}(0)$.

### §16. The eigenvalue problem.

1. INTRODUCTION. Let $\Omega$ be a domain of $n$-dimensional space bounded by a sufficiently smooth surface $S$.

The problem of the eignevalues of the equation

$$\Delta u+\lambda u=0 \tag{16.1}$$

for the conditions on the boundary $S$

$$\left.\frac{\partial u}{\partial n}\right|_S - \left.hu\right|_S = 0 \quad (n \text{— exterior normal}) \tag{16.2}$$

consists of the determination of all values $\lambda$ for which the equation (16.1) with the conditions (16.2) has a non-null solution. On the function $h$ defined on the surface, we shall impose the following restrictions: there exist suitable continuously differentiable functions $a_i$ bounded with bounded derivatives on the closed domain $\Omega+S$ ($|a_i|<M$; $|\partial a_i/\partial x|<M$), such that

$$|h| \leqslant \sum a_i|_S \cos nx_i. \tag{16.3}$$

If $S$ is a suitably smooth surface, one may construct a smooth function $C_i$, reducing to $\cos nx_i$ on the surface. Then it suffices to take $a_i = Mc_i$, where

$$M = \max_S |h|.$$

The variational method makes possible the solution of this problem, finding all its eigenfunctions and showing the orthonormality and closedness in $L_2$ of the system of such functions. We recall that an orthonormal system of functions is called closed if one can approximate with linear combinations of these functions with arbitrary precision to any function in $L_2$.

2. AUXILIARY INEQUALITIES. We turn to the solution of this problem. Consider the functionals on $W_2^{(1)}$:

$$D(v) = \int \underset{\Omega}{\cdots} \int \sum_{i=1}^{n} \left(\frac{\partial v}{\partial x_i}\right)^2 d\Omega - \int \underset{S}{\cdots} \int hv^2 dS;$$

$$J(v) = \int \underset{\Omega}{\cdots} \int \sum_{i=1}^{n} \left(\frac{\partial v}{\partial x_i}\right)^2 d\Omega;$$

$$H(v) = \int \underset{\Omega}{\cdots} \int |v|^2 \, d\Omega = \|v\|^2_{L_2}.$$

LEMMA. *The following inequalities hold*

$$J(v) \leqslant L_1 D(v) + L_2 H(v); \tag{16.4}$$

$$D(v) \leqslant K_1 H(v) + K_2 J(v). \tag{16.5}$$

We shall prove (16.5). We have, using (16.3):

$$\left|\int \underset{S}{\cdots} \int hv^2 \, dS\right| \leqslant \int \underset{S}{\cdots} \int v^2 \sum a_i \cos nx_i \, dS =$$

$$= \int \underset{\Omega}{\cdots} \int \sum \frac{\partial (a_i v^2)}{\partial x_i} \, d\Omega = 2 \int \underset{\Omega}{\cdots} \int v \sum a_i \frac{\partial v}{\partial x_i} \, d\Omega +$$

$$+ \int \underset{\Omega}{\cdots} \int v^2 \sum \frac{\partial a_i}{\partial x_i} \, d\Omega \leqslant M \int \underset{\Omega}{\cdots} \int 2|v| \sum \left|\frac{\partial v}{\partial x_i}\right| \, d\Omega +$$

$$+ MnH(v).$$

Using the fact that

$$2|v|\left|\frac{\partial v}{\partial x_i}\right| \leqslant Kv^2 + \frac{1}{K}\left|\frac{\partial v}{\partial x_i}\right|^2,$$

where $K>0$ is an arbitrary number, we obtain:

$$\left|\int_S \cdots \int hv^2 dS\right| \leqslant MnH(v) + KMnH(v) + \frac{M}{K}J(v) =$$

$$= Mn(1+K)\,H(v) + \frac{M}{K}J(v).$$

From the definition of $D(v)$ and $J(v)$ we have:

$$D(v) \leqslant J(v) + \left|\int_S \cdots \int hv^2 dS\right| \leqslant Mn(1+K)\,H(v) +$$

$$+\left(1+\frac{M}{K}\right)J(v) = K_1 H(v) + K_2 J(v);$$

$$K_1 = Mn(1+K);$$

$$K_2 = 1 + \frac{M}{K},$$

and the inequality (16.5) is established.

On the other hand, from the definitions of $D(v)$ and $J(v)$ follows

$$J(v) = D(v) + \int_S \cdots \int hv^2 dS \leqslant D(v) + Mn(1+K)\,H(v) + \frac{M}{K}J(v),$$

from which

$$J(v)\left(1-\frac{M}{K}\right) \leqslant D(v) + Mn(1+K)\,K(v).$$

If we chose $K>2M$, we obtain $(1-(M/K))>1/2$, and:

$$J(v) < 2D(v) + 2Mn(1+K)\,H(v) = L_1 D(v) + L_2 H(v),$$

and inequality (16.4) is proved and the lemma as well.

From (16.4) it follows that

$$D(v) \geqslant \frac{1}{L_1}J(v) - \frac{L_2}{L_1}H(v) \geqslant -\frac{L_2}{L_1}H(v),$$

and hence if $H(v)=1$, that $D(v) \geqq -(L_2/L_1)$.

It follows from the latter that there exists

$$\inf_{H(v)=1} D(v) = \lambda_1$$

Hence there exists a minimizing sequence:

$$\{v_k\},\ v_k \in W_2^{(1)},\ H(v_k)=1,\ \lim_{k\to\infty} D(v_k) = \lambda_1.$$

3. MINIMAL SEQUENCES AND THE EQUATION OF VARIATIONS.

THEOREM 1. *There exists a function* $u_1 \in W_2^{(1)}$ *such that*

$$H(u_1) = 1, \quad D(u_1) = \lambda_1.$$

*The function* $u_1$ *has continuous derivatives of all orders in* $\Omega$ *and satisfies the equation*

$$\Delta u_1 + \lambda_1 u_1 = 0.$$

PROOF. Let $\{v_k\}$ be a minimizing sequence. We shall show that $\|v_k\|_{W_2^{(1)}}$ is bounded. Indeed, if we put

$$(\rho, v) = \int \underset{\Omega}{\cdots} \int v \, d\Omega, \quad [(\rho, 1) \neq 0],$$

if we take into account $H(v_k) = 1$, we will have the boundedness of $D(v_k)$ from (16.4):

$$\|v_k\|_{W_2^{(1)}} = \{(\rho, v_k)^2 + J(v_k)\}^{\frac{1}{2}} \leqslant$$

$$\leqslant \{\Omega H(v_k) + L_1 D(v_k) + L_2 H(v_k)\}^{\frac{1}{2}} \leqslant \{(\Omega + L_2) + L_1 A\}^{\frac{1}{2}},$$

where $A = \sup_k |D(v_k)|$. Thus the boundedness of $\|v_k\|_{W_2^{(1)}}$ is established. From the complete continuity of the imbedding operator it follows that $\{v_k\}$ is strongly compact in $L_2$ on $\Omega$. From $\{v_k\}$ we may choose a subsequence converging in $L_2$. This subsequence will itself be a minimizing sequence. Hence there exists a minimizing sequence converging in $L_2$. We may assume from the beginning that our original sequence has this property. Furthermore, $\|v_k - v_l\| = H(v_k - v_l) < \epsilon$ as soon as $k, l > N(\epsilon)$. From the obvious equality,

$$H\left(\frac{v_k + v_l}{2}\right) = \frac{1}{2} H(v_k) + \frac{1}{2} H(v_l) - H\left(\frac{v_k - v_l}{2}\right)$$

we obtain, using $H(v_k) = H(v_l) = 1$, $H(v_k - v_l)/2 < \epsilon/4$ that

$$H\left(\frac{v_k + v_l}{2}\right) > 1 - \frac{\varepsilon}{4}.$$

The functionals $D(v)$ and $H(v)$ are homogeneous quadratic functionals and therefore their ratio $D(v)/H(v)$ does not change under the passage from $v$ to $cv$ ($c = \text{const.} \neq 0$, $H(v) \neq 0$), and hence

$$\inf_{v\in W_2^{(1)}} \frac{D(v)}{H(v)} = \inf_{\substack{H(v)=1\\ v\in W_2^{(1)}}} D(v) = \lambda_1.$$

Therefore $D(v) \geqq \lambda_1 H(v)$ for all $v \in W_2^{(1)}$, and in particular

$$D\left(\frac{v_k+v_l}{2}\right) > \lambda_1\left(1-\frac{\varepsilon}{4}\right).$$

Then, taking $k$ and $l$ sufficiently large, $D(v_k) < \lambda_1+\epsilon$, $D(v_l) < \lambda_1+\epsilon$, we obtain

$$D\left(\frac{v_k-v_l}{2}\right) = \frac{1}{2}D(v_k)+\frac{1}{2}D(v_l) - D\left(\frac{v_k+v_l}{2}\right) < \frac{\lambda_1+\varepsilon}{2} + \frac{\lambda_1+\varepsilon}{2} - \lambda_1 + \frac{\lambda_1\varepsilon}{4} = \varepsilon\left(1+\frac{\lambda_1}{4}\right),$$

i.e., $D(v_k - v_l) \to 0$ for $k, l \to \infty$.

From the inequality (16.4), there follows

$$0 < J(v_k - v_l) \leqslant L_1 D(v_k - v_l) + L_2 H(v_k - v_l),$$

i.e.,

$$J(v_k - v_l) \to 0 \quad k,\ l \to \infty.$$

But

$$\| v_k - v_l \|^2_{W_2^{(1)}} = (\rho,\ v_k - v_l)^2 + J(v_k - v_l) \leqslant \leqslant \Omega H(v_k - v_l) + J(v_k - v_l) \to 0,$$

i.e.,

$$\| v_k - v_l \|_{W_2^{(1)}} \to 0.$$

Hence $\{v_k\}$ converges in $W_2^{(l)}$ and as a result of the completeness of $W_2^{(l)}$ there exists a limit function $u_1 \in W_2^{(l)}$ such that

$$\| u_1 - v_k \|_{W_2^{(1)}} \to 0; \quad k \to \infty.$$

In addition,

$$|D(v_k) - D(u_1)| = |J(v_k) - J(u_1) + \int_S \cdots \int h(u_1^2 - v_k^2)\,dS| \leqslant \leqslant |\sqrt{J(v_k)} - \sqrt{J(u_1)}|\,(\sqrt{J(v_k)} + \sqrt{J(u_1)}) + + \sqrt{\int_S \cdots \int (u_1 - v_k)^2\,dS}\sqrt{\int_S \cdots \int h^2 (u_1 + v_k)^2\,ds} \to 0$$

for $k \to \infty$, since

$$|\sqrt{J(v_k)}-\sqrt{J(u_1)}|=|\|v_k\|_{L_2^{(1)}}-\|u_1\|_{L_2^{(1)}}|\leqslant\|v_k-u_1\|_{L_2^{(1)}}\leqslant$$
$$\leqslant\|v_k-u_1\|_{W_2^{(1)}}\to 0;$$

$$\left[\int_S\cdots\int(u_1-v_k)^2\,dS\right]^{\frac{1}{2}}=\|u_1-v_k\|_{L_2(S)}\leqslant N\|u_1-v_k\|_{W_2^{(1)}}\to 0$$

while the terms $(J(v_k))^{\frac{1}{2}}+(J(v_l))^{\frac{1}{2}}$ and

$$\sqrt{J(v_k)}+\sqrt{J(u_1)} \text{ and } \int_S\cdots\int h^2(u_1+v_k)^2\,dS$$

are bounded. Therefore $D(v_k)\to D(u_1)$ and, hence,

$$D(u_1)=\lambda_1. \tag{16.6}$$

Analogously one sees that

$$H(u_1)=1. \tag{16.7}$$

Suppose now that $\xi$ is some function from $W_2^{(1)}$. Consider the ratio

$$\frac{D(u_1+\mu\xi)}{H(u_1+\mu\xi)}=\frac{D(u_1)+2\mu D(u_1,\xi)+\mu^2D(\xi)}{H(u_1)+2\mu H(u_1,\xi)+\mu^2H(\xi)},$$

where

$$D(u,\xi)=\int_\Omega\cdots\int\sum\frac{\partial u}{\partial x_i}\frac{\partial\xi}{\partial x_i}\,d\Omega-\int_S\cdots\int hu\xi\,dS;$$
$$H(u,\xi)=\int_\Omega\cdots\int u\xi\,d\Omega.$$

It yields a continuously differentiable function of $\mu$ on some interval around the point $\mu=0$. This ratio has a minimum at $\mu=0$ equal to $\lambda_1$ and by the basic theorem of Fermat, we have

$$\left[\frac{D(u_1+\mu\xi)}{H(u_1+\mu\xi)}\right]'_{\mu=0}=\frac{2D(u_1,\xi)H(u_1)-2H(u_1,\xi)D(u_1)}{[H(u_1)]^2}=0,$$

which by virtue of (16.6) and (16.7) gives

$$D(u_1,\xi)-\lambda_1H(u_1,\xi)=0. \tag{16.8}$$

Equation (16.8) holds for any $\xi\in W_2^{(1)}$.

4. EXISTENCE OF THE FIRST EIGENFUNCTION. We shall show that $u_1$ has continuous derivatives of all orders and satisfies the equation

$$\Delta u_1+\lambda_1u_1=0. \tag{16.9}$$

Indeed, let $\delta$ be the distance between some point $\vec{P}\in\Omega$ and the boundary $S$. Set

$$\xi = \left[\psi\left(\frac{r}{h_1}\right) - \psi\left(\frac{r}{h_2}\right)\right] r^{-\frac{n-2}{2}} Y_{\frac{n-2}{2}}(\sqrt{\lambda_1} r) = \left[\psi\left(\frac{r}{h_1}\right) - \psi\left(\frac{r}{h_2}\right)\right] X(r),$$

where $h_1$ and $h_2 < \delta$, $Y_{(n-2)/2}$ is the Bessel function of second kind of order $(n-2)/2$, and $\psi(r/h_i)$ is the averaging function introduced earlier. $X(r)$ is a solution of the equation

$$\Delta X + \lambda_1 X = 0,$$

and for $r=0$ has a singularity of order $r^{-n+2}$, and $X'$ of order $r^{-n+1}$. Since $\psi(r/h_1) = \psi(r/h_2)$ on the sphere $r \leqq \frac{1}{2}\min(h_1, h_2)$, we have $\xi = 0$ on this sphere. Hence $\xi$ has no singularity for $r=0$ and has continuous derivatives of all orders. Therefore, $\xi \in W_2^{(1)}$. In addition, $\xi = 0$ outside the sphere $r \geqq \max(h_1, h_2)$ and since $\max(h_1, h_2) < \delta$, we have $\xi \equiv 0$ on $S$. Therefore (16.8) for $\xi$ takes the form

$$\int_\Omega \cdots \int \left(\sum \frac{\partial u_1}{\partial x_i}\frac{\partial \xi}{\partial x_i} - \lambda_1 u_1 \xi\right) d\Omega = 0.$$

Since

$$\int_\Omega \cdots \int \frac{\partial u_1}{\partial x_i}\frac{\partial \xi}{\partial x_i}\, d\Omega = -\int_\Omega \cdots \int u_1 \frac{\partial^2 \xi}{\partial x_i^2}\, d\Omega$$

(by the definition of the generalized derivative $\partial u_1/\partial x_i$), it follows that

$$\int_\Omega \cdots \int u_1 (\Delta \xi + \lambda_1 \xi)\, d\Omega = 0. \tag{16.10}$$

We set

$$\frac{1}{c(h)}\left\{\Delta\left[\psi\left(\frac{r}{h}\right) X(r)\right] + \lambda_1 \psi\left(\frac{r}{h}\right) X(r)\right\} = \omega_h(r),$$

where $c(h)$ is a constant to be defined later. It is obvious that $\omega_h(r) \equiv 0$ for $r \geqq h$. In addition, $\omega_h(r) \equiv 0$ for $r \leqq (h/2)$, since in this case $\psi(r/h) \equiv 1$, $\Delta X + \lambda_1 X = 0$. Hence, $\omega_h(r)$ has continuous derivatives of arbitrary order.

Since

$$\Delta \xi + \lambda_1 \xi = c(h_1)\, \omega_{h_1}(r) - c(h_2)\, \omega_{h_2}(r),$$

equation (16.8) takes the form

$$c(h_1) \int_\Omega \cdots \int u_1 \omega_{h_1}(r)\, d\Omega = c(h_2) \int_\Omega \cdots \int u_1 \omega_{h_2}(r)\, d\Omega. \tag{16.11}$$

We put

$$c(h) = \int_\Omega \cdots \int \left\{\Delta\left[\psi\left(\frac{r}{h}\right) X(r)\right] + \lambda_1 \psi\left(\frac{r}{h}\right) X(r)\right\} d\Omega.$$

Then

$$\int \cdots \int \omega_h(r)\, d\Omega = 1.$$

It is not difficult to show that there exists a limit different from zero for

$$\lim_{h \to 0} c\ (h) = c_0.$$

The function $\omega_h(r)$ may be considered as an averaging kernel. The equation (16.11) then means that the averaged functions for $u_1$ on the domain $\Omega_\delta$ for various $h<\delta$ differ only by constants. We obtain

$$u_{h_2} = \frac{c\,(h_1)}{c\,(h_2)}\, u_{h_1}.$$

By virtue of the fact that $u_1$ is the limit of its averaged functions, it differs merely by a constant from any one of its averaged functions. But the latter are continuously differentiable arbitrarily often and hence the same is true for $u_1$ on the domain $\Omega_\delta$. Since $\delta$ is arbitrary, $u_1$ is infinitely often differentiable at any interior point of $\Omega$. Suppose now that $\xi$ is continuously differentiable and vanishes on a boundary strip. Then from (16.8) we obtain by integrating by parts

$$\int \cdots \int_{\Omega} \xi (\Delta u_1 + \lambda_1 u_1)\, d\Omega = 0,$$

from which by virtue of the arbitrariness of $\xi$, there follows

$$\Delta u_1 + \lambda_1 u_1 = 0, \tag{16.12}$$

and the theorem is proved.

REMARKS. Consider a sequence of domains $\{\Omega'\}$ lying in the interior of $\Omega$ and converging to $\Omega$. Let the boundaries $S'$ of these domains be piecewise continuously differentiable. The equation (16.8) for $\xi$ takes the form

$$\int \cdots \int_{\Omega} \Big( \sum_{i=1}^{n} \frac{\partial u_1}{\partial x_i} \frac{\partial \xi}{\partial x_i} - \lambda_1 u_1 \xi \Big) d\Omega - \int \cdots \int_{S} h u_1 \xi\, dS = 0. \tag{16.13}$$

But

$$\int \cdots \int_{\Omega} \Big( \sum_{i=1}^{n} \frac{\partial u_1}{\partial x_i} \frac{\partial \xi}{\partial x_i} - \lambda_1 u_1 \xi \Big) d\Omega =$$

$$= \lim_{\Omega' \to \Omega} \int \cdots \int_{\Omega'} \Big( \sum_{i=1}^{n} \frac{\partial u_1}{\partial x_i} \frac{\partial \xi}{\partial x_i} - \lambda_1 u_1 \xi \Big) d\Omega =$$

$$= \lim_{\Omega' \to \Omega} \Big[ - \int \cdots \int_{\Omega'} \xi (\Delta u_1 + \lambda_1 u_1)\, d\Omega + \int \cdots \int_{S'} \xi \frac{\partial u_1}{\partial n}\, dS' \Big] =$$

$$= \lim_{\Omega' \to \Omega} \int \cdots \int_{S'} \xi \frac{\partial u_1}{\partial n} \, dS'.$$

Thus (16.13) takes the form

$$\lim_{\Omega' \to \Omega} \int \cdots \int_{S'} \xi \frac{\partial u_1}{\partial n} \, dS' - \int \cdots \int_{S} h u_1 \xi \, dS = 0. \qquad (16.14)$$

If in addition $S' \to S$ in the sense that not only the points of $S'$ converge to the points of $S$ but also the normals at these points converge to the corresponding normals of $S$, then

$$\int \cdots \int_{S} h u_1 \xi \, dS = \lim_{\Omega' \to \Omega} \int \cdots \int_{S'} h u_1 \xi \, dS',$$

if we assume that $h$ is the value on $S$ of some function given on $\Omega$.

Then condition (16.14) takes the form

$$\lim_{\Omega' \to \Omega} \int \cdots \int_{S'} \left( \frac{\partial u_1}{\partial n} - h u_1 \right) \xi \, d_1 S = 0. \qquad (16.15)$$

Thus, $u_1$ satisfies the condition (16.2) "in the mean".

We turn now to the search for the other eigenfunctions.

5. EXISTENCE OF THE LATER EIGENFUNCTIONS. Let us assume that we have already found $(m-1)$ functions $u_i \in W_2^{(l)}$ and $(m-1)$ numbers $\lambda_i$ $(i=1,2,\cdots,m-1)$ such that

$$D(u_i, \xi) - \lambda_i H(u_i, \xi) = 0; \qquad (16.16)$$

$$H(u_i) = 1 \quad H(u_i, u_k) = 0 \quad (i \neq k)$$
$$i, k = 1, 2, \ldots, m-1, \qquad (16.17)$$

where (16.16) holds for an arbitrary $\xi \in W_2^{(1)}$. Suppose that $v \in W_2^{(1)}$ satisfies the $m-1$ conditions

$$H(v, u_i) = 0 \quad (i = 1, 2, \ldots, m-1). \qquad (16.18)$$

The family of such functions $v \in W_2^{(1)}$ we denote by $W_2^{(1)}(u_1, \cdots, u_{m-1})$. Obviously $W_2^{(1)}(u_1, \cdots, u_{m-1})$ is a linear space since every linear combindation of its elements belongs to the set. We shall show that this set is closed. Indeed, suppose that a sequence $\{v_k\}$ converges to $v$ in $W_2^{(1)}$, i.e., $\|v - v_k\|_{W_2^{(1)}} \to 0$. By the Imbedding Theorem

$$\|v - v_k\|_{L_2} \leqslant C \|v - v_k\|_{W_2^{(1)}},$$

and hence

$$v - v_k\|_{L_2} \to 0.$$

Thus

$$|H(v, u_i) - H(v_k, u_i)| = \left| \int \cdots \int_{\Omega} u_i (v - v_k)\, d\Omega \right| \leqslant$$

$$\leqslant \sqrt{\int \cdots \int_{\Omega} u_i^2\, d\Omega} \sqrt{\int \cdots \int_{\Omega} |v - v_k|^2\, d\Omega} \doteq \|v - v_k\|_{L_2} \to 0,$$

i.e., $H(v_k, u_i) = \lim_{k\to\infty} H(v_k, u_i)$, and if $v_k \in W_2^{(1)}(u_1, \cdots, u_{m-1})$, then

$$v \in W_2^{(1)}(u_1, \cdots, u_{m-1}).$$

Hence $W_2^{(1)}(u_1, \cdots, u_{m-1})$ is closed. Since $W_2^{(1)}(u_1, \cdots, u_{m-1}) \subset W_2^{(1)}$, $D(v)$ is bounded from below for $v \in W_2^{(1)}(u_1, \cdots, u_{m-1})$ and $H(v) = 1$. The greatest lower bound of $D(v)$ for this family we denote by $\lambda_m$.

$$\left. \begin{aligned} &\inf D(v) = \lambda_m; \\ &v \in W_2^{(1)}(u_1, \ldots, u_{m-1}) \\ &H(v) = 1. \end{aligned} \right\} \qquad (16.19)$$

THEOREM 2. *There exists a function* $u_m \in W_2^{(1)}(u_1, \cdots, u_{m-1})$, $H(u_m) = 1$, *such that*

$$D(u_m, \xi) - \lambda_m H(u_m, \xi) = 0 \qquad (16.20)$$

*for arbitrary* $\xi \in W_2^{(1)}$. *In particular, for* $\xi = u_m$, (16.20) *takes the form:*

$$D(u_m) = \lambda_m. \qquad (16.21)$$

*In addition, on* $\Omega, u_m$ *has continuous derivatives of arbitrary order and satisfies the equation*

$$\Delta u_m + \lambda_m u = 0. \qquad (16.22)$$

PROOF. From the definition of the greatest lower bound, it follows that we can find a minimizing sequence $\{v_k\}$ in $W_2^{(1)}(u_1, \cdots, u_{m-1})$:

$$v_k \in W_2^{(1)}(u_1, \ldots, u_{m-1}),\ H(v_k) = 1,\ \lim D(v_k) = \lambda_m.$$

Repeating literally the proof of Theorem 1, we show that $\|v_k\|_{W_2^{(1)}}$ is bounded and hence that $\{v_k\}$ is strongly compact in $L_2$. We may assume that the same sequence $\{v_k\}$ converges in $L_2$. Then for sufficiently large $k$ and $l$; $H(v_k - v_l) < \epsilon$, and as in Theorem 1, $H(v_k + v_l)/2 > 1 - (\epsilon/4)$.

For all $v \in W_2^{(1)}(u_1, \cdots, u_{m-1})$, we have:

$$D(v) \geqslant \lambda_m H(v),$$

and since $(v_k + v_l)/2 \in W_2^{(1)}(u_1, \cdots, u_{m-1})$ simultaneously with $v_k$ and $v_l$,

$$D\left(\frac{v_k + v_l}{2}\right) \geqslant \lambda_m H\left(\frac{v_k + v_l}{2}\right) > \lambda_m \left(1 - \frac{\varepsilon}{4}\right).$$

From this, as in Theorem 1, we find that:

$$D(v_k - v_l) \to 0 \qquad k,\ l \to \infty$$

and further

$$\|v_k - v_l\|_{W_2^{(1)}} \to 0 \qquad k,\ l \to \infty,$$

i.e., $\{v_k\}$ converges in $W_2^{(l)}$. As a consequence of completeness, there exists a limit function $u_m \in W_2^{(1)}$. Since $v_k \in W_2^{(1)}(u_1,\cdots,u_{m-1})$ and $W_2^{(1)}(u_1,\cdots,u_{m-1})$ is closed, it follows that $u_m \in W_2^{(1)}(u_1,\cdots,u_{m-1})$. In addition, as in Theorem 1, we show that

$$D(u_m) = \lambda_m,$$

$$H(u_m) = 1.$$

Let $\eta \in W_2^{(1)}(u_1,\cdots,u_{m-1})$. Consider the ratio

$$\frac{D(u_m + \mu\eta)}{H(u_m + \mu\eta)} = \frac{D(u_m) + 2\mu D(u_m, \eta) + \mu^2 D(\eta)}{H(u_m) + 2\mu H(u_m, \eta) + \mu^2 H(\eta)}.$$

This ratio is a continuously differentiable function on some interval about the point $\mu = 0$, having for $\mu = 0$ a minimum equal to $\lambda_m$ (since $u_m + \mu\eta \in W_2^{(1)}(u_1,\cdots,u_{m-1})$. From this, it follows by Fermat's theorem that

$$D(u_m, \eta) - \lambda_m H(u_m, \eta) = 0. \qquad (16.23)$$

We shall show that (16.23) holds for all $\eta \in W_2^{(1)}$. Indeed, put $\xi = u_m$ in (16.16). Since

$$u_m \in W_2^{(1)}(u_1, \ldots, u_{m-1}), \text{ then } H(u_i, u_m) = 0 \ (i = 1, 2, \ldots, m-1),$$

and (16.16) takes the form

$$D(u_i, u_m) = 0.$$

Therefore (16.23) holds for $\eta = u_i$ $(i = 1,2,\cdots,m-1)$. Hence (16.23) holds for an arbitrary linear combination of the functions $u_i$ and $\eta \in W_2^{(1)}(u_1,\cdots, u_{m-1})$.

Let $\xi \in W_2^{(1)}$ be an arbitrary function. Then the function

$$\eta = \xi - \sum_{i=1}^{m-1} u_i H(u_i, \xi) \in W_2^{(1)}(u_1, \ldots, u_{m-1}),$$

as is easy to verify. Thus $\xi = \eta + \sum_{i=1}^{m-1} u_i H(u_i, \xi)$, i.e., an arbitrary $\xi \in W_2^{(1)}$ is a linear combination of the $u_i$ and of $\eta \in W_2^{(1)}(u_1,\cdots,u_{m-1})$. Thus (16.23) holds for $\xi$, i.e., the correctness of (16.20) has been established.

The proof of the fact that $u_m$ has continuous derivatives of all orders in $\Omega$ and satisfies (16.22) there is a literal repetition of the corresponding part of the proof of Theorem 1.

As far as the fulfillment by the function $u_m$ of the boundary conditions (16.22) is concerned, this may be dealt with as in the remarks to the proof of Theorem 1.

6. THE INFINITE SEQUENCE OF EIGENVALUES.

THEOREM 3. *There exists a nondecreasing infinite sequence of numbers* $\{\lambda_m\}$ *and a corresponding sequence of infinitely differentiable functions* $u_m$ *on* $\Omega$ *such that*

$$\left.\begin{aligned} &\Delta u_m + \lambda_m u_m = 0; \\ &H(u_i, u_j) = \delta_{ij}; \\ &D(u_i, u_j) = \delta_{ij}\lambda_i. \end{aligned}\right\} \tag{16.24}$$

*Each* $u_m$ *satisfies the boundary condition* (16.2) *in the sense of equation* (16.15).

PROOF. Theorem 1 asserts the existence of a number $\lambda_1$ and a function $u_1$ such that

$$\Delta u_1 + \lambda_1 u_1 = 0 \quad H(u_1) = 1 \quad D(u_1) = \lambda_1.$$

Consider the closed linear space $W_2^{(1)}(u_1)$. By Theorem 2, there exists a number $\lambda_2$ and a function $u_2$ such that

$$\Delta u_2 + \lambda_2 u_2 = 0; H(u_2) = 1; H(u_1, u_2) = 0; D(u_2) = \lambda_2; D(u_1, u_2) = 0.$$

Thus $\lambda_1, u_1; \lambda_2, u_2$ satisfy (16.24) for $i, j = 1, 2$. Since $W_2^{(1)}(u_1) \subset W_2^{(1)}$, we have

$$\inf_{v \in W_2^{(1)}(u_1)} D(v) \geqslant \inf_{v \in W_2^{(1)}} D(v), \; H(v) = 1,$$

and by the virtue of the fact that the greatest lower bound of $D(v)$ on a smaller set is not less than the greater lower bound of $D(v)$ on $W_2^{(1)}$, we know that $\lambda_2 \geqq \lambda_1$. Considering the closed linear space $W_2^{(1)}(u_1, u_2)$ and using Theorem 2, we obtain $\lambda_3$ and $u_3$. Using Theorem 2, this process for obtaining $\lambda_i$ and $u_i$ can be continued indefinitely. We obtain thereby $\lambda_m = \inf D(v)$ $(v \in W_2^{(1)}(u_1, \cdots, u_{m-1}),\; H(v) = 1)$. Since

$$W_2^{(1)}(u_1, u_2, \ldots, u_{m-1}) \supset W_2^{(1)}(u_1, \ldots, u_m)$$

(a function $v$ belonging to $W_2^{(1)}(u_1, \cdots, u_{m-1})$ satisfies $m-1$ conditions and at the same time $v \in W_2^{(1)}(u_1, \cdots, u_m)$ satisfies the additional condition $H(v, u_m) = 0$), we have $\lambda_m \leqq \lambda_{m+1}$.

The fulfillment of condition (16.24) follows from Theorem 2.

Since $H(u_i, u_j) = \delta_{ij}$, the sequence $\{u_m\}$ forms an orthogonal and normal system of functions.

THEOREM 4.

$$\lim_{m \to \infty} \lambda_m = +\infty.$$

*Indeed $\lambda_m$ is bounded from below and nondecreasing with $m$. Suppose that $\lambda_m$ does not converge to $\infty$. Then $\lambda_m$ would be bounded:*

$$|\lambda_m| < A \quad m = 1,2, \ldots$$

*From* (16.24), *it follows that*

$$|D(u_m)| < A \qquad H(u_m) = 1 \ (m = 1,2, \ldots).$$

*On the basis of the inequality* (16.4), *we have*

$$J(u_m) \leqslant L_1 D(u_m) + L_2 H(u_m) < L_1 A + L_2 = B,$$

*i.e., the sequence $\{u_m\}$ is uniformly bounded in the norm of $L_2^{(1)}$ by the number $\sqrt{B}$.*

We show then that the sequence $\{u_m\}$ would be bounded in the norm of $W_2^{(1)}$. Indeed, if we define the norm on $W_2^{(1)}$ by the equation

$$\|v\|^2_{W_2^{(1)}} = \left\{\left(\int \underset{\Omega}{\cdots} \int v\, d\Omega\right)^2 + \|v\|^2_{L_2^{(1)}}\right\},$$

we find

$$\|u_m\|^2_{W_2^{(1)}} = \left\{\left(\int \cdots \int u_m\, d\Omega\right)^2 + \|u_m\|^2_{L_2^{(1)}}\right\} <$$

$$< \left\{\Omega \int \cdots \int u_m^2\, d\Omega + B\right\} = \Omega + B.$$

As a consequence of the complete continuity of the imbedding operator, we would conclude that $\{u_m\}$ is strongly compact in $L_{q^*}$ on any hyperplane of dimension $s$ where $s > n-2$:

$$q^* < q = \frac{2s}{n-2}.$$

In particular, if we should take $s = n$, $q^* = 2$, we would find that $\{u_m\}$ is strongly compact in $L_2$ on $\Omega$. In other words, from $\{u_m\}$ we may choose a subsequence $\{u_{m_i}\}$ $(i = 1,2,\cdots)$ such that

$$H(u_{m_i} - u_{m_k}) \to 0, \quad \text{if} \quad i, k \to \infty.$$

The latter is impossible, since

$$H(u_{m_i}-u_{m_k})=H(u_{m_i})-2H(u_{m_i},u_{m_k})+H(u_{m_k})=2.$$

Thus, the assumption that $\lim \lambda_m \neq \infty$ leads to a contradiction, and the theorem is proved.

7. CLOSEDNESS OF THE SET OF EIGENFUNCTIONS.

THEOREM 5. *The system* $\{u_m\}$ *is closed in* $L_2$, i.e., *for every* $\phi \in L_2$ *we have the equality*

$$\int_\Omega \cdots \int \varphi^2 d\Omega = \sum_{m=1}^{\infty}\left(\int \cdots \int \varphi u_m \, d\Omega\right)^2.$$

Since $\lambda_m \to \infty$, it follows that, starting from some integer, $j: \lambda_m > 0$ $(m \geqq j)$. Let $K$ be any number, $k > j$. Let $v \in W_2^{(1)}$ be any function. Set

$$R_k = v - \sum_{m=1}^{k} u_m H(v, u_m).$$

It is obvious that $R_k \in W_2^{(1)}(u_1, \cdots, u_k)$ since

$$H(R_k, u_i) = 0 \quad (i = 1, 2, \ldots, k).$$

Hence

$$\frac{D(R_k)}{H(R_k)} \geqslant \lambda_{k+1},$$

from which it follows that

$$(1) \quad D(R_k) \geqslant 0$$

and

$$(2) \quad H(R_k) \leqslant \frac{D(R_k)}{\lambda_{k+1}}.$$

On the other hand,

$$\begin{aligned}
0 < D(R_k) &= D(v) - 2D\Big(v, \sum_{m=1}^{k} u_m H(v, u_m)\Big) + D\Big(\sum_{m=1}^{k} u_m H(v, u_m)\Big) = \\
&= D(v) - 2\sum_{m=1}^{k} H(v, u_m) D(v, u_m) + \sum_{m=1}^{k}\sum_{l=1}^{k} H(v, u_m) H(v, u_l) D(u_m u_l) = \\
&= D(v) - 2\sum_{m=1}^{k} H(v, u_m)\lambda_m H(v, u_m) + \sum_{m=1}^{k} [H(v, u_m)]^2 D(u_m) = \\
&= D(v) - \sum_{m=1}^{k} \lambda_m [H(v, u_m)]^2 = \Big\{ D(v) - \sum_{m=1}^{j} \lambda_m [H(v, u_m)]^2 \Big\} - \\
&- \sum_{m=j+1}^{k} \lambda_m [H(v, u_m)]^2 \leqslant D(v) - \sum_{m=1}^{j} \lambda_m [H(v, u_m)]^2.
\end{aligned}$$

Thus $D(R_k)$ is bounded, and from $H(R_k) \leqq D(R_k)/\lambda_{k+1}$ and $\lambda_{k+1} \to \infty$, it follows that $H(R_k) \to 0$, which means the closedness of $\{u_m\}$ in the class of functions $v \in W_2^{(1)}$. Let $\phi \in L_2$ be an arbitrary function. For any $\epsilon > 0$, no matter how small, we can find $v \in W_2^{(1)}$ such that

$$\|\varphi - v\|_{L_2} \leqslant \frac{\varepsilon}{2}. \tag{16.25}$$

Indeed, denoting by $\Omega_\delta$ the subset of points of $\Omega$ lying at distance more than $\delta$ from the boundary, we will have:

$$\int_\Omega \cdots \int \varphi^2 \, d\Omega = \lim_{\delta \to 0} \int_{\Omega_\delta} \cdots \int \varphi^2 \, d\Omega,$$

from which it follows that for given $\epsilon > 0$, we can find $\delta > 0$ such that

$$\left| \int_\Omega \cdots \int \varphi^2 d\Omega - \int_{\Omega_\delta} \cdots \int \varphi^2 \, d\Omega \right| < \frac{\varepsilon^2}{16}.$$

Introducing the function

$$\varphi_\delta = \begin{cases} \varphi & \text{within} \quad \Omega_\delta \\ 0 & \text{outside of} \quad \Omega_\delta \end{cases} \quad (\varphi_\delta \in L_2),$$

we obtain

$$\int_\Omega \cdots \int |\varphi - \varphi_\delta|^2 \, d\Omega < \frac{\varepsilon^2}{16}. \tag{16.26}$$

If we construct as $\phi_\delta$ the averaged function for radius $< \delta$, we obtain a function $v$ having continuous first derivatives on the closed domain $\Omega$. Indeed, $v \in W_2^{(1)}$, and choosing a sufficiently small radius for averaging, we will have

$$\int_\Omega \cdots \int |\varphi_\delta - v|^2 \, d\Omega < \frac{\varepsilon^2}{16}. \tag{16.27}$$

From (16.26) and (16.27) we conclude the correctness of (16.25). For the function $v$, as was shown, we can find such a linear combination of a finite number of the functions $u_m$ that

$$\sqrt{H\left(v - \sum_{m=1}^{k} a_m u_m\right)} = \left\| v - \sum_{m=1}^{k} a_m u_m \right\|_{L_2} < \frac{\varepsilon}{2},$$

which, together with (16.25), gives

$$\|\varphi - \sum_{m=1}^{k} a_m u_m\|_{L_2} < \varepsilon.$$

If the $a_m$ are replaced by $H(\phi, u_m)$, then it follows from the general theory of the orthogonal systems, the left side does not increase, and we know the following inequality

$$\left\| \varphi - \sum_{m=1}^{k} u_m H(\varphi, u_m) \right\|_{L_2} < \varepsilon,$$

which, in view of the arbitrariness of $\epsilon$, leads to the equality

$$\int \cdots \int_{\Omega} \varphi^2 \, d\Omega = \sum_{m=1}^{\infty} \left( \int \cdots \int_{\Omega} \varphi u_m \, d\Omega \right)^2.$$

The theorem is proved.

## CHAPTER III

# THE THEORY OF HYPERBOLIC PARTIAL DIFFERENTIAL EQUATIONS

In the present chapter we consider some problems arising from the theory of hyperbolic partial differential equations. We shall treat three topics:

(1) The solution of the wave equation with constant coefficients. Dependence of the solutions upon initial data, and generalized solutions.

(2) Wave equations with variable coefficients.

(3) The theory of nonlinear equations.

All these topics are analyzed by a general method of investigation which consists of seeking solutions in the spaces $L_p^{(l)}$ and $W_p^{(l)}$ considered in the first chapter. However, en route we shall have to solve an auxiliary problem in the integration of wave equations with sufficiently smooth variable coefficients. It is necessary for us to show the existence of a solution for the Cauchy problem for such equations with sufficiently smooth data. For this purpose, we employ the theory of characteristics in $(2k+1)$-dimensional space and the method of descent in the space of $2k$ dimensions.

### §17. Solution of the wave equation with smooth initial conditions.

1. DERIVATION OF THE BASIC INEQUALITY. Consider the wave operator

$$\square u = \Delta u - \frac{\partial^2 u}{\partial t^2} \tag{17.1}$$

on the domain $\Omega$ in the space of $n+1$ dimensions with coordinates $x_1, x_2, \cdots, x_n, t$, bounded by a smooth surface $S$. Let $u(x_1, x_2, \cdots, x_n, t)$ and $v(x_1, x_2, \cdots, x_n, t)$ be twice differentiable in $\Omega$ with their first derivatives continuous up to the surface $S$.

Let

$$\square u = f; \quad \square v = \varphi.$$

Consider the integral

$$J=\int \underset{S}{\cdots} \int\left\{-\sum_{i=1}^{n}\left(\frac{\partial u}{\partial x_i}\frac{\partial v}{\partial t}+\frac{\partial v}{\partial x_i}\frac{\partial u}{\partial t}\right)\cos \vec{n}x_i+\right.$$
$$\left.+\left(\frac{\partial u}{\partial t}\frac{\partial v}{\partial t}+\sum_{i=1}^{n}\frac{\partial u}{\partial x_i}\frac{\partial v}{\partial x_i}\right)\cos \vec{nt}\right\} dS,$$

where $\vec{n}$ is the interior normal to $S$.

A simple transformation yields:

$$J=-\int \underset{\Omega}{\cdots} \int\left\{\frac{\partial}{\partial t}\left(\frac{\partial u}{\partial t}\frac{\partial v}{\partial t}+\sum_{i=1}^{n}\frac{\partial u}{\partial x_i}\frac{\partial v}{\partial x_i}\right)-\sum_{i=1}^{n}\frac{\partial}{\partial x_i}\left(\frac{\partial u}{\partial x_i}\frac{\partial v}{\partial t}+\frac{\partial v}{\partial x_i}\frac{\partial u}{\partial t}\right)\right\}d\Omega=$$
$$=-\int \underset{\Omega}{\cdots} \int\left\{\frac{\partial u}{\partial t}\left[\frac{\partial^2 v}{\partial t^2}-\sum_{i=1}^{n}\frac{\partial^2 v}{\partial x_i^2}\right]+\frac{\partial v}{\partial t}\left[\frac{\partial^2 u}{\partial t^2}-\sum_{i=1}^{n}\frac{\partial^2 u}{\partial x_i^2}\right]\right\} d\Omega=$$
$$=\int \underset{\Omega}{\cdots} \int\left\{\frac{\partial u}{\partial t}\square v+\frac{\partial v}{\partial t}\square u\right\} d\Omega. \tag{17.2}$$

Replacing $\square u$ and $\square v$ by their values, we will have:

$$J=\int \underset{\Omega}{\cdots} \int\left(\frac{\partial u}{\partial t}\varphi+\frac{\partial v}{\partial t}f\right)d\Omega.$$

Consider the expression

$$\sum_{i=1}^{n}\left(\frac{\partial u}{\partial x_i}\cos \vec{nt}-\frac{\partial u}{\partial t}\cos \vec{n}x_i\right)\left(\frac{\partial v}{\partial x_i}\cos \vec{nt}-\frac{\partial v}{\partial t}\cos \vec{n}x_i\right)=$$
$$=\frac{\partial u}{\partial t}\frac{\partial v}{\partial t}\sum_{i=1}^{n}(\cos \vec{n}x_i)^2+\cos^2 \vec{nt}\sum_{i=1}^{n}\frac{\partial u}{\partial x_i}\frac{\partial v}{\partial x_i}-$$
$$-\sum_{i=1}^{n}\left(\frac{\partial u}{\partial t}\frac{\partial v}{\partial x_i}+\frac{\partial v}{\partial t}\frac{\partial u}{\partial x_i}\right)\cos \vec{n}x_i\cos \vec{nt}=\cos \vec{nt}\left[\frac{\partial u}{\partial t}\frac{\partial v}{\partial t}\cos \vec{nt}+\right.$$
$$\left.+\sum_{i=1}^{n}\frac{\partial u}{\partial x_i}\frac{\partial v}{\partial x_i}\cos \vec{nt}-\sum_{i=1}^{n}\left(\frac{\partial u}{\partial t}\frac{\partial v}{\partial x_i}+\frac{\partial v}{\partial t}\frac{\partial u}{\partial x_i}\right)\cos \vec{n}x_i\right]+$$
$$+\frac{\partial u}{\partial t}\frac{\partial v}{\partial t}\left\{\sum_{i=1}^{n}\cos^2 \vec{n}x_i-\cos^2 \vec{nt}\right\}.$$

Everywhere except at the points of the surface $S$ where $\cos \vec{n}t=0$

$$J=\int\underset{S}{\cdots}\int\Big[\frac{1}{\cos\vec{nt}}\sum_{i=1}^{n}\Big(\frac{\partial u}{\partial x_i}\cos\vec{nt}-\frac{\partial u}{\partial t}\cos\vec{nx}_i\Big)\Big(\frac{\partial v}{\partial x_i}\cos\vec{nt}-$$
$$-\frac{\partial v}{\partial t}\cos\vec{nx}_i\Big)-\frac{\partial u}{\partial t}\frac{\partial v}{\partial t}\;\frac{1-2\cos^2\vec{nt}}{\cos\vec{nt}}\Big]dS=\int\underset{S}{\cdots}\int\Phi\,dS, \tag{17.3}$$

where by $\phi$ we denote the whole integrand lying within $J$.

It is useful to note that if $u=v$, then at all points of the surface where $|\cos\vec{nt}|\geqq 1/\sqrt{2}$, the integrated expression always has the same sign, which is also the sign of $\cos\vec{nt}$.

$$\operatorname{sign}\Phi=\operatorname{sign}\cos\vec{nt}. \tag{17.4}$$

Wherever $\cos\vec{nt}=0$, we obtain:

$$\Phi=-\Big(\frac{\partial u}{\partial t}\frac{\partial v}{\partial\vec{n}}+\frac{\partial v}{\partial t}\frac{\partial u}{\partial\vec{n}}\Big).$$

Suppose that $u$ is a solution of the wave equation

$$\Box u=0 \tag{17.5}$$

in the homogeneous infinite medium.

Taking $v=u$, we apply the formula (17.2) derived above to the function $u$; taking for the domain $\Omega$ a truncated cone whose generator makes an angle of $\pi/4$ with the $ot$ axis (Figure 8). Then

$$\cos\vec{nt}=\begin{cases}-\dfrac{1}{\sqrt{2}} & \text{at } S_1,\\ -1 & \text{at } S_2,\\ +1 & \text{at } S_3,\end{cases}$$

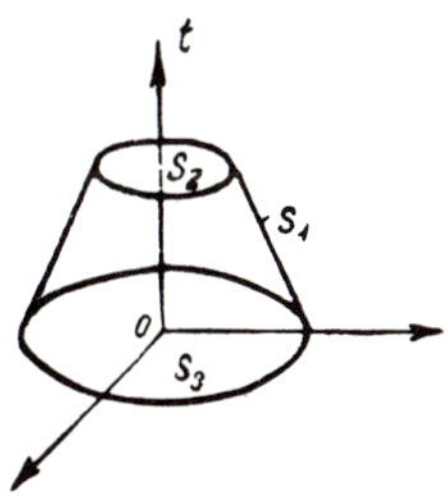


Figure 8

where $S_3$ is the lower base, $S_2$ the upper base and $S_1$ is the lateral surface of the truncated cone. Suppose, for the sake of definiteness, the quantity $t$ on $S_2$ is equal to $t_0$. By means of (17.2) and (17.5), we obtain

$$J(u,u)=\int\underset{S_1}{\cdots}\int\Phi\,dS+\int\underset{S_2}{\cdots}\int\Phi\,dS+$$
$$+\int\underset{S_3}{\cdots}\int\Big\{-2\sum_{i=1}^{n}\frac{\partial u}{\partial x_i}\frac{\partial u}{\partial t}\cos\vec{nx}_i+\Big[\Big(\frac{\partial u}{\partial t}\Big)^2+\sum_{i=1}^{n}\Big(\frac{\partial u}{\partial x_i}\Big)^2\Big]\cos\vec{nt}\Big\}\,dS=0. \tag{17.6}$$

By (17.4),

$$\int\underset{S_1}{\cdots}\int\Phi\,dS<0.$$

Thus

$$-\int_{S_2}\cdots\int\Phi\,dS=\int_{S_3}\cdots\int\Phi\,dS+\int_{S_1}\cdots\int\Phi\,dS<\int_{S_3}\cdots\int\Phi\,dS.$$

From this, since on $S_2$ and $S_3$: $\cos\vec{n}x_i=0$, we have the estimate

$$\int_{S_2}\cdots\int\left[\sum_{i=1}^{n}\left(\frac{\partial u}{\partial x_i}\right)^2+\left(\frac{\partial u}{\partial t}\right)^2\right]dS\leqslant$$

$$\leqslant\int_{S_3}\cdots\int\left[\sum_{i=1}^{n}\left(\frac{\partial u}{\partial x_i}\right)^2+\left(\frac{\partial u}{\partial t}\right)^2\right]dS. \tag{17.7}$$

2. ESTIMATES FOR THE GROWTH OF THE SOLUTION AND ITS DERIVATIVES. For our purposes, it is also necessary to estimate the integral

$$\int_{S_2}\cdots\int u^2\,dS.$$

From (17.7) it follows that

$$\int_{S_2}\cdots\int\left(\frac{\partial u}{\partial t}\right)^2dS$$

is bounded. Denoting by $y(t)$ the quantity

$$y(t)=\int_{\Sigma t}\cdots\int u^2\,dS,$$

where $\sum t$ is the surface where the coordinates $x_1,x_2,\cdots,x_n$ take the same values as on $S_2$ while the variable $t$ changes from 0 to $t_0$, we obtain

$$y'(t)=2\int_{\Sigma t}\cdots\int u(t)\,\frac{\partial u}{\partial t}\,dS.$$

Applying the Cauchy-Bunjakovsky inequality, we obtain

$$|y'(t)|\leqslant 2\left[\int_{\Sigma t}\cdots\int\left(\frac{\partial u}{\partial t}\right)^2dS\right]^{\frac{1}{2}}\left[\int_{\Sigma t}\cdots\int u(t)^2\,dS\right]^{\frac{1}{2}}$$

and by virtue of the inequality (17.7)

$$|y'(t)|\leqslant 2Ay(t)^{\frac{1}{2}}, \tag{17.8}$$

where

$$A=\left[\int_{S_3}\cdots\int\left\{\sum_{i=1}^{n}\left(\frac{\partial u}{\partial x_i}\right)^2+\left(\frac{\partial u}{\partial t}\right)^2\right\}dS_3\right]^{\frac{1}{2}}.$$

The inequality (17.8) implies that

$$\frac{d}{dt}\sqrt{y} \leqslant A$$

or:

$$\sqrt{y_1} \leqslant \sqrt{y_0} + At,$$

from which finally:

$$y \leqslant y_0 + 2A\sqrt{y_0}t + A^2t^2.$$

Setting also

$$\int_{S_3} \cdots \int u^2\, dS = B^2,$$

we have $y_0 \leqq B^2$, and hence

$$y \leqslant (B + At)^2.$$

We intersect our truncated cone with the plane $t =$ const. and let $\sum t$ be the resulting domain of $n$-dimensional space obtained as the intersection.

Arguing as before, we obtain:

$$\overbrace{\int_{\Sigma t} \cdots \int}^{n} \left\{\sum_{i=1}^{n}\left(\frac{\partial u}{\partial x_i}\right)^2 + \left(\frac{\partial u}{\partial t}\right)^2\right\} dS \leqslant \int_{S_3} \cdots \int \left\{\sum_{i=1}^{n}\left(\frac{\partial u}{\partial x_i}\right)^2 + \left(\frac{\partial u}{\partial t}\right)^2\right\} dS.$$

Integrating in $t$ from 0 to $t_0$, we obtain

$$\overbrace{\int_V \cdots \int}^{n+1} \left\{\sum_{i=1}^{n}\left(\frac{\partial u}{\partial x_i}\right)^2 + \left(\frac{\partial u}{\partial t}\right)^2\right\} dV \leqslant$$

$$\leqslant t_0 \int_{S_3} \cdots \int \left\{\sum_{i=1}^{n}\left(\frac{\partial u}{\partial x_i}\right)^2 + \left(\frac{\partial u}{\partial t}\right)^2\right\} dS \leqslant A^2 t_0, \qquad (17.9)$$

where $V$ is our truncated cone.

Analogously, using the fact that

$$y = \int_{\Sigma t} \cdots \int u^2\, dS \leqslant (B + At)^2 \qquad (17.10)$$

and integrating in $t$ from 0 to $t_0$, we obtain:

$$\overbrace{\int_V \cdots \int}^{n+1} u^2\, dV \leqslant \frac{1}{A}[(B + At_0)^3 - B^3] = 3B^2t_0 + 3ABt_0^2 + A^2t_0^3. \qquad (17.11)$$

The inequality (17.10) has a number of important corollaries.

COROLLARY 1. *Suppose that the initial values of $u$ and $\partial u/\partial t$ vanish in the interior of $S_3$. Then $A=B=0$ and as a consequence of* (17.10) $y\equiv 0$, *i.e.*, $u\equiv 0$ *in* $V$.

Thus, if on the basis of the truncated cone $u\equiv 0$, $\partial u/\partial t\equiv 0$, then $u\equiv 0$ in the interior of this cone.

COROLLARY 2. *The value of the function $u$, a solution of the given equation, at a point $x_1^0, x_2^0, \cdots, x_n^0, t^0$ is determined by the values of the initial data $u$ and $\partial u/\partial t$ on the "sphere" $\eta=(\sum_{i=1}^{n}(x_i-x_i^0)^2)^{\frac{1}{2}}\leqq t_0$, which is the intersection of the characteristic cone with vertex at the given point with the plane $t=0$.*

Indeed, if for any two solutions of the wave equation the initial data of $u$ and $\partial u/\partial t$ coincide on this domain, the data of their difference will vanish on this domain and by Corollary 1, the difference will be null at the vertex of the cone. Hence, at the vertex of the cone, i.e., at the point in question the two solutions will coincide, as was to be shown.

3. SOLUTIONS FOR SPECIAL INITIAL DATA.

THEOREM. *Let the function $u$ be a solution of the homogeneous wave equation. If the initial values $u|_{t=0}$ and $\partial u/\partial t|_{t=0}$ are infinitely differentiable on the whole of the space of the $x_2, \cdots, x_n$, then the function $u$ itself has all derivatives of all orders.*

PROOF. To begin with, we shall establish this theorem in a more special situation. We prove a lemma.

LEMMA. *Let the function $u$ satisfy the equation*

$$\Delta u - \frac{\partial^2 u}{\partial t^2} = 0 \tag{17.5}$$

*on the domain $-a\leqq x_i\leqq a$ $(i=1,2,\cdots,n)$ where $a$ is some constant, and suppose in addition that*

$$u\,|_{x_i=\pm a} = 0, \tag{17.12}$$

*i.e., the function $u$ vanishes on the boundary of this domain. Suppose in addition that for $t=0$*

$$\begin{aligned} u\,|_{t=0} &= \varphi_0 \\ \left.\frac{\partial u}{\partial t}\right|_{t=0} &= \varphi_1, \end{aligned} \tag{17.13}$$

*where the functions $\phi_0$ and $\phi_1$ have continuous derivatives of all orders and vanish on the boundary of the domain together with all their derivatives. Then the function $u$ has continuous derivatives of all orders.*

Indeed in this case, the solution can be written in an explicit form with the aid of Fourier series.

We expand the functions $\phi_0$ and $\phi_1$ in Fourier series

$$\varphi_0 = \sum_{j_k=1}^{\infty} b_{j_1, j_2, \ldots, j_n} \sin j_1 \frac{(x_1+a)\pi}{2a} \sin j_2 \frac{(x_2+a)\pi}{2a} \ldots \sin j_n \frac{(x_n+a)\pi}{2a}$$

$$\varphi_1 = \sum_{j_k=1}^{\infty} g_{j_1, j_2, \ldots, j_n} \sin j_1 \frac{(x_1+a)\pi}{2a} \times$$

$$\times \sin j_2 \frac{(x_2+a)\pi}{2a} \ldots \sin j_n \frac{(x_n+a)\pi}{2a}. \tag{17.14}$$

The functions $\phi_0$ and $\phi_1$ are continuous together with all their derivatives and may be extended in a periodic way to the whole space with preservation of the continuity of all their derivatives. It follows that the Fourier series for these functions will converge uniformly together with all their derivatives of arbitrary order.

Consider the partial sums of these series. Set:

$$\varphi_0^{(N)} = \sum_{j_k=1}^{N} b_{j_1, j_2, \ldots, j_n} \sin j_1 \frac{(x_1+a)\pi}{2a} \sin j_2 \frac{(x_2+a)\pi}{2a} \ldots \sin j_n \frac{(x_n+a)\pi}{2a};$$

$$\varphi_1^{(N)} = \sum_{j_k=1}^{N} g_{j_1, \ldots, j_n} \sin j_1 \frac{(x_1+a)\pi}{2a} \sin j_2 \frac{(x_2+a)\pi}{2a} \ldots \sin j_n \frac{(x_n+a)\pi}{2a}.$$

If in the initial data we replace $\phi_0$ and $\phi_1$ by $\phi_0^{(N)}$ and $\phi_1^{(N)}$, we obtain for the solution $u^{(N)}$ of the wave equation satisfying these initial conditions

$$u^{(N)} = \sum_{j_k=1}^{N} \Big\{ b_{j_1, j_2, \ldots, j_n} \cos \sqrt{j_1^2 + j_2^2 + \ldots + j_n^2}\, t +$$

$$+ \frac{g_{j_1, j_2, \ldots, j_n}}{\sqrt{j_1^2 + j_2^2 + \ldots + j_n^2}} \sin \sqrt{j_1^2 + j_2^2 + \ldots + j_n^2}\, t \Big\} \times$$

$$\times \sin j_1 \frac{(x_1+a)\pi}{2a} \sin j_2 \frac{(x_2+a)\pi}{2a} \ldots \sin j_n \frac{(x_n+a)\pi}{2a}.$$

This solution, obviously, is infinitely differentiable. We shall show that with increasing $N$ the sequence $u^{(N)}$ converges in every space $W_2^{(l)}$, where $l$ is an arbitrary number, to some function $u$. It follows already from this that the limit function $u$ is a solution of the wave equation satisfying the initial conditions (17.13) and is infinitely differentiable. Since this solution is unique, our lemma will follow.

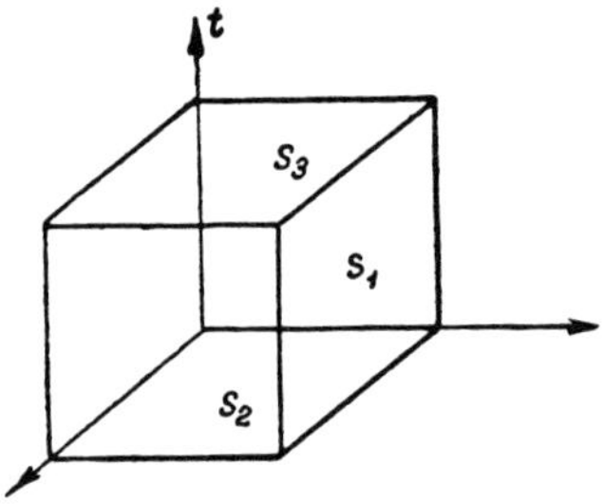

Figure 9

4. PROOF OF THE BASIC THEOREM. It remains for us to prove the convergence of the sequence $u^{(N)}$. Applying the integral equality (17.6) to the "parallelepiped" whose base $S_2$ is the "square" domain $\Omega$: $|x_i| \leq a$ in the plane $t=0$ and the upper base $S_3$ of which lies in the plane $t=t_0$ (Figure 9). If we use the fact that on the lateral surface $S_1$ of this domain ($|x_i|=a$) we have the equality $u^{(N)}= \partial u^{(N)}/\partial t=0$, we obtain

$$\int_{S_1}\cdots\int \frac{\partial u^{(N)}}{\partial t}\frac{\partial u^{(N)}}{\partial n}\,dS_1=0,$$

and hence

$$\int_{S_2}\cdots\int\left[\sum_{i=1}^{n}\left(\frac{\partial u^{(N)}}{\partial x_i}\right)^2+\left(\frac{\partial u^{(N)}}{\partial t}\right)^2\right]dS_2=$$

$$=\int_{S_3}\cdots\int\left[\sum_{i=1}^{n}\left(\frac{\partial u^{(N)}}{\partial x_i}\right)^2+\left(\frac{\partial u^{(N)}}{\partial t}\right)^2\right]dS_3. \qquad (17.15)$$

We consider also the functions

$$v^{(N)}_{\alpha_1\alpha_2,\ldots,\alpha_n}=\frac{\partial^\alpha u^{(N)}}{\partial x_1^{\alpha_1}\ldots\partial x_n^{\alpha_n}}.$$

These functions in their turn satisfy the wave equation.[1)] We note also that on the boundary of the parallelepiped for $|x_j|=a$, the function $v^{(N)}_{\alpha_1,\ldots,\alpha_n}$ will satisfy either the condition $v^{(N)}_{\alpha_1,\ldots,\alpha_n}=0$, if $\alpha_j$ is even or the condition $\partial v^{(N)}_{\alpha_1,\ldots,\alpha_n}/\partial n=0$, if $\alpha_j$ is odd. Applying formula (17.6) to these functions, we obtain in this way:

---

[1)] If the solution $u$ of the wave equation has continuous derivatives up to $(l+2)$nd order on the whole space, then obviously an arbitrary derivative of $u$ of order $l$ will also be a solution of the wave equation.

$$\int_{S_2}\cdots\int\left[\sum_{i=1}^{n}\left(\frac{\partial v^{(N)}_{\alpha_1,\ldots,\alpha_n}}{\partial x_i}\right)^2+\left(\frac{\partial v^{(N)}_{\alpha_1,\ldots,\alpha_n}}{\partial t}\right)^2\right]dS_2=$$

$$=\int_{S_3}\cdots\int\left[\sum_{i=1}^{N}\left(\frac{\partial v^{(N)}_{\alpha_1,\ldots,\alpha_n}}{\partial x_i}\right)^2+\left(\frac{\partial v^{(N)}_{\alpha_1,\ldots,\alpha_n}}{\partial t}\right)^2\right]dS_3. \tag{17.16}$$

On the initial plane $S_2$ all the integrals for given $\alpha_1,\alpha_2,\cdots,\alpha_n$ are bounded numbers which do not depend on $N$.

We consider also the functions

$$w^{(k,r)}_{\alpha_1,\ldots,\alpha_n}=v^{(k)}_{\alpha_1,\ldots\alpha_n}-v^{(r)}_{\alpha_1,\ldots,\alpha_n}.$$

For these functions we obtain as before:

$$\int_{S_3}\cdots\int\left[\sum_{i=1}^{n}\left(\frac{\partial w^{(k,r)}_{\alpha_1,\ldots,\alpha_n}}{\partial x_i}\right)^2+\left(\frac{\partial w^{(k,r)}_{\alpha_1,\ldots,\alpha_n}}{\partial t}\right)^2\right]dS_3=$$

$$=\int_{S_2}\cdots\int\left[\sum_{i=1}^{n}\left(\frac{\partial w^{(k,r)}_{\alpha_1,\ldots,\alpha_n}}{\partial x_i}\right)^2+\left(\frac{\partial w^{(k,r)}_{\alpha_1,\ldots,\alpha_n}}{\partial t}\right)^2\right]dS_2. \tag{17.17}$$

It is not hard to establish that for sufficiently large $k$ and $r$, the integral on the right hand side of the last equation will be as small as one pleases. This follows immediately from the convergence with all derivatives of the Fourier series for $\phi_0$ and $\phi_1$. From this it follows that the quantity on the left side will be arbitrarily small

$$\int_{S_3}\cdots\int\left[\sum_{i=1}^{n}\left(\frac{\partial w^{(k,r)}_{\alpha_1,\ldots,\alpha_n}}{\partial x_i}\right)^2+\left(\frac{\partial w^{(k,r)}_{\alpha_1,\ldots,\alpha_n}}{\partial t}\right)^2\right]dS_3<\varepsilon.$$

Integrating this last inequality in the variable $t$ between the limits 0 and $T$. We will have:

$$\int_{\Omega}\cdots\int\left[\sum_{i=1}^{n}\left(\frac{\partial w^{(k,r)}_{\alpha_1,\ldots,\alpha_n}}{\partial x_i}\right)^2+\left(\frac{\partial w^{(k,r)}_{\alpha_1,\ldots,\alpha_n}}{\partial t}\right)^2\right]d\Omega<T\varepsilon,$$

where, by $\Omega$, we denote the domain $0\leqq t\leqq T$; $0\leqq|x_i|\leqq a$. Using the same argument as we employed to derive (17.11), we show that

$$\int_{\Omega}\cdots\int(w^{(k,r)}_{\alpha_1,\ldots,\alpha_n})^2\,d\Omega\leqslant(\varepsilon_0+\varepsilon_1 T)^2\leqslant\varepsilon.$$

By virtue of the completeness of the space $W_2^{(l)}$, we conclude that the sequence $v^{(N)}_{\alpha_1,\alpha_2,\cdots,\alpha_n}$ which satisfies the Cauchy convergence criterion must converge in this space. The convergence of all the derivatives of $u^{(N)}$ in $W_2^{(l)}$ implies the uniform convergence of all the derivatives of these functions by virtue of the Imbedding Theorem. The lemma is proved.

If we use the lemma, it is easy to prove our theorem. Indeed the values of the unknown function in the interior of the pyramid

$$0 \leqslant |x_i| + t \leqslant \frac{a}{2} \tag{17.18}$$

depend only on the values of the functions $\phi_0$ and $\phi_1$ within the domain $0 \leqq x_1 \leqq a/2$ for $t=0$.

We construct the functions

$$\varphi_0^{(a)} = \varphi_0 \prod_{i=1}^{n} \psi\left(\left|\frac{x_i}{a}\right|\right);$$

$$\varphi_i^{(a)} = \varphi_1 \prod_{i=1}^{n} \psi\left(\left|\frac{x_i}{a}\right|\right),$$

where $\psi(\xi)$ is a function equal to one for $\xi < \frac{1}{2}$ and zero for $\xi > 1$, and infinitely differentiable. We will seek a solution $u^{(a)}$ of the wave equation satisfying the conditions

$$u^{(a)}|_{t=0} = \varphi_0^{(a)};$$

$$\left.\frac{\partial u^{(a)}}{\partial t}\right|_{t=0} = \varphi_1^{(a)}.$$

On the basis of the lemma, we see that $u^{(a)}$ is infinitely differentiable. But by what was shown earlier, within the pyramid (17.18) this solution coincides with $u$.

Hence $u$ in its turn will be infinitely differentiable, as was to be proved.

## §18. The generalized Cauchy problem for the wave equation.

1. TWICE DIFFERENTIABLE SOLUTIONS. We pose the following problem: to find a solution of the wave equation

$$\Box u = \Delta u - \frac{\partial^2 u}{\partial t^2} = 0 \tag{17.5}$$

in the whole space satisfying the initial conditions

$$\left.\begin{aligned} u|_{t=0} &= u_0, \\ \left.\frac{\partial u}{\partial t}\right|_{t=0} &= u_1. \end{aligned}\right\} \tag{17.13}$$

It has been shown that if $u_0$ and $u_1$ have derivatives of every order, the solution of the problem exists and is infinitely differentiable. But of course, there is no necessity for infinite differentiability of the data to obtain solutions, especially since the equation involves only derivatives of second order.

We consider first of all the following problem.

**PROBLEM** 1. To determine what conditions imposed upon $u_0$ and $u_1$ will assure the existence of twice differentiable solutions.

In the previous chapter we became acquainted with generalized derivatives, and may immediately introduce the generalized wave operator.

Let $u(t, x_1, \cdots, x_n)$ be a summable function on some domain $\Omega$ of the $(n+1)$-dimensional space. If there exists a summable function $f(x_1, \cdots, x_n, t)$ such that

$$\overbrace{\int \cdots \int_{\Omega}}^{n+1} u \,\Box\, \psi \, dv = \int \cdots \int_{\Omega} \psi f \, dv,$$

for any twice differentiable function $\psi(t, x_1, \cdots, x_n)$ vanishing outside of some closed subdomain of $\Omega$, then $f$ is called the generalized wave operator of $u$ and we shall write $\Box u = f$.

**PROBLEM** 2. A function which has a generalized wave operator equal to zero will be called a generalized solution of the wave equation. It is natural to pose the Cauchy problem for generalized solutions. We shall consider both these problems.

**THEOREM** 1. *If $u_0$ has generalized derivatives up to order $[n/2]+3$ square-integrable on every bounded domain and $u_1$ has similar generalized derivatives up to order $[n/2]+2$, then the equation* (17.5) *has a twice differentiable solution satisfying the conditions* (17.13).

**PROOF**. We construct the sequences of averaged functions $\{u_{0h}\}$ and $\{u_{1h}\}$. By the theorem of §17, item 4, there exists solutions of equation (17.5) satisfying the initial conditions

$$\left. \begin{aligned} u\,|_{t=0} &= u_{0h} \\ \left.\frac{\partial u}{\partial t}\right|_{t=0} &= u_{1h} \end{aligned} \right\} \tag{18.1}$$

and having derivatives of arbitrary order.

Consider the function: $v_{p,q} = v_{hp} - v_{hq}$; $v_{p,q}$ is a solution of equation (17.5) with the boundary conditions

$$v_{p,q}\,|_{t=0} = u_{0h_p} - u_{0h_q}; \quad \left.\frac{\partial v_{p,q}}{\partial t}\right|_{t=0} = u_{1h_p} - u_{1h_q}.$$

By inequality (17.7) we have (Figure 10)

$$\int \underset{S_3}{\cdots} \int \left[ \sum_{i=1}^{n} \left( \frac{\partial v_{p,q}}{\partial x_i} \right)^2 + \left( \frac{\partial v_{p,q}}{\partial t} \right)^2 \right] dS \leqslant \int \underset{S_2}{\cdots} \int \left[ \sum_{i=1}^{n} \left( \frac{\partial v_{p,q}}{\partial x_i} \right)^2 + \right.$$

$$\left. + \left( \frac{\partial v_{p,q}}{\partial t} \right)^2 \right] dS \tag{18.2}$$

and analogously for any derivative of $v_{pq}$

$$\int \underset{S_3}{\cdots} \int \left[ \sum_{i=1}^{n} \left( \frac{\partial}{\partial x_i} \frac{\partial^\alpha v_{p,q}}{\partial x_1^{\alpha_1} \ldots \partial x_n^{\alpha_n}} \right)^2 + \left( \frac{\partial}{\partial t} \frac{\partial^\alpha v_{p,q}}{\partial x_1^{\alpha_1} \ldots \partial x_n^{\alpha_n}} \right)^2 \right] dS \leqslant$$

$$\leqslant \int \underset{S_2}{\cdots} \int \left[ \sum_{i=1}^{n} \left( \frac{\partial}{\partial x_i} \frac{\partial^\alpha v_{p,q}}{\partial x_1^{\alpha_1} \ldots \partial x_n^{\alpha_n}} \right)^2 + \left( \frac{\partial}{\partial t} \frac{\partial^\alpha v_{p,q}}{\partial x_1^{\alpha_1} \ldots \partial x_n^{\alpha_n}} \right)^2 \right] dS \tag{18.3}$$

and by (17.10)

$$\int \underset{S_3}{\cdots} \int (v_{p,q})^2 dS \leqslant \left\{ \left[ \int \underset{S_2}{\cdots} \int v_{p,q}^2 dS \right]^{\frac{1}{2}} + \right.$$

$$\left. + \left[ \int \underset{S_2}{\cdots} \int \sum_{i=1}^{n} \left( \frac{\partial v_{p,q}}{\partial x_i} \right)^2 dS + \int \underset{S_2}{\cdots} \int \left( \frac{\partial v_{p,q}}{\partial t} \right)^2 dS \right]^{\frac{1}{2}} t \right\}^2. \tag{18.4}$$

By a property of averaged functions (§5, item 2) $u_{0h} \Rightarrow u_0$ in $W_2^{[n/2]+3}$ and $u_{1h} \Rightarrow u_1$ in $W_2^{[n/2]+2}$ and consequently the right sides of inequalities (18.2), (18.3) and (18.4) may be made arbitrarily small for sufficiently small $h_p$ and $h_q$ and $\alpha \leqq [n/2]+3$, and thereby the left sides also, i.e., for an arbitrary domain in the plane $t=$ const. the sequence $\{u_h\}$ converges strongly in the sense of $W_2^{[n/2]+3}$. However, there follows from this by the Imbedding Theorem the convergence of the functions $u_h$ in $C^2$ (in the $n$-dimensional space of $x_1, \cdots, x_n$). By analogous estimates, we show that $\partial u/\partial t \in C^1$ and $\partial^2 u/\partial t^2 \in C$, i.e., $u$ is twice continuously differentiable in the $(n+1)$-dimensional space and is a solution of the wave equation.

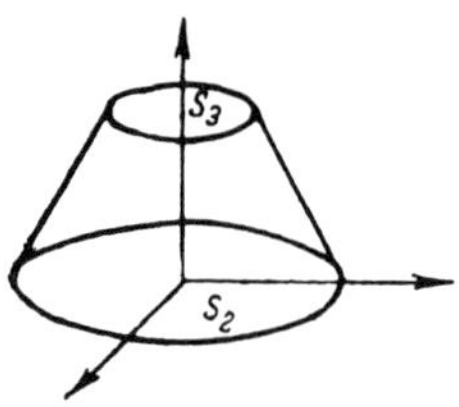


Figure 10

2. EXAMPLE. We consider the wave equation in a 5-dimensional space. We saw above that it sufficed to demand by way of the initial conditions the existence of square integrable fifth derivatives for $u_0$ and fourth derivatives for $u_1$. We shall show that if the square integrable derivatives exist up to lower orders only (for $u_0$ derivatives of the fourth order, for $u_1$ derivatives of the third order), the solutions may not have continuous derivatives of second order.

Let $r=(\sum_{i=1}^{5} x_i^2)^{1/2}$; then if $f(y)$ is continuously differentiable $m$ times $(m \geqq 5)$,

$$u=\frac{f(t-r)-f(t+r)}{r^3}+\frac{f'(t-r)+f'(t+r)}{r^2} \tag{18.5}$$

will be a solution of the wave equation with initial data

$$u_0=-\frac{2f_1(r)}{r^3}+\frac{2f_1'(r)}{r^2}, \text{ where } 2f_1(r)=f(-r)-f(r);$$

$$u_1=-\frac{2f_2'(r)}{r^3}+\frac{2f_2''(r)}{r^2} \qquad 2f_2(r)=f(r)+f(-r).$$

We have

$$\frac{\partial^2 u}{\partial t^2}=\frac{f''(t-r)-f''(t+r)}{r^3}+\frac{f'''(t-r)+f'''(t+r)}{r^2}=$$
$$=\frac{[f''(t-r)-f''(t+r)]+r[f'''(t-r)+f'''(t+r)]}{r^3}.$$

If $m\geqq 5$, it is not difficult to verify that $\partial^2 u/\partial t^2$ and $\partial^2 u/\partial r^2$ are continuous, and an application of l'Hospital's rule gives

$$\left.\frac{\partial u^2}{\partial t^2}\right|_{r=0}=\lim_{r\to 0}\frac{[f''(t-r)-f''(t+r)]+r[f'''(t-r)+f'''(t+r)]}{r^3}=\frac{2}{3}f_1^{(V)}(t).$$

Let

$$f(y)=\begin{cases} 0, & y<1,\\ (y-1)^\alpha, & y>1.\end{cases}$$

Then

$$u_0=\begin{cases} 0, & r<1;\\ -\dfrac{(r-1)^\alpha}{r^3}+\alpha\dfrac{(r-1)^{\alpha-1}}{r^2}, & r>1;\end{cases}$$

$$u_1=\begin{cases} 0, & r<1;\\ -\alpha\dfrac{(r-1)^{\alpha-1}}{r^3}+\alpha(\alpha-1)\dfrac{(r-1)^{\alpha-2}}{r^2}, & r>1;\end{cases}$$

and it is not difficult to see that $u_0\in W_2^{(4)}$, $u_1\in W_2^{(3)}$ but $u_0\overline{\in} W_2^{(5)}$, $u_1\overline{\in} W_2^{(4)}$ if $9/2<\alpha<11/2$. Thereby we obtain:

$$\left.\frac{\partial^2 u}{\partial t^2}\right|_{r=0}=\begin{cases} 0, & t<1\\ A(t-1)^{\alpha-5}, & t>1 \quad (A=\text{const.}\neq 0),\end{cases}$$

and the function $u(r,t)$ given by (18.5) turns out not to be twice continuously differentiable if $9/2<\alpha\leqq 5$. As is easy to see, $u(r,t)$ is a generalized solution of the wave equation for the data $u_0$, $u_1$ and moreover the unique solution (cf. Theorem 3). If $5<\alpha<11/2$, it is still true that $u_0\in W_2^{(5)}$, $u_1\in W_2^{(4)}$ but the corresponding solution is twice continuously differentiable and, as a result, the conditions $u_0\in W_2^{[n/2]+3}$, $u_1\in W_2^{[n/2]+2}$ are not necessary. Analogous examples may easily be constructed for $n=(2k+1)\,(k=1,2,\cdots)$.

3. GENERALIZED SOLUTIONS. We saw above that a solution of the wave equation having continuous derivatives possesses the property that its norm in an arbitrary space $W_2^{(l)}$ (if we consider it as an element of this space) remains uniformly bounded. These norms may be estimated if the initial values of the function are known, more precisely, if we know the norm of $u_0$ in the same space $W_2^{(l)}$ and $u_1$ in $W_2^{(l-1)}$.

The invariance of the norm in Hilbert space yields the possibility of proceeding to a rather convenient generalized interpretation of the solutions of the wave equation.

We shall say that the function $u$ is a generalized solution of the wave equation in $W_2^{(1)}$:

$$\Delta u-\frac{\partial^2 u}{\partial t^2}=0,$$

if the function lies in $W_2^{(1)}$ on every bounded domain and in addition it has the generalized wave operator

$$\left(\Delta-\frac{\partial^2}{\partial t^2}\right)u=\Box\, u,$$

equal to zero.

Obviously, every generalized solution of the wave equation in $W_2^{(1)}$ is the limit of a sequence of functions $u_k$ converging in $W_2^{(1)}$ on every bounded domain and themselves continuous solutions of the wave equation

$$\Box\, u_k=0.$$

Indeed, by what was shown above, $u_k$ will be the limit in $W_2^{(1)}$ on bounded sets of the sequence of averaged functions $u_k$

$$u_{h_k}\Rightarrow u_k.$$

Furthermore, each averaged function obviously satisfies the wave equation since the wave operator commutes with the operation of averaging. This is shown by the same means which we used earlier for the proof of the fact that the derivatives of averaged functions are the averages of their derivatives (cf. Chapter I, §5, item 2).

We shall prove the converse theorem.

THEOREM 1. *If the function $u$ is the limit of a sequence $u_k$ of twice continuously differentiable solutions of the wave equation having uniformly bounded $W_2^{(1)}$ norms with the sequence converging weakly in $L_1$, then $u$ is a generalized solution in $W_2^{(1)}$.*

PROOF. On the basis of the theorem on the existence of generalized derivatives, we conclude that $u$ belongs to $W_2^{(1)}$. The fact that $u$ has the generalized wave operator zero follows from

$$\int \cdots \int u \,\Box\, \psi \, dv = \lim_{k\to\infty} \int \cdots \int u_k \,\Box\, \psi \, dv,$$

where $\psi$ is an arbitrary function having continuous derivatives. If we choose $\psi$, in particular, to be a function vanishing outside of some domain $V_\psi$ and remark that

$$\int \cdots \int u_k \,\Box\, \psi \, dv = 0,$$

we obtain

$$\int \cdots \int u \,\Box\, \psi \, dv = 0, \tag{18.6}$$

as was to be shown.

4. EXISTENCE OF INITIAL DATA.

THEOREM 2. *Every generalized solution in $W_2^{(1)}$ has on the plane $t=0$ limiting values $u_0 \in W_2^{(1)}$ and $u_1 \in L_2$ corresponding to n-dimensional space. Thus on every domain $S_1$ in the space of $x_1, \cdots, x_n$, we will have*

$$\left.\begin{aligned} &\lim_{k\to 0} \| u\,(k) - u_0 \|_{W_2^{(1)}} = 0; \\ &\lim_{k\to 0} \left\| \frac{\partial u\,(k)}{\partial t} - u_1 \right\|_{L_2} = 0. \end{aligned}\right\} \tag{18.7}$$

The plane $t=0$, of course, was chosen completely arbitrarily. Our theorem leads to the conclusion that a solution of the wave equation in $W_2^{(1)}$ can have the plane $t=$ const. only in the role of a removable singularity.

We recall that from the Imbedding Theorem there follows in this case a somewhat weaker result, namely that every solution $u$ of the wave equation in $W_2^{(1)}$ can have the plane $t=$ const. only as a removable singularity if we consider the values of $u$ in $L_2$ (without derivatives).

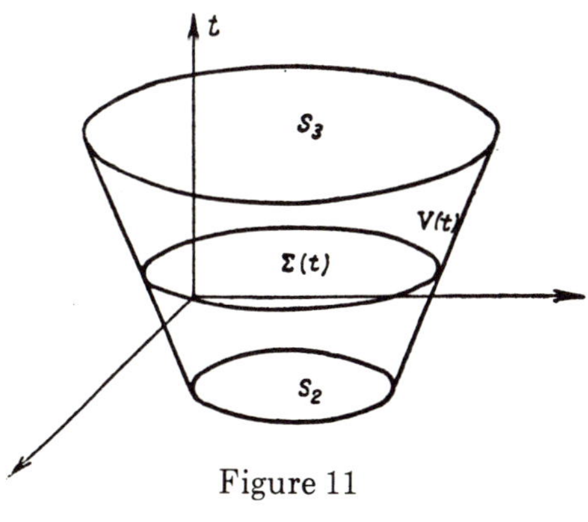

Figure 11

We proceed to the proof of our theorem.

We consider in the plane $t=0$ an arbitrary $n$-dimensional sphere $S_2$ and construct through its boundary the reverse characteristic cone (Figure 11).

Denote by $\sum(t)$ the sphere obtained from the intersection of this cone and the plane $t=\text{const}$. The value of $\sum(t)$ for $t=T$ will be denoted by $S_3$. Let $V(t)$ be the conical piece bounded by the lateral surface of the cone and by the planes $S_3$ and $\sum(t)$.

Suppose that $v$ is some twice differentiable solution of the wave equation. For it we have the inequality:

$$\int_{\Sigma(t)}\!\!\cdots\!\int\left[\sum_{i=1}^{n}\left(\frac{\partial v}{\partial x_i}\right)^2+\left(\frac{\partial v}{\partial t}\right)^2\right]dS \geqslant \int_{\Sigma(t_1)}\!\!\cdots\!\int\left[\sum_{i=1}^{n}\left(\frac{\partial v}{\partial x_i}\right)^2+\left(\frac{\partial v}{\partial t}\right)^2\right]dS,$$

if $t>t_1$, obtained from (17.7) by replacing $t$ by $T-t$.

Integrating in $t$ from $t_1$ to $T$, we obtain

$$\int_{V(t_1)}\!\!\cdots\!\int\left[\sum_{i=1}^{n}\left(\frac{\partial v}{\partial x_i}\right)^2+\left(\frac{\partial v}{\partial t}\right)^2\right]dV \geqslant (T-t_1)\int_{\Sigma(t_1)}\!\!\cdots\!\int\left[\sum_{i=1}^{n}\left(\frac{\partial v}{\partial x_i}\right)^2+\left(\frac{\partial v}{\partial t}\right)^2\right]dS. \tag{18.8}$$

Let $v_h=u_h(t+k)-u_h(t)$, where $u_h$ is an averaged function for $u$ and $k>0$.

Applying (18.8) to $v_h$ for a fixed value of $k$ and substituting its value for $v_h$, we will have:

$$\frac{1}{T-t_1}\int_{V(t_1)}\!\!\cdots\!\int\left[\sum_{i=1}^{n}\left(\frac{\partial u_h(t+k)}{\partial x_i}-\frac{\partial u_h(t)}{\partial x_i}\right)^2+\left(\frac{\partial u_h(t+k)}{\partial t}-\frac{\partial u_h(t)}{\partial t}\right)^2\right]dV \geqslant$$
$$\geqslant \int_{\Sigma(t_1)}\!\!\cdots\!\int\left[\sum_{i=1}^{n}\left(\frac{\partial u_h(t+k)}{\partial x_i}-\frac{\partial u_h(t)}{\partial x_i}\right)^2+\left(\frac{\partial u_h(t+k)}{\partial t}-\frac{\partial u_h(t)}{\partial t}\right)^2\right]dS. \tag{18.9}$$

In formula (18.9) we may pass to the limit as $h\to 0$ since the convergence of $u_h$ to $u$ takes place in $W_2^{(1)}$, i.e., together with the derivatives of first order. From this, we have

$$\int_{\Sigma(t_1)}\!\!\cdots\!\int\left\{\sum_{i=1}^{n}\left[\frac{\partial}{\partial x_i}(u(t+k)-u(t))\right]^2+\left[\frac{\partial}{\partial t}(u(t+k)-u(t))\right]^2\right\}dS \leqslant$$
$$\leqslant \frac{1}{T-t_1}\int_{V(t_1)}\!\!\cdots\!\int\left\{\sum_{i=1}^{n}\left[\frac{\partial}{\partial x_i}(u(t+k)-u(t))\right]^2+\right.$$
$$\left.+\left[\frac{\partial}{\partial t}(u(t+k)-u(t))\right]^2\right\}dS. \tag{18.10}$$

The functions $\partial u/\partial x_i$ and $\partial u/\partial t$ belong to $L_2$ on $V$, and, as a result, are continuous in the large on the domain $V$. Let $t_1 < T/2$. We may then choose $\delta(\epsilon)$ so small that for $k < \delta(\epsilon)$ the right side of the inequality (18.10) will be less than $\epsilon$ independently of $t_1$.

Thereby we obtain:

$$\int_{\Sigma(t_1)} \cdots \int \left\{ \sum_{i=1}^{n} \left[ \frac{\partial}{\partial x_i} (u(t_1+k) - u(t_1)) \right]^2 + \right.$$
$$\left. + \left[ \frac{\partial}{\partial t} (u(t_1+k) - u(t_1)) \right]^2 \right\} dS < \varepsilon.$$

We put $t_1 + k = t_2$ and obtain in the $n$-dimensional space:

$$\| u(t_2) - u(t_1) \|_{L_2^{(1)}} < \varepsilon \quad \left\| \frac{\partial u}{\partial t}(t_2) - \frac{\partial u}{\partial t}(t_1) \right\|_{L_2} < \varepsilon$$

for $t_1$ and $t_2$ sufficiently close and in particular for arbitrarily small $t_1$ and $t_2$.

By virtue of the completeness of the spaces $L_2$ and $L_2^{(1)}$, there follows from this the existence of a limiting value for $t \to 0$ in $L_2$ for $\partial u/\partial t$ and a limiting value for $t \to 0$ in $L_2^{(1)}$ for $u$. If we note also that a limiting value for $u$ in $L_2$ exists by virtue of the Imbedding Theorem, we arrive at the conclusion that there exist values for $u$ and $\partial u/\partial t$ for $t \to 0$ belonging respectively to $W_2^{(1)}$ and $L_2$ respectively such that

$$\| u(k) - u(0) \|_{W_2^{(1)}} \to 0, \quad \left\| \frac{\partial u}{\partial t}(k) - \frac{\partial u}{\partial t}(0) \right\|_{L_2} \to 0.$$

The theorem is proved.

5. SOLUTIONS OF THE GENERALIZED CAUCHY PROBLEM.

THEOREM 3. *For any functions* $u_0(x_1, x_2, \cdots, x_n) \in W_2^{(1)}$ *and* $u_1(x_1, \cdots, x_n) \in L_2$, *there exists a unique generalized solution $u$ of the wave equation in* $W_2^{(1)}$ *satisfying the conditions*:

$$\left. \begin{aligned} u \big|_{t=0} &= u_0, \\ \frac{\partial u}{\partial t} \Big|_{t=0} &= u_1, \end{aligned} \right\} \tag{17.13}$$

*where for $t \to 0$, $u(t)$ converges to $u_0$ in the metric of $W_2^{(1)}$ while $\partial u/\partial t$ converges to $u_1$ in the metric of $L_2$.*

PROOF. Consider the averaged functions for the initial data: $u_{0h}$ and $u_{1h}$. If we put them in place of $u_0$ and $u_1$ as initial data, we may find solutions of the Cauchy problem since these new initial data are infinitely differentiable.

Suppose the new solutions of the wave equation are $u_h$. It is easy to show that the sequence $u_h$ converges in $W_2^{(1)}$ in the $(n+1)$-dimensional space of $x_1, x_2, \cdots, x_n, t$ and in the $n$-dimensional space of $x_1, \cdots, x_n$ for an arbitrary fixed value of $t$. Indeed, put $u_{h_p} - u_{h_q} = v_{p,q}$.

The function $v_{p,q}$ is a solution of the wave equation, for which the quantities $A$ and $B$ will be as small as we please for sufficiently small $h_p$ and $h_q$. If we use inequalities (17.9) and (17.11), we see that for sufficiently small $h_p$ and $h_q$ we obtain:

$$\int_V \cdots \int v_{p,q}^2 \, dV < \varepsilon; \quad \int_V \cdots \int \left[ \sum_{i=1}^{n} \left( \frac{\partial v_{p,q}}{\partial x_i} \right)^2 + \left( \frac{\partial v_{p,q}}{\partial t} \right)^2 \right] dS < \varepsilon,$$

from which it follows directly that the sequence $u_h$ converges by virtue of the completeness of $W_2^{(1)}$ and $L_2$ in the space of $x_1, x_2, \cdots, x_n, t$. The limit function $u$ by virtue of Theorem 1 will be a generalized solution of the wave equation.

Furthermore, if we employ inequalities (17.7) and (17.10) replacing in them $S_2$ by $S_3$, we will have

$$\int_{S_3} \cdots \int \left[ \sum_{i=1}^{n} \left( \frac{\partial v_{p,q}}{\partial x_i} \right)^2 + \left( \frac{\partial v_{p,q}}{\partial t} \right)^2 \right] dS \leqslant \varepsilon$$

$$\int_{S_3} \cdots \int (v_{p,q})^2 \, dS \leqslant \varepsilon.$$

These inequalities mean that the sequence $u_h$ converges on an arbitrary plane $t = \text{const.}$ in the sense of $W_2^{(1)}$ with the convergence uniform with respect to $t$.

One easily concludes from this that the limit function $u$, which was shown in the second theorem to have initial values for $u$ and $\partial u / \partial t$ respectively in $W_2^{(1)}$ and $L_2$, assumes on the initial planes the given values $u_0$ and $u_1$. In fact,

$$\| u(k) - u(0) \|_{W_2^{(1)}} \leqslant \| u_h(k) - u_h(0) \|_{W_2^{(1)}} + $$
$$+ \| u_h(0) - u(0) \|_{W_2^{(1)}} + \| u_h(k) - u(k) \|_{W_2^{(1)}}.$$

for sufficiently small $h$ we have:

$$\|u_h(k) - u(k)\|_{W_2^{(1)}} \leqslant \frac{\varepsilon}{3};$$

$$\|u_h(0) - u(0)\|_{W_2^{(1)}} \leqslant \frac{\varepsilon}{3}.$$

Choosing $k$ sufficiently small for fixed $h$, we will have

$$\|u_h(k) - u_h(0)\|_{W_2^{(1)}} \leqslant \frac{\varepsilon}{3},$$

whereby

$$\|u(k) - u(0)\|_{W_2^{(1)}} \leqslant \varepsilon.$$

Analogously, it is easily shown that

$$\left\|\frac{\partial u}{\partial t}(k) - \frac{\partial u}{\partial t}(0)\right\|_{L_2} \leqslant \varepsilon.$$

Thus Theorem 3 is proved.

§19. **Linear equations of normal hyperbolic type with variable coefficients (basic properties).**

1. CHARACTERISTICS AND BICHARACTERISTICS. In the present section we shall show the existence of solutions of the Cauchy problem for linear equations of hyperbolic type with sufficiently smooth coefficients for sufficiently smooth initial data.

We consider the equation:

$$Lu \equiv \sum_{i=0}^{2k+1} \sum_{j=0}^{2k+1} A_{ij} \frac{\partial^2 u}{\partial x_i \partial x_j} + \sum_{i=0}^{2k+1} B_i \frac{\partial u}{\partial x_i} + Cu = F, \qquad (19.1)$$

where $A_{ij}$ $(A_{ij} = A_{ji})$, $B_i$, $C$, and $F$ are functions of the variables $x_0, x_1, \cdots \cdots, x_{2k+1}$ continuous together with their derivatives up to order $K+1$ inclusive, where $K$ is a sufficiently large number.

We assume that for each point of the space, the quadratic form

$$A(p) = \sum_{i=0}^{2k+1} \sum_{j=0}^{2k+1} A_{ij} p_i p_j \qquad (19.2)$$

may be brought into the form

$$A(p) = -\sum_{i=1}^{2k+1} q_i^2 + q_0^2 \qquad (19.3)$$

with the aid of a linear change of variables.

The equation (19.1) is called in that case an equation of normal hyperbolic type.

A characteristic surface or characteristic for equation (19.1) is a surface

$$G(x_0, x_1, \ldots, x_{2k+1}) = 0, \tag{19.4}$$

for which

$$A\left(\frac{\partial G}{\partial x_i}\right) = 0, \tag{19.5}$$

while

$$\sum_{i=0}^{2k+1} \left(\frac{\partial G}{\partial x_i}\right)^2 > 0 \tag{19.6}$$

at each point of the surface.

An equation of normal hyperbolic type has, among others, a characteristic surface with a conical point at an arbitrary given point of the space $x_0^0, x_1^0, x_2^0, \cdots, x_{2k+1}^0$. This surface is called the characteristic conoid. In the case of an equation with constant coefficients the characteristic conoid reduces to the characteristic cone.

We recall some of the simplest properties of the characteristic conoids and of their construction.

We set:

$$\frac{\partial G}{\partial x_i} = p_i. \tag{19.7}$$

As is known from a standard course on the theory of partial differential equations of first order a surface (19.4) satisfying the equation (19.5) is obtained as a manifold built up out of bicharacteristics, i.e., solutions of the system of ordinary differential equations:

$$\frac{dx_i}{\frac{1}{2}\frac{\partial A}{\partial p_i}} = \frac{-dp_i}{\frac{1}{2}\frac{\partial A}{\partial x_i}} = ds. \tag{19.8}$$

More precisely: a parametrized equation for the surface (19.4) is obtained in the form

$$x_i = \xi_i(s, x_0^{(0)}, x_1^{(0)}, \ldots, x_{2k+1}^{(0)}, p_0^{(0)}, p_1^{(0)}, \ldots, p_{2k+1}^{(0)}), \tag{19.9}$$

where $x_0^{(0)}, x_1^{(0)}, \cdots, x_{2k+1}^{(0)}, p_0^{(0)}, p_1^{(0)}, \cdots, p_{2k+1}^{(0)}$ are functions of $2k$ independent variables $v_1, v_2, \cdots, v_{2k}$. For these:

(a) The functions (19.9) together with

$$p_i = \frac{1}{s}\pi_i(s, x_0^{(0)}, \ldots, x_{2k+1}^{(0)}, p_0^{(0)}, p_1^{(0)}, \ldots, p_{2k+1}^{(0)}) \tag{19.10}$$

should represent the general solution of the system (19.8) depending on

$4k+4$ arbitrary constants. In (19.9) and (19.10) we set

$$x_i\big|_{s=0} = x_i^{(0)} \quad p_i\big|_{s=0} = p_i^{(0)}; \tag{19.11}$$

(b) The functions $x_i^{(0)}(v_1,\cdots,v_{2k})$, $p_i^{(0)}(v_1,\cdots,v_{2k})$ should satisfy the conditions:

$$A(p)\big|_{s=0} = 0, \tag{19.12}$$

$$\sum_{i=0}^{2k+1} p_i^{(0)} \frac{\partial x_i^{(0)}}{\partial v_j} = 0; \ (i=0,1,2,\ldots, 2k). \tag{19.13}$$

In the condition (19.13) the quantities $x_i$ and $p_i$ should be understood as expressed in $s$ by formulas (19.9) and (19.10);

(c) The equation (19.9) should give a parametric representation for a manifold of dimension $2k+1$ (and not lower).

We shall show how to construct by means of this theory the characteristic conoid with vertex at the point $x_0^{(0)}, x_1^{(0)},\cdots,x_{2k+1}^{(0)}$.

We assume that we have constructed the solutions (19.9) and (19.10) of the system (19.8). In them we shall consider $p_0^{(0)},p_1^{(0)},\cdots,p_{2k+1}^{(0)}$ as independent parameters, upon which we shall impose two conditions: the equation (19.12) and the normative condition:

$$\sum_{i=0}^{2k+1} p_i^{(0)2} = 1. \tag{19.14}$$

The quantities $x_0^{(0)}, x_1^{(0)},\cdots,x_{2k+1}^{(0)}$ will be taken as constants, not depending on $p_0^{(0)},p_1^{(0)},\cdots,p_{2k+1}^{(0)}$.

It is not difficult to verify that the equation (19.9) gives for this parametric equation a surface satisfying the equation (19.5). Condition (19.12) is satisfied by the choice of the $p_i^{(0)}$. The condition (19.13) is also satisfied since $\partial x_i^{(0)}/\partial p_j^{(0)}=0$.

As to the fact that in this case we will have from (19.9) the parametric equation of a manifold of dimension $2k+1$, we shall give a brief proof of it later.

We note first of all some important properties of the equations (19.9) and (19.10).

Set

$$\left.\begin{aligned} sp_i^{(0)} &= y_i, \\ sp_i &= \pi_i. \end{aligned}\right\} \qquad 19.15)$$

We show that the functions $\xi_i$ and $\pi_i$ depend on $s$ and on $p_i^{(0)}$ only through $y_i$, i.e., that

$$\left.\begin{array}{l}\xi_i\left(s,\ x_0^{(0)},\ x_1^{(0)},\ \ldots,\ x_{2k+1}^{(0)},\ \frac{y_0}{s},\ \frac{y_1}{s},\ \ldots,\ \frac{y_{2k+1}}{s}\right),\\ \pi_i\left(s,\ x_0^{(0)},\ x_1^{(0)},\ \ldots,\ x_{2k+1}^{(0)},\ \frac{y_0}{s},\ \frac{y_1}{s},\ \ldots,\ \frac{y_{2k+1}}{s}\right),\end{array}\right\} \tag{19.16}$$

do not depend on $s$ for given $y_i$ and $x_i^{(0)}$.

Indeed, in place of $s$ and $p_i$ consider the new variables $s_1$ and $p_i^{(1)}$, setting:

$$s=\alpha s_1;\quad p_i=\frac{p_i^{(1)}}{\alpha}. \tag{19.17}$$

Putting these new variables into the system (19.8), we obtain a system of equations for $p_i^{(1)}$ and $x_i$ with independent variable $s_1$. This system turns out to coincide with the system (19.8) and may be obtained from it by a single change of independent variable and of the functions $p_i$. It follows from this that the functions

$$\left.\begin{array}{l}x_i=\xi_i(\alpha s_1,\ x_0^{(0)},\ \ldots,\ x_{2k+1}^{(0)},\ p_0^{(0)},\ \ldots,\ p_{2k+1}^{(0)}),\\ p_i^{(1)}=\alpha p_i=\frac{1}{s_1}\pi_i(\alpha s_1,\ x_0^{(0)},\ x_1^{(0)},\ \ldots\ x_{2k+1}^{(0)},\ p_0^{(0)},\ \ldots\ p_{2k+1}^{(0)})\end{array}\right\} \tag{19.18}$$

also satisfy the system (19.8).

Set

$$p_i^{(1)}\big|_{s_1=0}=p_i^{(0)(1)} \tag{19.19}$$

where $p_i^{(0)(1)}=\alpha p_i^{(0)}$ or $p_i^{(0)}=p_i^{(0)(1)}/\alpha$.

We have obviously:

$$x_i\big|_{s_1=0}=x_i^{(0)}. \tag{19.20}$$

The equation (19.18) may therefore be rewritten as:

$$\left.\begin{array}{l}x_i=\xi_i\left(\alpha s_1,\ x_0^{(0)},\ \ldots,\ x_{2k+1}^{(0)},\ \frac{p_0^{(0)(1)}}{\alpha},\ \ldots,\ \frac{p_{2k+1}^{(0)(1)}}{\alpha}\right),\\ p_i^{(1)}=\frac{1}{s_1}\pi_i\left(\alpha s_1,\ x_0^{(0)},\ \ldots,\ x_{2k+1}^{(0)},\ \frac{p_0^{(0)(1)}}{\alpha},\ \ldots,\ \frac{p_{2k+1}^{(0)(1)}}{\alpha}\right).\end{array}\right\} \tag{19.21}$$

On the other hand, for the same functions $x_i$ and $p_i$ as solutions of the system (19.8) satisfying conditions (19.19) and (19.11), on the basis of the uniqueness theorem we will have:

$$\left.\begin{array}{l}x_i=\xi_i(s_1,\ x_0^{(0)},\ \ldots,\ x_{2k+1}^{(0)},\ p_0^{(0)(1)},\ \ldots,\ p_{2k+1}^{(0)(1)}),\\ p_i^{(1)}=\frac{1}{s_1}\pi_i(s_1,\ x_0^{(0)},\ \ldots,\ x_{2k+1}^{(0)},\ p_0^{(0)(1)},\ \ldots,\ p_{2k+1}^{(0)(1)}).\end{array}\right\} \tag{19.22}$$

The right hand sides of (19.21) and (19.22) are identical for any $\alpha$. Setting $s_1=1$; $\alpha=s$, we obtain our assertion.

Set

$$\xi_i(s, x_0^{(0)}, \ldots, x_{2k+1}^{(0)}, p_0^{(0)}, \ldots, p_{2k+1}^{(0)}) =$$
$$= X_i(x_0^{(0)}, \ldots, x_{2k+1}^{(0)}, y_0, y_1, \ldots, y_{2k+1})$$
$$\pi_i(s, x_0^{(0)}, \ldots, x_{2k+1}^{(0)}, p_0^{(0)}, \ldots, p_{2k+1}^{(0)}) =$$
$$= \Pi_i(x_0^{(0)}, \ldots, x_{2k+1}^{(0)}, y_0, y_1, \ldots, y_{2k+1})$$

We shall show that the equations

$$x_i = X_i(x_0^{(0)}, \ldots, x_{2k+1}^{(0)}, y_0, \ldots, y_{2k+1}) \tag{19.23}$$

express a change of coordinates from $x_i$ to $y_i$ in our space which carries the point $y_i=0$ into $x_i=x_i^{(0)}$, has close to this point a Jacobian different from zero, and is continuous together with derivatives up to order $K$, where $K$ is a sufficiently large number.

By known theorems of the theory of ordinary differential equations, the functions $\pi_i$ and $\xi_i$ have continuous derivatives with respect $x_j^{(0)}$ and $p_j^{(0)}$ up to order $K+1$ inclusive.

We shall consider the functions $x_i$ and $p_i$ as functions of the variable $s$ and the parameters $x_0^{(0)}, x_1^{(0)}, x_2^{(0)}, \cdots, x_{2k+1}^{(0)}$, $p_0^{(0)}, p_1^{(0)}, \cdots, p_{2k+1}^{(0)}$ and consider the derivatives of $x_i$ and $p_j$ with respect to $s$. We shall show that $\lim_{s\to 0} (d^\alpha x_i/ds^\alpha)$ is a homogeneous polynomial of degree $\alpha$ and $\lim_{s\to\infty} (d^\alpha p_i/ds^\alpha)$ is one of degree $\alpha+1$ in the $p_i^{(0)}$.

For the proof we employ once more the system of equations (19.8). Differentiating the equations of the system with respect to $s$ a sequence of times and removing each time from the right hand side the first derivatives, we may express $d^\alpha x_i/ds^\alpha$ and $d^\alpha p_i/ds^\alpha$ in terms of the quantities $x_i$ and $p_i$. We shall show that we shall obtain each time

$$\left.\begin{aligned} \frac{d^\alpha x_i}{ds^\alpha} &= X_i^{(\alpha)}(x_0, x_1, \ldots, x_{2k+1}, p_0, p_1, \ldots, p_{2k+1}), \\ \frac{d^\alpha p_i}{ds^\alpha} &= P_i^{(\alpha)}(x_0, x_1, \ldots, x_{2k+1}, p_0, p_1, \ldots, p_{2k+1}), \end{aligned}\right\} \tag{19.24}$$

where $X_i^{(\alpha)}$ and $P_i^{(\alpha)}$ are polynomials of degree $\alpha$ or $\alpha+1$ in $p_i$.

In order to prove this, we apply the principle of mathematical induction. For $\alpha=1$, our assertion is obvious. Suppose that for some $\alpha$ our assertion has been proved.

Differentiating (19.24) with respect to $s$, we obtain:

$$\frac{d}{ds}\left(\frac{d^\alpha x_i}{ds^\alpha}\right)=\frac{dX_i^{(\alpha)}}{ds}=\sum_{j=0}^{2k+1}\frac{\partial X_i^{(\alpha)}}{\partial x_j}\frac{dx_j}{ds}+\sum_{j=0}^{2k+1}\frac{\partial X_i^{(\alpha)}}{\partial p_j}\frac{dp_j}{ds}=$$

$$=\sum_{j=0}^{2k+1}\frac{\partial X_i^{(\alpha)}}{\partial x_j}\sum_{l=0}^{2k+1}A_{jl}p_l+\sum_{j=0}^{2k+1}\frac{\partial X_i^{(\alpha)}}{\partial p_j}\cdot\frac{1}{2}\sum_{l=0}^{2k+1}\sum_{m=0}^{2k+1}\frac{\partial A_{lm}}{\partial x_j}p_lp_m,$$

from which it is clear that our assertion is still valid for $\alpha+1$.

Passing to the limit for $s\to 0$, we obtain the desired assertion. Analogously we prove the assertion for $p_i$. Since the derivatives in $s$ exist up to order $K+1$, applying Taylor's formula, we obtain:

$$\left.\begin{aligned}x_i&=x_i^{(0)}+s\sum_{j=0}^{2k+1}A_{ij}^{(0)}p_j^{(0)}+\sum_{\alpha=2}^{K}s^\alpha X_i^{(\alpha)}(p_j^{(0)})+R_K^{(i)},\\ p_i&=p_i^{(0)}+\frac{1}{s}\sum_{\alpha=2}^{K}s^{\alpha+1}\prod\nolimits_i^{(\alpha+1)}(p_j^{(0)})+R_K^{1(i)},\end{aligned}\right\}\tag{19.25}$$

where $X_i^{(\alpha)}$, $\prod_i^{(\alpha+1)}$ are polynomials of degree $\alpha$ and $\alpha+1$ in $p_j^{(0)}$. From this it follows that for any $r<K+1$, we may express $x_i$ and $p_i$ in terms of $y_j$ by

$$x_i-x_i^{(0)}=\sum_{j=0}^{2k+1}A_{ij}^{(0)}y_j+\sum_{n=2}^{r+1}X_i^{(n)}(y_j)+R_{r+1}^{(i)},$$

$$p_i=\frac{1}{s}\left\{y_i+\sum_{n=2}^{r+1}\prod\nolimits_i^{(n)}(y_j)+R_{r+1}^{1(i)}\right\},$$

where the derivatives of order $\alpha$ of $R_{r+1}^{(i)}$ and $R_{r+1}^{1(i)}$ taken in $p$, vanish for $s=0$ like $s^{k+2-\alpha}$. A simple calculation shows that thereby the derivatives of $R_{r+1}^{(i)}$ and $R_{r+1}^{1(i)}$ in $y_j$ up to order $r+1$ inclusive vanish at the origin of coordinates. As a result, the derivatives of $x_i$ with respect to $y_j$ up to the order of $r+1$ exist and are everywhere continuous.

It is obvious that the Jacobian

$$\left.\frac{D(x_0,x_1,\dots,x_{2k+1})}{D(y_0,y_1,\dots,y_{2k+1})}\right|_{s=0}\neq 0,$$

and on the basis of the implicit function theorem, there exists a domain with center at $x_0^{(0)},\cdots,x_{2k+1}^{(0)}$ on which an inverse is defined: $y_0,y_1,\cdots,y_{2k+1}$ are one-valued functions of the variables $x_0,x_1,\cdots,x_{2k+1}$.

On this domain, we obtain

$$y_i=\sum_{j=0}^{2k+1}H_{ij}^{(0)}(x_j-x_j^{(0)})+\sum_{n=2}^{r+1}Y_i^{(n)}(x_j-x_j^{(0)})+R,\tag{19.26}$$

where $H_{ij}^{(0)}$ is the inverse matrix to $A_{ij}^{(0)}$ and $Y_i^{(n)}$ is a polynomial of degree $n$ in $(x_j - x_j^{(0)})$.

Thus it is established that the $(2k+2)$-dimensional manifold of the $y_i$ corresponds to the $(2k+2)$-dimensional manifold of all possible values of $s$ and of values $p_0^{(0)}, p_1^{(0)}, \cdots, p_{2k+1}^{(0)}$, bound by conditions (19.12), whereby to each value of $y_i$ corresponds a value of $(s, p_j^{(0)})$ and conversely.

The equation $A(p_i^{(0)}) = 0$ in the variables $y_i$ may be written in the form

$$A_0(y_i) = 0 \tag{19.27}$$

and consequently represents the equation of a cone, i.e., a manifold of $2k+1$ dimensions.

Thus we have proved that conditions (a), (b), and (c) are satisfied as given above and consequently (19.27) is the equation of the characteristic conoid.

By our assumption that the equation is of normal hyperbolic type, the characteristic cone will divide the whole space into three parts: the exterior, the upper interior, and the lower interior of the cone. Any direction at an arbitrary point of the space will either point into the interior of the cone and will by analogy with the ordinary case be called time-like or it will point into the exterior of the cone and will in that case be called space-like.

In the analytic definition, those directions $\vec{l}$ will be space-like for which $A(l_0, \cdots, l_{2k+1}) < 0$ and time-like for which $A(l_0, \cdots, l_{2k+1}) > 0$.

On a circle around the given point $(x_0^{(0)}, \cdots, x_{2k+1}^{(0)})$ we may make a transformation of coordinates carrying the matrix $\| A_{ij}^{(0)} \|$ into canonical form. Suppose that this transformation is

$$x_i - x_i^{(0)} = \sum_j \alpha_{ij} z_j; \quad z_j = \sum_j \gamma_{ij} (x_i - x_i^{(0)}),$$

where $\alpha_{ij}$ and $\gamma_{ij}$ are bounded functions continuously differentiable $K+1$ times of the variables $x_0^{(0)}, x_1^{(0)}, \cdots, x_{2k+1}^{(0)}$ for which the determinant

$$|\alpha_{ij}| > h,$$

where $h$ is a number not depending on $x_0^{(0)}, x_1^{(0)}, \cdots, x_{2k+1}^{(0)}$ and $K$ is a sufficiently large number.

Then

$$\sum_{i=0}^{2k+1} \sum_{j=0}^{2k+1} A_{ij} p_i p_j \Big|_{x_0^{(0)}, x_1^{(0)}, \cdots, x_{2k+1}^{(0)}} =$$

$$= \sum_{l=0}^{2k+1} \sum_{m=0}^{2k+1} \Big\{ \sum_{i=0}^{2k+1} \sum_{j=0}^{2k+1} A_{ij} \gamma_{il} \gamma_{jm} \Big\} q_l q_m,$$

and, as a result, by the assumption on the canonical transformation, we must have

$$\sum_{i=0}^{2k+1} \sum_{j=0}^{2k+1} A_{ij} \gamma_{il} \gamma_{jm} = \begin{cases} 0 & l \neq m \\ -1 & l = m \neq 0 \\ 1 & l = m = 0. \end{cases}$$

We shall assume that such a transformation has already been carried through, and that $y_i$ is the corresponding local system of coordinates, giving (19.27) the form

$$y_0^2 - \sum_{i=1}^{2k+1} y_i^2 = 0 \tag{19.28}$$

or if we set

$$\rho = \sqrt{\sum_{i=1}^{2k+1} y_i^2},$$

$$y_0^2 - \rho^2 = 0, \tag{19.29}$$

i.e., in the coordinates $y_i$ the characteristic conoid turns into a right circular cone. Such a transformation can be made at each point of the space.

In the case of variable coefficients it is necessary, on the other hand, to note the following important circumstance. In the solution of the Cauchy problem for the wave equation, the bicharacteristics were straight lines and therefore our solutions of the characteristic equation could be extended to arbitrary time $t$. In the case of an equation with variable coefficients, the field of bicharacteristics may have some kind of singularity (for example a focus) and therefore we may not obtain the right formula for the conoid in the whole space, but obtain it only in some band not containing singularities of the field of bicharacteristics. The width of this band may be estimated from the coefficients of the derivatives of second order, but this estimate is not of interest to us.

2. THE CHARACTERISTIC CONOID. We transform the equation (19.1) into the coordinates $y_i$. In the new coordinates our equation takes the form:

$$\sum \sum \tilde{A}_{ij} \frac{\partial^2 u}{\partial y_i \partial y_j} + \sum \tilde{B}_i \frac{\partial u}{\partial y_i} + \tilde{C} u = \tilde{F}. \tag{19.30}$$

For this, the cone with the equation:

$$G(y_0, y_1, \ldots, y_{2k+1}) \equiv y_0 + \rho = 0, \tag{19.31}$$

where

$$\rho = \sqrt{\sum_{i=1}^{2k+1} y_i^2}$$

will be the characteristic cone, and the lines

$$y_i = \alpha_i y_0,$$

where

$$\sum \alpha_i^2 = 1$$

will be the bicharacteristics. It is useful to consider what consequences follow from these facts. On the characteristic cone:

$$\left.\begin{aligned} q_i &= \frac{\partial G}{\partial y_i} = \frac{y_i}{\rho} = -\frac{y_i}{y_0} = -\alpha_i \quad (i \neq 0), \\ q_0 &= \frac{\partial G}{\partial y_0} = 1. \end{aligned}\right\} \tag{19.32}$$

We substitute the solutions of the system of equations:

$$\frac{dy_i}{\frac{1}{2}\frac{\partial \widetilde{A}}{\partial q_i}} = \frac{-dq_i}{\frac{1}{2}\frac{\partial \widetilde{A}}{\partial y_i}} = ds,$$

where

$$\widetilde{A} = \sum_{i=0}^{2k+1} \sum_{j=0}^{2k+1} \widetilde{A}_{ij} q_i q_j.$$

If we set:

$$\frac{dy_0}{ds} = \varphi(s), \tag{19.33}$$

we write our system in the form:

$$\begin{aligned} &\frac{-\alpha_l}{\frac{1}{2}\sum_{i=0}^{2k+1}\sum_{j=0}^{2k+1}\frac{\partial \widetilde{A}_{ij}}{\partial y_l} q_i q_j} = \frac{\alpha_i \varphi(s)}{A_{0i} - \sum_{j=1}^{2k+1} \widetilde{A}_{ij}\alpha_j} = \\ &= \frac{\varphi(s)}{A_{00} - \sum_{j=1}^{2k+1} \widetilde{A}_{j0}\alpha_j} = 1. \end{aligned} \tag{19.34}$$

Thus,

$$\widetilde{A}_{00} - \sum_{i=1}^{2k+1} \widetilde{A}_{i0}\alpha_i = \varphi(s), \tag{19.35}$$

$$\widetilde{A}_{0i}-\sum_{j=1}^{2k+1}\widetilde{A}_{ij}\alpha_j=\alpha_i\varphi(s). \tag{19.36}$$

Multiplying the second equality by $\alpha_i$ and adding, we will have:

$$\sum_{i=1}^{2k+1}\widetilde{A}_{0i}\alpha_i-\sum_{i=1}^{2k+1}\sum_{j=1}^{2k+1}\widetilde{A}_{ij}\alpha_i\alpha_j=\sum_{i=1}^{n}\alpha_i^2\varphi(s)=\varphi(s). \tag{19.37}$$

From this, among other things, it follows that on the surface of the characteristic cone

$$\widetilde{A}_{00}-\sum_{i=1}^{2k+1}\sum_{j=1}^{2k+1}\widetilde{A}_{ij}\alpha_i\alpha_j=2\varphi(s) \tag{19.38}$$

or

$$\sum_{i=1}^{2k+1}\sum_{j=1}^{2k+1}\widetilde{A}_{ij}\alpha_i\alpha_j=\widetilde{A}_{00}-2\varphi(s). \tag{19.39}$$

3. EQUATIONS IN CANONICAL COORDINATES. We pass to the study of the Cauchy problem for the equation (19.1). Let it be required to find solutions of that equation which satisfy the conditions:

$$\left.\begin{aligned} u\big|_{x_0=0}&=u_0(x_1,\ldots,x_{2k+1}),\\ \frac{\partial u}{\partial x_0}\Big|_{x_0=0}&=u_1(x_1,\ldots,x_{2k+1}).\end{aligned}\right\} \tag{19.40}$$

We make another important assumption. We assume that at each point of the part of space being considered, $A_{00}\geqq m>0$, $A_{ii}\leqq -m<0$ $(i\neq 0)$. This is equivalent to the requirement that the direction of the $x_0$ axis should be time-like, and that all the remaining coordinate axes should be space-like.

We now introduce a new variable, setting:

$$t=y_0+\rho.$$

For simplicity in the argument, we shall in the following designate by $\partial/\partial y_i$ the partial derivative with respect to $y_i$ taken on the surface $y_0=$ const., i.e., for constant $y_0$, and by $D/Dy_i$ the partial derivative with respect to $y_i$ taken on the surface $t=$const. Then we obtain:

$$\frac{\partial}{\partial y_i}=\frac{D}{Dy_i}+\frac{y_i}{\rho}\frac{D}{Dt};\quad \frac{\partial}{\partial y_0}=\frac{D}{Dt};\quad \frac{\partial^2}{\partial y_0^2}=\frac{D^2}{Dt^2};$$

$$\frac{\partial^2}{\partial y_i\,\partial t}=\frac{D^2}{Dy_i\,Dt}+\frac{y_i}{\rho}\frac{D^2}{Dt^2};$$

$$\frac{\partial^2}{\partial y_i\,\partial y_j} = \frac{D^2}{Dy_i\,Dy_j} - \frac{y_i y_j}{\rho^3}\frac{D}{Dt} + \frac{y_i}{\rho}\frac{D^2}{Dy_i\,Dt} + \\ + \frac{y_j}{\rho}\frac{D^2}{Dy_j\,Dt} + \frac{y_i y_j}{\rho^2}\frac{D^2}{Dt^2}; \quad (i \neq j);$$

$$\frac{\partial^2}{\partial y_i^2} = \frac{D^2}{Dy_i^2} + \left(\frac{1}{\rho} - \frac{y_i^2}{\rho^3}\right)\frac{D}{Dt} + \frac{2y_i}{\rho}\frac{D^2}{Dy_i\,Dt} + \frac{y_i^2}{\rho^2}\frac{D^2}{Dt^2}.$$

Setting these values in our equation, we will have:

$$\sum_{i=1}^{2k+1}\sum_{j=1}^{2k+1}\tilde{A}_{ij}\frac{D^2u}{Dy_i\,Dy_j} + 2\sum_{i=1}^{2k+1}\sum_{j=1}^{2k+1}\tilde{A}_{ij}\frac{y_i}{\rho}\frac{D^2u}{Dy_j\,Dt} + \\ + \sum_{i=1}^{2k+1}\sum_{j=1}^{2k+1}\tilde{A}_{ij}\frac{y_i y_j}{\rho^2}\frac{D^2u}{Dt^2} + 2\sum_{i=1}^{2k+1}\tilde{A}_{i0}\frac{D^2u}{Dy_i\,Dt} + \\ + 2\sum_{i=1}^{2k+1}\tilde{A}_{0i}\frac{y_i}{\rho}\frac{D^2u}{Dt^2} + \tilde{A}_{00}\frac{D^2u}{Dt^2} + \left[\sum_{i=1}^{2k+1}\sum_{j=1}^{2k+1}\left(-\tilde{A}_{ij}\frac{y_i y_j}{\rho^3}\right) + \right. \\ \left. + \sum_{i=1}^{2k+1}\tilde{A}_{ii}\frac{1}{\rho} + \sum_{i=1}^{2k+1}\tilde{B}_i\frac{y_i}{\rho} + \tilde{B}_0\right]\frac{Du}{Dt} + \sum_{i=1}^{2k+1}\tilde{B}_i\frac{Du}{Dy_i} + \tilde{C}u = \tilde{F}$$

or

$$\left\{\sum_{i=1}^{2k+1}\sum_{j=1}^{2k+1}\tilde{A}_{ij}\frac{D^2u}{Dy_i\,Dy_j} + \sum_{i=1}^{2k+1}\tilde{B}_i\frac{Du}{Dy_i} + \tilde{C}u\right\} + \\ + \left\{2\sum_{i=1}^{2k+1}\left[\sum_{j=1}^{2k+1}\tilde{A}_{ij}\frac{y_i}{\rho} + \tilde{A}_{i0}\right]\frac{D}{Dy_i}\left(\frac{Du}{Dt}\right) + \right. \\ \left. + \left[\sum_{i=1}^{2k+1}\tilde{A}_{ii}\frac{1}{\rho} - \sum_{i=1}^{2k+1}\sum_{j=1}^{2k+1}\tilde{A}_{ij}\frac{y_i y_j}{\rho^3} + \sum_{i=1}^{2k+1}\tilde{B}_i\frac{y_i}{\rho} + \tilde{B}_0\right]\frac{Du}{Dt}\right\} + \\ + \left\{\sum_{i=1}^{2k+1}\sum_{j=1}^{2k+1}\tilde{A}_{ij}\frac{y_i y_j}{\rho^2} + 2\sum_{i=1}^{2k+1}\tilde{A}_{0i}\frac{y_i}{\rho} + \tilde{A}_{00}\right\}\frac{D^2u}{Dt^2} = \tilde{F}, \qquad (19.41)$$

which we write more briefly in the form

$$L^{(0)}u + M^{(0)}\frac{Du}{Dt} + J\frac{D^2u}{Dt^2} = \tilde{F}, \qquad (19.42)$$

where by $L^{(0)}$ and $M^{(0)}$ we denote the operators appearing, respectively, in the first and second curly brackets on the left side of (19.41), and by $J$ the third curly bracket in (19.41). As is easily seen by virtue of (19.32),

(19.35), and (19.39), on the characteristic cone $t=0$, we have $J|_{t=0}=0$, and (19.42) takes the form

$$L^{(0)}u+M^{(0)}\frac{Du}{Dt}=\tilde{F}.$$

We carry out still one more change of coordinates, introducing polar coordinates instead of $y_1, y_2, \cdots, y_{2k+1}$.

4. THE BASIC OPERATORS $M^{(0)}$ AND $L^{(0)}$ IN POLAR COORDINATES. We will define the position of a point by the coordinates $\rho$, where

$$\rho=\sqrt{\sum_{i=1}^{2k+1} y_i^2},$$

and the unit vector $\vec{\lambda}$.

It is necessary for us to investigate more closely differential operators in these variables.

In order to calculate any differential operator on the unit sphere, we choose some coordinate system on the surface of this sphere. For us the choice of the system is indifferent, and therefore we may, for example, choose for each point of the sphere its own coordinate system regular in a neighborhood of this point.

We may, for example, consider for this purpose polar coordinates.

The system of polar coordinates on the unit sphere is given by the equations:

$$\left.\begin{aligned} z_{2k+1} &= \rho\cos\vartheta_{2k},\\ z_{2k} &= \rho\sin\vartheta_{2k}\cos\vartheta_{2k-1}\\ &\cdots\cdots\cdots\cdots\\ z_3 &= \rho\sin\vartheta_{2k}\sin\vartheta_{2k-1}\ \ldots\ \sin\vartheta_3\cos\vartheta_2,\\ z_2 &= \rho\sin\vartheta_{2k}\sin\vartheta_{2k-1}\ \ldots\ \sin\vartheta_3\sin\vartheta_2\cos\vartheta_1,\\ z_1 &= \rho\sin\vartheta_{2k}\sin\vartheta_{2k-1}\ \ldots\ \sin\vartheta_3\sin\vartheta_2\sin\vartheta_1, \end{aligned}\right\} \tag{19.43}$$

where $-\pi<\vartheta_1<\pi$; $0<\vartheta_i<\pi$; $i=1,2,3,\cdots,2k$, and $z_i$ is an arbitrary Cartesian coordinate system obtained for the $y_j$ by a rotation of axes, i.e., by the affine orthogonal transformation:

$$y_i=\sum_{j=1}^{2k+1}\gamma_{ij}z_j.$$

It is useful to note the formula

$$\frac{D(y_1, y_2, \ldots, y_{2k+1})}{D(\rho, \vartheta_1, \vartheta_2, \ldots, \vartheta_{2k})} = \frac{D(z_1, z_2, \ldots, z_{2k+1})}{D(\rho, \vartheta_1, \vartheta_2, \ldots, \vartheta_{2k})} =$$
$$= \rho^{2k} \sin^{2k-1} \vartheta_{2k} \sin^{2k-2} \vartheta_{2k-1}, \ldots, \sin^2 \vartheta_3 \sin \vartheta_2. \tag{19.44}$$

From this, there follows immediately

$$d\Omega = \rho^{2k} \sin^{2k-1} \vartheta_{2k} \sin^{2k-2} \vartheta_{2k-1} \cdots$$
$$\sin^2 \vartheta_3 \sin \vartheta_2 \, d\rho \, d\vartheta_1 \ldots d\vartheta_{2k}. \tag{19.45}$$

If we take into consideration that the surface $\rho = \text{const.}$ is a sphere, the line $\vartheta_1 = \text{const.}, \cdots, \vartheta_{2k} = \text{const.}$ is a radius of this sphere, and that $\partial\rho/\partial n = 1$, where $\vec{n}$ is the exterior normal to the sphere, we conclude that the surface element of the sphere is:

$$dS = \rho^{2k} \sin^{2k-1} \vartheta_{2k} \sin^{2k-2} \vartheta_{2k-1}, \ldots, \sin \vartheta_2 \, d\vartheta_1 \, d\vartheta_2, \ldots, d\vartheta_{2k}. \tag{19.46}$$

The system of equations (19.43) in the form when they are solved for the $\vartheta_i$, will be:

$$z_1 = \rho_1; \quad \sqrt{z_1^2 + z_2^2} = \rho_2; \quad \ldots, \quad \sqrt{\sum_{i=1}^{8} z_i^2} = \rho_8; \quad \ldots,$$
$$\sqrt{\sum_{i=1}^{2k} z_i^2} = \rho_{2k}; \quad \sqrt{\sum_{i=1}^{2k+1} z_i^2} = \rho; \tag{19.47}$$
$$\vartheta_1 = (\operatorname{sign} z_1) \arccos \frac{z_2}{\rho_2}, \quad \vartheta_2 = \arccos \frac{z_3}{\rho_3}, \quad \ldots, \quad \vartheta_{2k-1} = \arccos \frac{z_{2k}}{\rho_{2k}},$$
$$\vartheta_{2k} = \arccos \frac{z_{2k+1}}{\rho}.$$

In the following, for convenience in calculating any operator at the point $\vec{\lambda}_0$ on the sphere, we shall choose the axes $z_1, z_2, \cdots, z_{2k+1}$ in such a fashion that the point $\vec{\lambda}_0$ under consideration falls on the $z_1$ axis, i.e., that we have at this point $z_1 = \rho$, $z_2 = z_3 = \cdots = z_{2k} = 0$.

A function $\Psi$ given on the sphere is said to be $s$ times continuously differentiable at the point $\vec{\lambda}_0$ if it has at that point continuous derivatives up to order $s$, inclusive, with respect to all the coordinates $\vartheta_1, \vartheta_2, \cdots, \vartheta_{2k}$ for the choice of coordinate system as given above.

A function which is continuously differentiable $s$ times at every point of the sphere is said to be continuously differentiable on the sphere.

A linear differential operator of order $s$ on the sphere is said to be continuous at the point $\vec{\lambda}_0$ of the sphere if it consists of a linear combination of derivatives up to order $s$ inclusive with respect to the variables $\vartheta_1, \vartheta_2, \cdots, \vartheta_{2k}$ with coefficients continuous on a neighborhood of $\vec{\lambda}_0$. If it is continuous on a neighborhood of each point, it is said to be continuous on the sphere.

An operator of order $s$ may be applied to any $s$-times continuously differentiable function.

We will call an operator $Lv$ defined on the unit sphere $r$-times differentiable if its coefficients at any point are $r$-times continuously differentiable. If an operator of order $l$ which is $r$-times differentiable is applied to a function which is $m$-times differentiable, where $l<m<l+r$, then the result is a function which is $(m-l)$-times differentiable.

After these remarks, we pass to the calculation of the operators $M^{(0)}$ and $L^{(0)}$ in polar coordinates.

To begin, we calculate the form of the operator $M^{(0)}$.

It is easy to see that

$$\frac{Dv}{Dy_i}=\sum_{j=1}^{2k+1}\varepsilon_{ij}\frac{Dv}{Dz_j}. \tag{19.48}$$

In its turn

$$\frac{Dv}{Dz_j}=\sum_{s=j-1}^{2k+1}\frac{Dv}{D\vartheta_s}\frac{D\vartheta_s}{Dz_j}+\frac{Dv}{D\rho}\frac{D\rho}{Dz_j}=\frac{z_j}{\rho}\frac{\partial v}{\partial\rho}+$$

$$+\frac{\partial v}{\partial\vartheta_{j-1}}\frac{1}{1+\left(\frac{\rho_{j-1}}{z_j}\right)^2}\left(-\frac{\rho_{j-1}}{z_j^2}\right)+\sum_{s=j}^{2k+1}\frac{\partial v}{\partial\vartheta_s}\frac{1}{1+\left(\frac{\rho_s}{z_{s+1}}\right)^2}\left[\frac{z_j}{z_{s+1}\rho_s}\right]=$$

$$=\frac{z_j}{\rho}\frac{\partial v}{\partial\rho}-\frac{\partial v}{\partial\vartheta_{j-1}}\frac{\rho_{j-1}}{\rho_j^2}+\sum_{s=j}^{2k+1}\frac{\partial v}{\partial v_s}\frac{z_{s+1}z_j}{\rho_s\rho_{s+1}^2}.$$

Replacing all the $z$'s and $\rho$ by their expressions and taking into consideration that at the point concerned all the $\rho_k$ equal $\rho$, we may give $M^{(0)}$ the form:

$$M^{(0)}=G^{(0)}\frac{\partial v}{\partial\rho}+\frac{1}{\rho}\Lambda^{(0)}v, \tag{19.49}$$

where $G^{(0)}$ is a function of the variables $t,\rho,\vartheta_1,\cdots,\vartheta_{2k}$ defined on the whole $\vartheta$-sphere and for all $\rho$ in the interval $0<\rho<M$, continuously differentiable with respect to all its variables, while $\Lambda$ is a differential operator of first order on the unit sphere which is sufficiently often continuously differentiable.

We also examine the value of the operator $M^{(0)}$ for $t=0$, i.e., on the surface of the characteristic cone. For this, we have $y_j/\rho=-\alpha_j$, and by virtue of the equations (19.36):

$$2\sum_{i=1}^{2k+1}\Big(\sum_{j=1}^{2k+1}\widetilde{A}_{ij}\frac{y_j}{\rho}+\widetilde{A}_{i0}\Big)\frac{Dv}{Dy_i}=2\varphi(s)\sum_{i=1}^{2k+1}\alpha_i\frac{Dv}{Dy_i}=$$

$$=-2\varphi(s)\sum_{i=1}^{2k+1}\frac{y_i}{\rho}\frac{Dv}{Dy_i}=-2\varphi(s)\frac{\partial v}{\partial\rho},$$

and by (19.39)

$$\sum_{i=1}^{2k+1}\widetilde{A}_{ij}\frac{1}{\rho}-\sum_{i=1}^{2k+1}\sum_{j=i}^{2k+1}\widetilde{A}_{ij}\frac{y_iy_j}{\rho^3}+\sum_{i=1}^{2k+1}\widetilde{B}_i\frac{y_i}{\rho}+\widetilde{B}_0=$$

$$=\Big[\sum_{i=1}^{2k+1}\widetilde{A}_{ii}-\widetilde{A}_{00}+2\varphi(s)\Big]\frac{1}{\rho}+\sum_{i=1}^{2k+1}\widetilde{B}_i\frac{y_i}{\rho}+\widetilde{B}_0.$$

Therefore

$$M^0v\,|_{t=0}=$$

$$=-2\varphi(s)\left\{\frac{\partial v}{\partial\rho}+\left[\frac{1}{\rho}\left(\frac{\widetilde{A}_{00}-\sum\limits_{i=1}^{2k+1}A_{ii}}{2\varphi(s)}-1\right)-\frac{\Big(\widetilde{B}_0+\sum\limits_{i=1}^{2k+1}\widetilde{B}_i\frac{y_i}{\rho}\Big)}{2\varphi(s)}\right]v\right\}.$$

We show now that

$$[\widetilde{A}_{00}-\sum_{i=1}^{2k+1}\widetilde{A}_{ii}]_{\rho=0}=(2k+2)\,\varphi(0). \tag{19.50}$$

We consider for this purpose the values

$$\widetilde{A}_{ij}\,|_{\rho=0}.$$

Obviously, the following equalities hold:

$$\widetilde{A}_{00}\,|_{\rho=0}-\sum_{i=1}^{2k+1}\alpha_i\widetilde{A}_{i0}\,|_{\rho=0}=\varphi(0);$$

$$\widetilde{A}_{0i}\,|_{\rho=0}-\sum_{j=1}^{2k+1}\alpha_j\widetilde{A}_{ij}\,|_{\rho=0}=\alpha_i\varphi(0),$$

where $\alpha_i$ are completely arbitrary numbers such that $\sum\alpha_i^2=1$. If we set

$$\alpha_l=\pm 1;\quad \alpha_i=0;\quad i\neq l,$$

we see that

$$\widetilde{A}_{00}\,|_{\rho=0}=\pm\widetilde{A}_{0l}\,|_{\rho=0}+\varphi(0),$$

from which it follows that

$$\tilde{A}_{00}|_{\rho=0} = \varphi(0); \quad \tilde{A}_{0l}|_{\rho=0} = 0.$$

Further, putting

$$\alpha_l = \sin\omega$$
$$\alpha_m = \pm\cos\omega, \text{ where } l \neq m,$$
$$\alpha_i = 0, \qquad i \neq l, \quad i \neq m,$$

we will have

$$\sin\omega\,[\tilde{A}_{ll}|_{\rho=0} + \varphi(0)] \pm \cos\omega\,\tilde{A}_{lm}|_{\rho=0} = 0,$$

from which we obtain, using the arbitrariness of $\omega$,

$$\tilde{A}_{lm}|_{\rho=0} = 0; \quad \tilde{A}_{ll}|_{\rho=0} = -\varphi(0).$$

From these equalities, formula (19.50) follows.

Furthermore from the existence of continuous derivatives with respect to the variables $y_0, y_1, \cdots, y_{2k-1}$ for the coefficients $\tilde{A}_{ij}$, it follows that

$$\tilde{A}_{ij} = \tilde{A}_{ij}|_{\rho=0} + \rho A_{ij}^{(1)} + \frac{\rho^2}{2} A_{ij}^{(2)} + \cdots + \frac{\rho^k}{k!} A_{ij}^{(k)} + R_{ij}^{(k)},$$

where $R_{ij}^{(k)}$ the remainder vanishes together with its derivatives up to order $K$ inclusive for $\rho=0$.

Analogously

$$\frac{\tilde{A}_{ij}}{\varphi(s)} = \frac{\tilde{A}_{ij}|_{\rho=0}}{\varphi(0)} + \sum_{k=1}^{k} \rho^k \tilde{A}_{ij}^{(k)} + P_{ij}^{(k)}.$$

Considering this decomposition, we finally obtain for the operator $M^{(0)}|_{t=0}$:

$$M^{(0)}v|_{t=0} = -2\varphi(s)\left[\frac{\partial v}{\partial \rho} + \left(\frac{k}{\rho} + G\right)v\right], \tag{19.51}$$

where the function $G(\vec{\lambda}, \rho)$ is bounded and has bounded derivatives of sufficiently high order.

We shall now express the operator $L^{(0)}$ in the same coordinates.

We will have after an obvious computation:

$$L^{(0)}v = Q^{(0)} \frac{\partial^2 v}{\partial \rho^2} + \frac{1}{\rho} R^{(0)} \frac{\partial v}{\partial \rho} + \frac{1}{\rho^2} S^{(0)} v, \tag{19.52}$$

where $Q$ is a function sufficiently often continuously differentiable on the sphere, $R$ is an operator of first order on the sphere, and $S$ is an operator of second order on the sphere which is differentiable sufficiently many times.

It is of interest to calculate the value of the operator $L^{(0)}v$ for $t=0$. For this purpose, we represent the operator $L^{(0)}v|_{t=0}$ in the form:

$$L^{(0)}v|_{t=0}=\left\{\sum_{i=1}^{2k+1}\frac{D^2v}{Dy_i^2}+\left[\rho\sum_{i=1}^{2k+1}\sum_{j=1}^{2k+1}\tilde{C}_{ij}\frac{D^2v}{Dy_iDy_j}+\right.\right.$$
$$\left.\left.+\sum_{i=1}^{2k+1}\tilde{B}_i\frac{Dv}{Dy_i}+\tilde{C}v\right]\right\}\varphi(s). \tag{19.53}$$

Such a representation obviously will be regular, where $\tilde{A}_{ij}$ will be functions continuously differentiable in $\rho$ and $\vartheta$.

We obtain, obviously:

$$\sum_{i=1}^{2k+1}\frac{D^2v}{Dy_i^2}=\sum_{i=1}^{2k+1}\frac{\partial^2v}{\partial\rho^2}+\frac{2k}{\rho}\frac{\partial v}{\partial\rho}+\frac{1}{\rho^2}\Omega v,$$

where $\Omega$ is the Laplace operator on the surface. We have:

$$L^{(0)}v|_{t=0}=\left\{\frac{\partial^2v}{\partial\rho^2}[1+\rho Q_1(\rho,\bar{\lambda})]+\frac{1}{\rho}[2k+\rho Q_2(\rho,\bar{\lambda})]\frac{\partial v}{\partial\rho}+\right.$$
$$\left.+\frac{1}{\rho^2}[\Omega+\rho Q_3(\rho,\vec{\lambda})]v\right\}\varphi(s), \tag{19.54}$$

5. THE SYSTEM OF BASIC RELATIONS ON THE CONE. Let $\Lambda^{(0)}$ be an operator of first order in the variables $y_1,\cdots,y_{2k+1}$, or, equivalently, in the variables $\rho,\vartheta_1,\cdots,\vartheta_{2k}$, the coefficients of which depend on the variables $t$:

$$\Lambda^{(0)}v=\sum_{i=1}^{2k+1}\sum_{j=1}^{2k+1}\tilde{A}_{ij}\frac{D^2v}{Dy_iDy_j}+\sum_{i=1}^{2k+1}\tilde{B}_i\frac{Dv}{Dy_i}+\tilde{C}v.$$

We shall denote by the symbol $d^l\Lambda/dt^l=\Lambda^{(l)}$ the operator of the form:

$$\Lambda^{(l)}v=\sum_{i=1}^{2k+1}\sum_{j=1}^{2k+1}\frac{D^l\tilde{A}_{ij}}{Dt^l}\frac{D^2v}{Dy_iDy_j}+\sum_{i=1}^{2k+1}\frac{D^l\tilde{B}_i}{Dt^l}\frac{Dv}{Dy_i}+\frac{d^l\tilde{C}}{dt^l}v, \tag{19.55}$$

the coefficients of which are the derivatives of order $l$ of the coefficients of the operator $\Lambda^{(0)}$.

We form the operators $M^{(l)}$ and $L^{(l)}$ by the rule written above and construct their expressions in polar coordinates.

By calculations analogous to those already made, we obtain:

$$\left.\begin{aligned} M^{(l)} &= G^{(l)} \frac{\partial v}{\partial \rho} + \frac{1}{\rho} \Lambda^{(l)} v; \\ L^{(l)} &= Q^{(l)} \frac{\partial^2 v}{\partial \rho^2} + \frac{1}{\rho} R^{(l)} \frac{\partial v}{\partial \rho} + \frac{1}{\rho^2} S^{(l)} v. \end{aligned}\right\} \tag{19.56}$$

The coefficients $G^{(l)}$ and $Q^{(l)}$ are functions continuously differentiable on the sphere, $\Lambda^{(l)}$ and $R^{(l)}$ are operators of first order on the sphere, and $S^{(l)}$ is an operator of second order on the sphere, differentiable sufficiently many times.

We return now to equation (19.42).

Differentiating it $l$-times with respect to $t$ and setting $t=0$, we obtain (if we take into account that $J|_{t=0}=0$):

$$\sum_{r=0}^{l} \frac{l!}{r!\,(l-r)!} L^{(l-r)} \frac{D^r u}{Dt^r} + \sum_{r=0}^{l} \frac{l!}{r!\,(l-r)!} M^{(l-r)} \frac{D^{r+1} u}{Dt^{r+1}} + \\ + \sum_{r=0}^{l-1} \frac{l!}{r!\,(l-r)!} \frac{D^{l-r} J}{Dt^{l-r}} \frac{D^{r+2} u}{Dt^{r+2}} = \frac{D^l \widetilde{F}}{Dt^l} \tag{19.57}$$

and introducing the notation

$$\left.\begin{aligned} D_0^{(l)} v &= L^{(l)} v; \quad D_1^{(l)} v = l L^{(l-1)} v + M^{(l)} v; \\ D_{l+1}^{(l)} v &= M^{(0)} v + \frac{DJ}{Dt} v; \\ D_r^{(l)} v &= \frac{l!}{r!\,(l-r)!} L^{(l-r)} v + \frac{l!}{(r-1)!\,(l-r+1)!} M^{(l-r+1)} v + \\ &+ \frac{l!}{(r-2)!\,(l-r+2)!} \frac{D^{l-r+2} J}{Dt^{l-r+2}}; \qquad 2 \leqslant r \leqslant l \end{aligned}\right\} \tag{19.58}$$

we rewrite equation (19.57) in the form:

$$\sum_{r=0}^{l+1} D_r^{(l)} \frac{D^r u}{Dt^r} \equiv M_{k-l} \left( u, \frac{Du}{Dt}, \ldots, \frac{D^{l+1} u}{Dt^{l+1}} \right) = \frac{D^l \widetilde{F}}{Dt^l}. \tag{19.59}$$

We set now

$$\left. \frac{D^r u}{Dt^r} \right|_{t=0} = u_r. \tag{19.60}$$

In this notation equation (19.59) may be rewritten in the form:

$$M_{k-l}(u_0, u_1, \ldots, u_{l+1}) \equiv \sum_{r=0}^{l+1} D_r^{(l)} u_r = \frac{\partial^l \widetilde{F}}{\partial t^l}. \tag{19.61}$$

On the right hand side of (19.61), there obviously appear known quantities.

§20. **The Cauchy problem for linear equations with smooth coefficients.**

1. THE OPERATORS ADJOINT TO THE OPERATORS OF THE BASIC SYSTEM. Given some system of $m$ linear operators on $l$ functions in the $(2k+1)$-dimensional space of $\rho, \vartheta_1, \cdots, \vartheta_{2k}$ with the volume element

$$d\Omega = \rho^{2k}\, d\rho\, dS = \rho^{2k}\varkappa\, d\rho\, d\vartheta_1\ \ d\vartheta_2,\ \ldots,\ \ d\vartheta_{2k}, \tag{20.1}$$

where $dS$ is the surface element on the unit sphere and $\kappa$ is a variable factor equal to

$$\varkappa = \sin^{2k-1}\vartheta_{2k}\sin^{2k-2}\vartheta_{2k-1}\ \ldots\ \sin^2\vartheta_3\sin\vartheta_2. \tag{20.2}$$

The operator $M_j^{(s)}$ may be displayed in the form of a rectangular matrix $M$ having $m$ rows and $l$ columns.

$$M = \begin{Vmatrix} M_1^{(1)}, & M_1^{(2)}, & \ldots, & M_1^{(s)}, & \ldots, & M_1^{(l)} \\ M_2^{(1)}, & M_2^{(2)}, & \ldots, & M_2^{(s)}, & \ldots, & M_2^{(l)} \\ \cdot & \cdot & \cdot & \cdot & \cdot & \cdot \\ M_j^{(1)}, & M_j^{(2)}, & \ldots, & M_j^{(s)}, & \ldots, & M_j^{(l)} \\ \cdot & \cdot & \cdot & \cdot & \cdot & \cdot \\ M_m^{(1)}, & M_m^{(2)}, & \ldots, & M_m^{(s)}, & \ldots, & M_m^{(l)} \end{Vmatrix}.$$

Let $M_j^{(s)*}$ denote the operator adjoint to the operator $M_j^{(s)}$, i.e., such that

$$\varkappa\rho^{2k}\left(w_j M_j^{(s)} v_s - v_s M_j^{(s)*} w_j\right) = \sum_{i=1}^{2k} \frac{\partial P_{ij}^{(s)}}{\partial \vartheta_i} + \frac{\partial P_{j0}^{(s)}}{\partial \rho}, \tag{20.3}$$

where $P_{ij}^{(s)}$ denotes the components of some vector.[2)]

We shall define the system adjoint to the system $M_j$ as the system $N_s$ consisting of $l$ linear operators on $m$ functions $w_1, w_2, \cdots, w_m$ of the form:

$$N_s(w_1,\ w_2,\ \ldots,\ w_m) = \sum_{j=1}^{m} M_j^{(s)*} w_j \equiv \sum_{j=1}^{m} N_s^{(j)} w_j.$$

The matrix of the operators $N_s^{(j)}$ will have the form

2) A simple calculation shows that the operator $M_j^{(s)*}$ is defined uniquely in this way.

$$N=\begin{Vmatrix} N_1^{(1)}, & N_1^{(2)}, & \ldots, & N_1^{(j)}, & \ldots, & N_1^{(m)} \\ N_2^{(1)}, & N_2^{(2)}, & \ldots, & N_2^{(j)}, & \ldots, & N_2^{(m)} \\ \cdot & \cdot & \cdot & \cdot & \cdot & \cdot \\ N_s^{(1)}, & N_s^{(2)}, & \ldots, & N_s^{(j)}, & \ldots, & N_s^{(m)} \\ \cdot & \cdot & \cdot & \cdot & \cdot & \cdot \\ N_l^{(1)}, & N_l^{(2)}, & \ldots, & N_l^{(j)}, & \ldots, & N_l^{(m)} \end{Vmatrix}.$$

This matrix is obtained from the matrix of the operator $M_j^{(s)}$ if in the latter one replaces rows by columns and takes the adjoint of each operator $N_s^{(j)}=M_j^{(s)*}$.

The system of adjoint operators satisfies the following relations:

$$\varkappa\rho^{2k}\left[\sum_{j=1}^{m} w_j M_j(v_1, \ldots, v_l)-\sum_{s=1}^{l} v_s N_s(w_1, w_2, \ldots, w_m)\right]=$$

$$=\sum_{i=1}^{2k}\frac{\partial P_i}{\partial\vartheta_i}+\frac{\partial P_0}{\partial\rho}, \text{ where } P_i=\sum_{j=1}^{m}\sum_{s=1}^{l}P_{ji}^{(s)}. \tag{20.4}$$

We construct the system of operators $N_j$ $(j=0,1,\cdots,k)$ on $k$ functions $\sigma_1,\sigma_2,\cdots,\sigma_k$ adjoint to the system (19.61).

The matrix $M$ for the system (19.61) has the form:

$$M=\begin{Vmatrix} D_0^{(k-1)}, & D_1^{(k-1)}, & D_2^{(k-1)}, & \ldots, & D_{k-3}^{(k-1)}, & D_{k-2}^{(k-1)}, & D_{k-1}^{(k-1)}, & D_k^{(k-1)} \\ D_0^{(k-2)}, & D_1^{(k-2)}, & D_2^{(k-2)}, & \ldots, & D_{k-3}^{(k-2)}, & D_{k-2}^{(k-2)}, & D_{k-1}^{(k-2)}, & 0 \\ D_0^{(k-3)}, & D_1^{(k-3)}, & D_2^{(k-3)}, & \ldots, & D_{k-3}^{(k-3)}, & D_{k-2}^{(k-3)}, & 0 & 0 \\ D_0^{(k-4)}, & D_1^{(k-4)}, & D_2^{(k-4)}, & \ldots, & D_{k-3}^{(k-4)}, & 0 & 0 & 0 \\ \cdot & \cdot & \cdot & \cdot & \cdot & \cdot & \cdot & \cdot \\ D_0^{(3)}, & D_1^{(3)}, & D_2^{(3)}, & \ldots, & 0 & 0 & 0 & 0 \\ D_0^{(2)}, & D_1^{(2)}, & D_2^{(2)}, & \ldots, & 0 & 0 & 0 & 0 \\ D_0^{(1)}, & D_1^{(1)}, & D_2^{(1)}, & \ldots, & 0 & 0 & 0 & 0 \\ D_0^{(0)}, & D_1^{(0)}, & 0 & \ldots, & 0 & 0 & 0 & 0 \end{Vmatrix}.$$

The corresponding matrix $N$ will be:

$$N = \begin{vmatrix} D_0^{(k-1)*}, & D_0^{(k-2)*}, & D_0^{(k-3)*}, & D_0^{(k-4)*}, & \dots, & D_0^{(3)*}, & D_0^{(2)*}, & D_0^{(1)*}, & D_0^{(0)*} \\ D_1^{(k-1)*}, & D_1^{(k-2)*}, & D_1^{(k-3)*}, & D_1^{(k-4)*}, & \dots, & D_1^{(3)*}, & D_1^{(2)*}, & D_1^{(1)*}, & D_1^{(0)*} \\ D_2^{(k-1)*}, & D_2^{(k-2)*}, & D_2^{(k-3)*}, & D_2^{(k-4)*}, & \dots, & D_2^{(3)*}, & D_2^{(2)*}, & D_2^{(1)*}, & 0 \\ D_3^{(k-1)*}, & D_3^{(k-2)*}, & D_3^{(k-3)*}, & D_3^{(k-4)*}, & \dots, & D_3^{(3)*}, & D_3^{(2)*}, & 0, & 0 \\ \cdots & \cdots & \cdots & \cdots & \cdots & \cdots & \cdots & \cdots & \cdots \\ D_{k-3}^{(k-1)*}, & D_{k-3}^{(k-2)*}, & D_{k-3}^{(k-3)*}, & D_{k-3}^{(k-4)*}, & \dots, & 0, & 0, & 0, & 0 \\ D_{k-2}^{(k-1)*}, & D_{k-2}^{(k-2)*}, & D_{k-2}^{(k-3)*}, & 0, & \dots, & 0, & 0, & 0, & 0 \\ D_{k-1}^{(k-1)*}, & D_{k-1}^{(k-2)*}, & 0, & 0, & \dots, & 0, & 0, & 0, & 0 \\ D_k^{(k-1)*}, & 0, & 0, & 0, & \dots, & 0, & 0, & 0, & 0 \end{vmatrix}.$$

2. THE CONSTRUCTION OF THE FUNCTIONS $\sigma$. We now construct the system of functions $\sigma_1, \sigma_2, \cdots, \sigma_k$ which are solutions of the equations:

$$\left.\begin{array}{l} D_k^{(k-1)*}\sigma_1 = 0; \\ D_{k-1}^{(k-2)*}\sigma_2 + D_{k-1}^{(k-1)*}\sigma_1 = 0; \\ D_{k-2}^{(k-3)*}\sigma_3 + D_{k-2}^{(k-2)*}\sigma_2 + D_{k-2}^{(k-1)*}\sigma_1 = 0; \\ \cdots\cdots\cdots\cdots\cdots\cdots \\ D_1^{(0)*}\sigma_k + D_1^{(1)*}\sigma_{k-1} + D_1^{(2)*}\sigma_{k-2} + \dots + D_1^{(k-1)*}\sigma_1 = 0. \end{array}\right\} \tag{20.5}$$

This system turns out to be recurrent, the functions $\sigma_1, \sigma_2, \cdots, \sigma_k$ being defined in order. one from another.

Further, let

$$N_0(\sigma_1, \dots \sigma_k) = D_0^{(0)*}\sigma_k + D_0^{(1)*}\sigma_{k-1} + D_0^{(2)*}\sigma_{k-2} + \\ + \dots + D_0^{(k-1)*}\sigma_1. \tag{20.6}$$

Obviously there holds the relation obtained from (20.4) and (20.5)

$$\varkappa\rho^{2k}\left[\sum_{l=0}^{k-1}\sigma_{k-l}\frac{\partial^l\widetilde{F}}{\partial t^l}-u_0N_0(\sigma_1,\ldots,\sigma_k)\right]=$$

$$=\varkappa\rho^{2k}\sum_{l=0}^{k-1}\sigma_{k-l}\left[\frac{\partial^l\widetilde{F}}{\partial t^l}-M_{k-l}(u_0,\ldots,u_{l+1})\right]+$$

$$+\varkappa\rho^{2k}\sum_{l=0}^{k-1}\sum_{r=0}^{l+1}\left\{\sigma_{k-l}D_r^{(l)}u_r-u_rD_r^{(l)*}\sigma_{k-l}\right\}=$$

$$=\varkappa\rho^{2k}\sum_{l=0}^{k-1}\sigma_{k-l}\left[\frac{\partial^l\widetilde{F}}{\partial t^l}-M_{k-l}(u_0,\ldots,u_{l+1})\right]+\sum_{m=1}^{2k}\frac{\partial P_m}{\partial\vartheta_m}+\frac{\partial P_0}{\partial\rho}, \tag{20.7}$$

where $P_m=\sum_{r=1}^{2k+1}\sum_{l=1}^{2k+1}P_m^{(r,l)}$.

We calculate the quantities $P_m^{(r,l)}$ in the usual way.

From the expression for $J(t,\rho,\vec{\lambda})$ it is easy to see that $J(t,\rho.\vec{\lambda})$ and its derivatives in $t$ up to order $k+1$ are continuous for $\rho=0$. Therefore, taking into account formulas (19.58) and (19.56) for $L^{(l)}$ and $M^{(l)}$ for $t=0$, we find:

$$\left.\begin{aligned}
&D_{l+1}^{(l)}v=M^{(0)}v+\frac{DJ}{Dt}v=-2\varphi(s)\left\{\frac{\partial v}{\partial\rho}+\frac{1}{\rho}\left[k+G(\rho,\vec{\lambda})\right]v\right\};\\
&D_l^{(l)}v=L^{(0)}v+lM^{(1)}v+\frac{l!}{2!\,(l-2)!}\frac{D^2J}{Dt^2}v=\varphi(s)\left\{\frac{\partial^2v}{\partial\rho^2}\times\right.\\
&\left.\times[1+\rho Q_1]+\frac{1}{\rho}[2k+\rho Q_2]\frac{\partial v}{\partial\rho}+\frac{1}{\rho^2}[\Omega+\rho Q_3]v\right\};\\
&D_r^{(l)}=\varphi(s)\left\{A\frac{\partial^2v}{\partial\rho^2}+\frac{1}{\rho}B\frac{\partial v}{\partial\rho}+\frac{1}{\rho^2}Cv\right\},\\
&\qquad\qquad 0\leqslant r\leqslant l-1,
\end{aligned}\right\} \tag{20.8}$$

where $G$, $Q_1$ and $A$ are functions, $Q_2$ and $B$ are operators of first order, $\Omega$, $Q_3$, and $C$ are operators of second order, sufficiently often differentiable.

We pass to the calculation of the adjoint operators. We note first of all for this purpose that the operators $B^*$ and $Q_2^*$, adjoint to the operators of first order $B$ and $Q_2^*$ defined on the unit sphere, will be themselves operators of first order defined on the unit sphere. The operators $\Omega^*$, $Q_3^*$, and $C^*$, adjoint to operators of second order on the sphere, will also be operators of second order on the sphere. The operator $\Omega^*$ will coincide with the operator $\Omega$, since the Laplace operator on the sphere is a self-adjoint operator.

Further, the following formulas hold:

$$\left(\frac{\partial}{\partial\rho}\right)^* = -\left(\frac{1}{\rho^{2k}}\frac{\partial}{\partial\rho}\rho^{2k}\right);\quad \left(\frac{\partial^2}{\partial\rho^2}\right)^* = \left[\frac{1}{\rho^{2k}}\frac{\partial^2}{\partial\rho^2}(\rho^{2k})\right]; \tag{20.9}$$

$$\left(Q_2\frac{\partial}{\partial\rho}\right)^* = -\left(\frac{1}{\rho^{2k}}\frac{\partial}{\partial\rho}\rho^{2k}Q_2^*\right).$$

Similarly:

$$\rho^{2k}\left[v\frac{\partial u}{\partial\rho} + u\frac{1}{\rho^{2k}}\frac{\partial}{\partial\rho}(\rho^{2k}v)\right] = (\rho^{2k}v)\frac{\partial u}{\partial\rho} + u\frac{\partial}{\partial\rho}(\rho^{2k}v) = \frac{\partial}{\partial\rho}(uv\rho^{2k});$$

$$\rho^{2k}\left[v\frac{\partial^2 u}{\partial\rho^2} - u\frac{1}{\rho^{2k}}\frac{\partial^2}{\partial\rho^2}(\rho^{2k}v)\right] = (\rho^{2k}v)\frac{\partial^2 u}{\partial\rho^2} - u\frac{\partial^2}{\partial\rho^2}(\rho^{2k}v) =$$

$$= \frac{\partial}{\partial\rho}\left[(\rho^{2k}v)\frac{\partial u}{\partial\rho} - u\frac{\partial}{\partial\rho}(\rho^{2k}v)\right];$$

$$\rho^{2k}\left\{vQ_2\frac{\partial u}{\partial\rho} + u\frac{1}{\rho^{2k}}\left[\left(\frac{\partial}{\partial\rho}\rho^{2k}Q_2^*\right)v\right]\right\} =$$

$$= \left[\rho^{2k}vQ_2\left(\frac{\partial u}{\partial\rho}\right) - Q_2^*(\rho^{2k}v)\frac{\partial u}{\partial\rho}\right] + \left[Q_2^*(\rho^{2k}v)\frac{\partial u}{\partial\rho} + u\frac{\partial}{\partial\rho}\left(Q_2^*\rho^{2k}u\right)\right] =$$

$$= \frac{\partial}{\partial\rho}\left[u\,(Q_2^*\rho^{2k}v)\right] + \sum_{i=1}^{2k}\frac{\partial P_i}{\partial\vartheta_i}$$

where $P_i$ are some expressions given on the surface of the unit sphere.

If we use the fact that the adjoint operator is unique, we see that our assertion is established. By virtue of formulas (20.8), we obtain:

$$D_{l+1}^{(l)*}\sigma = \frac{2}{\rho^{2k}}\frac{\partial}{\partial\rho}[\varphi(s)\rho^{2k}\sigma] - \frac{2k\varphi(s)}{\rho}\sigma - G(\rho,\vec{\lambda})[2\varphi(s)\sigma] =$$

$$= 2\varphi(s)\frac{\partial\sigma}{\partial\rho} + \frac{1}{\rho}(k + \rho H(\rho,\vec{\lambda}))\sigma\Big], \tag{20.10}$$

where $H$ is a continuously differentiable, bounded function.

Carrying through analogous calculations for $D_l^{(l)}$ and $D_r^{(l)}$, where $r \leqq l-1$, we have finally

$$D_{l+1}^{(l)*}\sigma = 2\varphi(s)\left\{\frac{\partial\sigma}{\partial\rho} + \frac{1}{\rho}[k + \rho H)\sigma\right\} \tag{20.11}$$

$$D_l^{(l)*}\sigma = \varphi(s)\left\{\frac{\partial^2\sigma}{\partial\rho^2}[1 + \rho T_1^{(l)}] + \frac{1}{\rho}[2k + \rho T_2^{(l)}]\frac{\partial\sigma}{\partial\rho} + \right.$$

$$\left. + \frac{1}{\rho^2}[\Omega + \rho T_3^{(l)}]\sigma\right\} \tag{20.12}$$

$$D_r^{(l)*}\sigma = \varphi(s)\left\{\Delta_r^{(s)}\frac{\partial^2\sigma}{\partial\rho^2} + \frac{1}{\rho}M_r^{(l)}\frac{\partial\sigma}{\partial\rho} + \frac{1}{\rho^2}N_r^{(l)}\sigma\right\},\ r < l-1. \tag{20.13}$$

3. INVESTIGATION OF THE PROPERTIES OF THE FUNCTIONS $\sigma$. We pass to the study of the character of the functions $\sigma$, defined by the system (20.5).

We show that with the appropriate choice, each of them can be represented in the form:

$$\sigma_{k-l}=\frac{(-2)^{l}\Gamma(2k-l-1)\Gamma(k)}{\Gamma(l+1)\Gamma(2k-1)\Gamma(k-l)}\frac{1}{\rho^{2k-l-1}}[1+\rho\Psi_{l}], \tag{20.14}$$

where $\Psi_l$ is a bounded function of the variables $\rho$ and $\vartheta_i$, differentiable sufficiently many times.

First of all, we remark that $\sigma_1$ is a solution of the equation

$$\frac{\partial\sigma_1}{\partial\rho}+\frac{1}{\rho}[k+\rho H]\sigma_1=0, \tag{20.15}$$

and consequently

$$\sigma_1=\chi(\vartheta_1,\ldots,\vartheta_{2k})e^{-\int\frac{1}{\rho_1}[k+\rho_1 H(\rho_1,\vartheta)]d\rho_1}=$$

$$=\chi(\vartheta_1,\vartheta_2,\ldots,\vartheta_{2k})e^{-\int\frac{k}{\rho_1}d\rho_1-\int_0^\rho H(\rho_1,\vartheta)d\rho_1}=$$

$$=\chi(\vartheta_1,\vartheta_2,\ldots,\vartheta_{2k-1})\frac{1}{\rho^k}e^{-\int_0^\rho H(\rho_1,\vartheta)d\rho_1} \tag{20.16}$$

We set

$$\chi(\vartheta_1,\vartheta_2,\ldots,\vartheta_{2k-1})=\frac{(-2)^{k-1}\Gamma(k)}{\Gamma(2k-1)}.$$

Our assertion for $\sigma_1$ is established.

The property for the function $\sigma_{k-l}$ for arbitrary $l$ is proved by induction. The equation for $\sigma_{k-l}$ has the form:

$$2\left[\frac{\partial\sigma_{k-l}}{\partial\rho}+\frac{1}{\rho}(k+\rho H)\sigma_{k-l}\right]=$$

$$=\chi_{k-l}(\rho,\vartheta_1,\ldots,\vartheta_{2k})$$

where

$$\chi_{k-l}=-\frac{1}{\varphi(s)}D_{l+1}^{(l+1)*}\sigma_{k-l-1}-\frac{1}{\varphi(s)}D_{l+1}^{(l+2)*}\sigma_{k-l-2}-$$

$$-\ldots-\frac{1}{\varphi(s)}D_{l+1}^{(k-1)*}\sigma_1. \tag{20.17}$$

Substituting in the right hand side of the last equality the values of $\sigma_1,\sigma_2,\cdots,\sigma_{k-l-1}$ which we assume to be known and to satisfy the condition asserted above, we see that the main term of the right hand side is obtained from substituting the function $\sigma_{k-l-1}$ in the operator $D_{l+1}^{(l+1)*}$ and has the form:

$$-(-2)^{l+1}\frac{\Gamma(2k-l-2)\,\Gamma(k)}{\Gamma(l+2)\,\Gamma(2k-1)\,\Gamma(k-l-1)}\left[\frac{\partial^2}{\partial\rho^2}\frac{1}{\rho^{2k-l-2}}+\frac{2k}{\rho}\frac{\partial}{\partial\rho}\frac{1}{\rho^{2k-l-2}}\right]=$$

$$=-(-2)^{l+1}\frac{\Gamma(2k-l-2)\,\Gamma(k)}{\Gamma(l+2)\,\Gamma(2k-1)\,\Gamma(k-l-1)}\cdot\frac{(2k-l-2)}{\rho^{2k-l}}\Big[(2k-l-1)-2k\Big]=$$

$$=(-2)^{l+1}\frac{\Gamma(2k-l-1)\,\Gamma(k)}{\Gamma(l+1)\,\Gamma(2k-1)\,\Gamma(k-l-1)}\,\frac{1}{\rho^{2k-l}}.$$

A partial solution of equation (20.17) is easy to obtain by standard methods. We will have:

$$\sigma_{k-l}=\sigma_1\int\frac{\chi_{k-l}}{\sigma_1}\,d\rho_1. \tag{20.18}$$

From this the value of the principal term in the decomposition of $\sigma_{k-l}$ is obtained by elementary means.

Our formula for $\sigma_{k-l}$ is proved.

We consider one further question, namely the estimate of the principal term in the operator $N_0\,(\sigma_1,\sigma_2,\cdots,\sigma_k)$ for $\rho\to 0$.

Obviously this principal term is obtained by substituting the function $\sigma_k$ in the operator $D_0^{(0)*}$.

This principal term is clearly equal to

$$\left(\frac{\partial^2}{\partial\rho^2}+\frac{2k}{\rho}\,\frac{\partial}{\partial\rho}+\frac{1}{\rho^2}\,\Omega\right)\frac{C}{\rho^{2k-1}}=0.$$

Thus, $N_0\,(\sigma_1,\cdots,\sigma_k)$ begins with terms of order $\rho^{-2k}$, since the terms of order $\rho^{-(2k+1)}$ are missing.

4. DERIVATION OF THE BASIC INTEGRAL IDENTITY $Bu=SF$. We now return to the identity (20.7).

We consider for $t=0$ the domain $\Omega$ in the space of $y_1,\cdots,y_{2k+1}$ containing the origin of coordinates in its interior. In the variables $x_1, x_2,\cdots,x_{2k+1}$, it will be a domain in the characteristic cone containing the vertex of the cone. We remove from this domain a sphere $\rho\leqq\epsilon$ around the origin of coordinates and denote the remaining domain by $\Omega'$. The boundary of $\Omega'$ consists of the exterior bounding surface $S$ and the surface $\rho=\epsilon$. Integrating both parts of the identity (20.7) over the volume $\Omega'$, we obtain:

$$\int\underset{\Omega'}{\cdots}\int\left\{\rho^{2k}\sum_{l=0}^{k-1}\sigma_{k-l}\left[\frac{\partial^l\widetilde{F}}{\partial t^l}-M_{k-l}(u_0,u_1,\ldots,u_{l+1})\right]+\sum_{m=1}^{2k}\frac{\partial P_m}{\partial\vartheta_m}+\right.$$

$$\left.+\frac{\partial P_0}{\partial\rho}\right\}d\rho\,dS=\int\underset{\Omega'}{\cdots}\int\rho^{2k}\left[\sum_{l=0}^{k-1}\sigma_{k-l}\frac{\partial^l\widetilde{F}}{\partial t^l}-\right.$$

$$\left.-u_0N_0(\sigma_1,\ldots\sigma_k)\right]d\rho\,dS. \tag{20.19}$$

Obviously $[(\partial^l \tilde{F}/\partial t^l) - M_{k-l}(u_0, u_1, \cdots, u_{l+1})] = 0$ by equation (19.61).

The left side may be transformed into an integral over the boundary of $\Omega'$. We obtain:

$$\int_{\Omega'} \cdots \int \rho^{2k}\left[\sum_{l=0}^{k-1} \sigma_{k-l} \frac{\partial^l \tilde{F}}{\partial t^l} - u_0 N_0(\sigma_1, \ldots, \sigma_k)\right] d\rho\, dS =$$

$$= \int_S \cdots \int \Xi\, dS + \int_{\rho=\varepsilon} \cdots \int P_0\, dS, \tag{20.20}$$

where $\Xi$ is defined by the equality

$$\Xi\, dS = \sum_{m=1}^{2k} P_m \frac{d\vartheta_1\, d\vartheta_2 \ldots d\vartheta_{2k}}{d\vartheta_m} + P_0\, d\vartheta_1 \ldots d\vartheta_{2k}. \tag{20.21}$$

We now take formula (20.20) to the limit as $\epsilon \to 0$. On the left side we will have a convergent integral since $N_0$ has a singularity of order not greater than $\rho^{-2k}$ and each of the functions $\sigma$ has a singularity of order not higher than $\rho^{-2k+1}$. We consider the limit of

$$\int_{\rho=\varepsilon} \cdots \int P_0\, dS.$$

We show that this limit will equal $Cu_0|_{\rho=0}$, where $C$ is a nonzero constant.

Indeed, there can be nonzero limits only from those terms arising from the decomposition of

$$\int \cdots \int P_0\, dS,$$

which do not contain the factor $\rho$. Such terms can be obtained from no other $\sigma_{k-s}$ than $\sigma_k$ since even $\sigma_{k-1}$ has a singularity of order $\rho^{-(2k-2)}$ and the derivatives of $\rho^{2k}\sigma_{k-1}$ with respect to $\rho$ for $\rho = 0$ must vanish.

It is also obvious that such a term cannot be obtained from terms containing $u_1$, since $u_1$ and $\sigma_k$ appear in equation (20.7) only in the form:

$$\sigma_k D_1^{(0)} u_1 - u_1 D_1^{(0)*} \sigma_k,$$

where $D_1^{(0)}$ is an operator of first order and as a result the terms $P_{1,i}^{(1)}$ can contain only derivatives of $\sigma_k$ with respect to $u_1$, and these derivatives do not vanish at the singularity to the order needed.

Thus, there remains only to calculate the boundary term $P_{00}^{(1)}$. We have for this part of the boundary

$$u_0 D_0^{(0)*}\sigma_k - \sigma_k D_0^{(0)} u_0 \cong u_0 \frac{1}{\rho^{2k}} \frac{\partial}{\partial \rho}\left(\rho^{2k} \frac{\partial \sigma_k}{\partial \rho}\right) - \sigma_k \frac{1}{\rho^{2k}} \frac{\partial}{\partial \rho}\left(\rho^{2k} \frac{\partial u_0}{\partial \rho}\right) =$$

$$= \frac{1}{\rho^{2k}} \frac{\partial}{\partial \rho}\left(u_0 \rho^{2k} \frac{\partial \sigma_k}{\partial \rho} - \sigma_k \rho^{2k} \frac{\partial u_0}{\partial \rho}\right) \cong \frac{C}{\rho^{2k}} \frac{\partial}{\partial \rho}(u_0 \rho^{2k}),$$

where $C$ is some constant. From this, there follows:

$$\lim_{\varepsilon \to 0} \int \underset{\rho = \varepsilon}{\cdots} \int P_0 dS = C_2 u_0 \big|_{\rho=0}.$$

Using this, we finally obtain after substituting this result in (20.20) and dividing by the corresponding constant:

$$u_0 \big|_{\rho=0} = \frac{1}{C}\Big\{\int \underset{\Omega}{\cdots} \int u_0 N_0(\sigma_1, \ldots, \sigma_k)\, d\Omega -$$

$$- \sum_{s=0}^{k-1} \int \underset{\Omega}{\cdots} \int \sigma_{k-l} \frac{\partial^l \widetilde{F}}{\partial t^l}\, d\Omega + \int \underset{S}{\cdots} \int \Xi dS\Big\}. \qquad (20.22)$$

5. THE INVERSE INTEGRAL OPERATOR $B^{-1}$ AND THE METHOD OF SUCCESSIVE APPROXIMATIONS. The relations we have obtained yield the possibility of constructing the solution of the Cauchy problem for equations with sufficiently smooth coefficients in the cases when those solutions exist. The existence of such solutions we shall establish later. We consider the problem of finding the solution of equation (19.1) for the conditions (19.40). Suppose that we need to find the value of the unknown function $u$ at the point $P_0(x_0, x_1, \cdots, x_{2k+1})$.

We construct the characteristic cone with vertex at $P_0$. This cone is in its lower part will intersect the plane $x_0=0$ in a set $S$ bounded by a part $\Omega$ of the surface of this cone which contains the point $P_0$ in its interior. Indeed, by virtue of the fact that the direction of the $x_0$-axis is always time-like, the transformation of coordinates from $x_0, x_1, \cdots, x_{2k+1}$ to $y_0, y_1, \cdots, y_{2k+1}$ will transform $x_0$ into $y_0$ in such a way that $dx_0/dy_0$ and $dy_0/dx_0$ will both be bounded and different from zero. This follows from the boundedness of scale and finiteness of angle between these two directions. Therefore, moving down the characteristic cone on any generator, we will always have $\partial x_0/\partial y_0 > h > 0$ and cannot reach the point $x_0=0$. On the set $S$ the quantity $\Xi$ will be known. Indeed this set lies in the plane $x_0=0$ where all the derivatives of the continuous function $u$ are obtainable from the equation (19.1) and the conditions (19.40). All derivatives containing at most one differentiation with respect to $x_0$ are given directly by the conditions (19.40) differentiated with respect to $x_1, \cdots, x_{2k+1}$, the derivative $\partial^2 u/\partial x_0^2|_{x_0=0}$ are determined from the equation (19.1) in which one finds only derivatives having not more than two differentiations with respect to $x_0$. Differentiating (19.1) with respect to $x_0$, we obtain equations for finding $\partial^3 u/\partial x_0^3$, $\partial^4 u/\partial x_0^4$, etc. An arbitrary derivative $\partial^s u/\partial x_0^s$ is determined by differentiating equation (19.1) and substitution of initial data for all known derivatives. The remaining derivatives are obtained by direct differentiation with respect to $x_1, \cdots, x_{2k+1}$ of the derivatives of the form $\partial^s u/\partial x_0^s$.

We introduce some notation. Set:

$$u(x_0, x_1, \ldots, x_{2k+1}) - \frac{1}{C}\int \cdots \int_\Omega u(x_0, \ldots, x_{2k+1}) N_0 d\Omega \equiv Bu. \quad (20.23)$$

The operator $B$ carries the function $u$ defined on the domain $x_0 > 0$ into a new function $u$ defined on the same domain, and can be applied to an arbitrary continuous function.

Let also

$$-\frac{1}{C}\int \cdots \int_\Omega \sum_{s=0}^{k-1} \sigma_{k-s} \frac{\widetilde{\partial}^s F}{\partial t^s} d\Omega \equiv SF. \quad (20.24)$$

The operator $S$ can be applied to an arbitrary function having continuous derivatives of order $k$ and defined for $x_0 > 0$ and carries it into a function defined on the same domain.

The equality (20.22) can be written in the form of the equation:

$$Bu = SF + f_1. \quad (20.25)$$

By the definition, $F$ is a known function. If we set $SF = f_2$, $f_1 + f_2 = f$, we will have

$$Bu = f. \quad (20.26)$$

The equation (20.26) is an integral equation of Volterra type. A little later, we shall establish its solvability and find its solution. We investigate a particular form of the equation (20.26).

We suppose that the function $F$ vanishes together with all its derivatives of order up to $k$ inclusive for $x_0 < 0$, and that the function $u$ also vanishes for $x_0 < 0$. Then in the equality (20.22), the integral

$$\int \cdots \int \Xi \, dS$$

vanishes since it depends only on the initial values of the function $u$ and its derivatives of order not higher than $k$ and these initial values are all zero, as follows easily from our assumption.

We obtain in this case:

$$Bu = SF \tag{20.27}$$

or

$$Bu = SLu, \tag{20.28}$$

where by $Lu$ we mean the hyperbolic differential operator standing on the left side of the equation (cf. (19.1)).

We study the properties of the operators $B$ and $S$.

THEOREM 1. *Let $T_1$ and $T_2$ be two values of the variable $x_0$. Let $T_2 > T_1$, and suppose that corresponding to the variable point $P_0\,(x_0^{(0)}, x_1^{(0)}, \cdots, x_{2k+1}^{(0)})$ we have constructed the characteristic cone $\sum$. Denote by $\Omega_1$ and $\Omega_2$ the domains cut out of the characteristic cone $\sum$ by the surfaces $x_0 = T_1$ and $x_0 = T_2$. Obviously $\Omega_1 \supset \Omega_2$. Then we have the estimate*:

$$\int_{\Omega_1 - \Omega_2} \cdots \int |N_0|\, d\Omega \leqslant K(T_2 - T_1),$$

*where $K$ is a constant not depending upon $P_0$.*

In order to carry through this estimate more easily, we replace in the expression $d\Omega$ the independent variables $\rho, \vartheta_1, \cdots, \vartheta_{2k}$ by the variables $x_0$, $\vartheta_1, \vartheta_2, \cdots, \vartheta_{2k}$. For these

$$\frac{D(\rho, \vartheta_1, \ldots, \vartheta_{2k})}{D(x_0, \vartheta_1, \ldots, \vartheta_{2k})} = \frac{\partial \rho}{\partial x_0}.$$

But $\rho = -y_0$ by our construction. Therefore $\partial\rho/\partial y_0 < M$, where $M$ is a constant. Furthermore $|N_0| \leq K_1 \rho^{-2k}$, and therefore for an arbitrary domain $\Omega'$,

$$\int_{\Omega'} \cdots \int |N_0|\, d\Omega \leqslant K_1 \int_{\Omega'} \cdots \int \varkappa \rho^{2k} \rho^{-2k}\, d\rho\, d\vartheta_1 \ldots d\vartheta_{2k} =$$

$$= K_1 \int_{\Omega'} \cdots \int d\rho\, d\vartheta_1 \ldots d\vartheta_{2k} \leqslant MK_2 \int_{\Omega'} \cdots \int dx_0\, d\vartheta_1 \ldots d\vartheta_{2k}.$$

From this, our inequality follows immediately.

THEOREM 2. *The operator $Bu = f$ has an inverse.*

We shall establish that the equation $Bu = f$ is solvable by the method of successive approximations. From this the existence of an inverse operator will follow without difficulty.

We set:

$$u^{(0)} = f;\ u^{(n)}(x_0^{(0)}, \dots, x_{2k+1}^{(0)}) = \int_\Omega \cdots \int u^{(n-1)} N_0 d\Omega + $$
$$+ f(x_0^{(0)}, \dots, x_{2k+1}^{(0)}).$$

We show that the sequence $u^{(n)}$ converges uniformly. From this, it will follow obviously that its limit will be a solution of the integral equation.

As usual, it suffices to consider $u^{(n+1)} - u^{(n)} = v^{(n)}$, where the $v^{(n)}$ are related by the homogeneous identity:

$$v^{(n)} = \int \cdots \int v^{(n-1)} N_0 \, d\Omega. \tag{20.29}$$

Suppose that $|v^{(0)}| \le M$.

We show that then

$$|v^{(n)}(x_0^{(0)}, \dots, x_{2k+1}^{(0)})| \leqslant MK^n \frac{x_0^{(0)n}}{n!}. \tag{20.30}$$

The proof proceeds by induction. Suppose that this inequality has been established for some $n$; we show that it remains valid for $n+1$. We have:

$$|v^{(n+1)}| \leqslant \int_\Omega \cdots \int |v^{(n)}| |N_0| \, d\Omega \leqslant \int_\Omega \cdots \int MK^n \frac{x_0^{(1)n}}{n!} |N_0| \, d\Omega =$$
$$= \int_0^{x_0^{(0)}} MK^n \frac{x_0^{(1)n}}{n!} \left( \frac{d}{dx_0^{(1)}} \int_{x_j \geqslant x_0^{(1)}} \cdots \int |N_0| \, d\Omega \right) dx_0^{(1)} \leqslant$$
$$\leqslant MK^{n+1} \int_0^{x_0^{(0)}} \frac{x_0^{(1)n}}{n!} dx_0^{(1)} \leqslant MK^{n+1} \frac{x_0^{(0)n+1}}{(n+1)!},$$

as was to be shown.

Having constructed the solution of the equation $Bu = f$, we find the solution of the Cauchy problem for the given equation of hyperbolic type if that solution exists.

It remains for us to show the existence of such a solution. As a preliminary, we shall prove some further theorems.

6. THE ADJOINT INTEGRAL OPERATOR $B^*$. Let $v(x_0, x_1, \cdots, x_{2k+1})$ be a function of the variables $x_0, \cdots, x_{2k+1}$ vanishing for $x_0 > T_0 > 0$ and for $|x_i| > T_0$, where $T_0$ is some constant. The function $u$ will be assumed to vanish for $x_0 < 0$. We form the integral:

$$\int \cdots \int v(x_0, x_1, \ldots, x_{2k+1}) \times$$
$$\times Bu(x_0, x_1, \ldots, x_{2k+1})\, dx_0, \ldots, dx_{2k+1} \tag{20.31}$$

over the whole space. This integral may be transformed into the integral

$$\int \cdots \int u(x_0, x_1, \ldots, x_{2k+1}) \times$$
$$\times B^* v(x_0, x_1, \ldots, x_{2k+1})\, dx_0, dx_1, \ldots, dx_{2k+1}.$$

The operator $B^*$ which appears as the multiplier of $u$ under the integral sign is called the adjoint operator of the operator $B$.

THEOREM. *Both operators $B$ and $B^*$ can be applied to arbitrary continuous functions, satisfying the corresponding conditions of vanishing in corresponding domains, and carry them again into continuous functions.*

*Both operators have continuous inverse operators, i.e., $B^{-1}$ and $B^{*-1}$ exists. In other words, the equations $Bu=f$ and $B^*w=\psi$ have unique continuous solutions for continuous right hand sides.*

The existence of a continuous inverse for $B$ was shown above. In order to prove the theorem, it is necessary for us first of all to construct the operator $B^*$ in an explicit form.

Transform the integral (20.31). We will have:

$$\overbrace{\int \cdots \int}^{2k+2} v(x_0, x_1, \ldots, x_{2k+1}) \, [u(x_0, x_1, \ldots, x_{2k+1}) -$$
$$- \int \underset{\Omega}{\cdots} \int u(x_0', x_1', \ldots, x_{2k+1}') \times$$
$$\times N_0(\rho, \vartheta_1, \ldots, \vartheta_{2k})\, \rho^{2k} \varkappa d\rho\, d\vartheta_1, \ldots, d\vartheta_{2k}]\, dx_0\, dx_1, \ldots, dx_{2k+1} =$$
$$= \overbrace{\int \cdots \int}^{2k+2} v(x_0, x_1, \ldots, x_{2k+1})\, u(x_0, x_1, \ldots, x_{2k+1}) \times$$
$$\times dx_0\, dx_1, \ldots, dx_{2k+1} + \overbrace{\int \cdots \int}^{4k+3} v(x_0, \ldots, x_{2k+1}) \times$$
$$\times N_0(x_0, x_1, \ldots, x_{2k+1}, \rho, \vartheta_1, \ldots, \vartheta_{2k})\, u(x_0'(x_0, x_1, \ldots, x_{2k+1},$$
$$\rho, \vartheta_1, \ldots, \vartheta_{2k}),\ x_1'(x_0, x_1, \ldots, x_{2k+1}, \rho, \vartheta_1, \ldots, \vartheta_{2k}), \ldots,$$
$$\ldots, x_{2k+1}'(x_0, x_1, \ldots, x_{2k+1}, \rho, \vartheta_1, \ldots, \vartheta_{2k})) \times$$
$$\times \rho^{2k} \varkappa d\rho\, d\vartheta_1, \ldots, d\vartheta_{2k}\, dx, \ldots, dx_{2k+1}. \tag{20.32}$$

In the last integral the function $N_0$ depends upon the coordinates of the vertex $x_0, x_1, \cdots, x_{2k+1}$ and polar coordinates on the cone. On these same coordinates depend the variables $x_0', \cdots, x_{2k+1}'$ appearing beyond the sign $N_0$ and calculated on the surface of the cone. We carry through a change of independent variables, taking as the new variables the coordinates $x_0', \cdots, x_{2k+1}'$ and leaving as before the independent variables $\rho, \vartheta_1, \cdots, \vartheta_{2k}$.

Thereby, the second integral in the equality (20.32) can be rewritten in the form:

$$\overbrace{\int \cdots \int}^{4k+3} N_0(\rho, \vartheta_1, \ldots, \vartheta_{2k}, x_0(x_0', \ldots, x_{2k+1}', \rho, \vartheta_1, \ldots, \vartheta_{2k}),$$

$$x_1(\ldots), \ldots x_{2k+1}(\ldots)) \cdot v(x_0(x_0', x_1', \ldots, x_{2k+1}', \rho, \vartheta_1, \ldots, \vartheta_{2k}),$$

$$x_1(\ldots), \ldots, x_{2k+1}(\ldots)) \times$$

$$\times u(x_0', x_1', \ldots, x_{2k+1}') \frac{D(x_0, x_1, \ldots, x_{2k+1}, \rho, \vartheta_1, \ldots, \vartheta_{2k})}{D(x_0', x_1', \ldots, x_{2k+1}', \rho, \vartheta_1, \ldots, \vartheta_{2k})} \times$$

$$\times dx_0' \, dx_1' \ldots dx_{2k+1}' \rho^{2k} \varkappa \, d\rho \, d\vartheta_1 \ldots d\vartheta_{2k} =$$

$$= \overbrace{\int \cdots \int}^{2k+2} u(x_0', \ldots, x_{2k+1}') \Big\{ \overbrace{\int \cdots \int}^{2k+1} N_0(\rho, \vartheta_1, \ldots, \vartheta_{2k},$$

$$x_0(x_1', x_2', \ldots, x_{2k+1}', \rho, \ldots, \vartheta_{2k}), x_1(x_1' \ldots) \ldots$$

$$\ldots x_{2k+1}(x_1', x_2' \ldots x_{2k+1}', \rho \ldots \vartheta_{2k}) \cdot v(x_0(\ldots), \ldots,$$

$$x_{2k+1}(\ldots)) \frac{D(x_0, x_1, \ldots, x_{2k+1}, \rho, \ldots, \vartheta_{2k})}{D(x_0' x_1', \ldots, x_{2k+1}', \rho, \ldots, \vartheta_{2k})} \rho^{2k} \varkappa \, d\rho \, d\vartheta_1 \ldots d\vartheta_{2k} \Big\} \times$$

$$\times dx_0' \ldots dx_{2k+1}'.$$

From this, it is evident that we can put:

$$B^* v \Big|_{x_0', \ldots x_{2k+1}'} = v(x_0', x_1', \ldots, x_{2k+1}') -$$

$$- \overbrace{\int \cdots \int}^{2k+1} N_0(\rho, \vartheta_1, \ldots, \vartheta_{2k}, x_0(x_0', \ldots, x_{2k+1}', \rho, \ldots, \vartheta_{2k}),$$

$$x_1(\ldots) \ldots x_{2k+1}(\ldots)).$$

$$\frac{D(x_0, x_1, \ldots x_{2k+1}, \rho, \vartheta_1, \ldots, \vartheta_{2k})}{D(x_0', x_1', \ldots, x_{2k+1}', \rho, \vartheta_1, \ldots, \vartheta_{2k})} v(x_0(\ldots),$$

$$x_1(\ldots), \ldots x_{2k+1}(\ldots)) \cdot \rho^{2k} \sin^{2k-1} \vartheta_{2k} \sin^{2k-2} \vartheta_{2k-1} \ldots$$

$$\ldots \sin \vartheta_2 \, d\rho \, d\vartheta_1 \, d\vartheta_2, \ldots d\vartheta_{2k}. \tag{20.33}$$

In this form, there appears a resemblance between the operators $B$ and $B^*$. This resemblance extends not just to the external form of their representations but also to many properties.

We investigate first of all the character of the surface in $(2k+2)$-dimensional space given by the parametric equations

$$\left.\begin{array}{l} x_0 = x_0(x_0', x_1', \ldots, x_{2k+1}', \rho, \vartheta_1, \ldots, \vartheta_{2k}); \\ \cdots\cdots\cdots\cdots\cdots\cdots\cdots\cdots \\ x_{2k+1} = x_{2k+1}(x_0', x_1', \ldots, x_{2k+1}', \rho, \vartheta_1, \ldots, \vartheta_{2k}). \end{array}\right\} \quad (20.34)$$

We shall show that the surface (20.34) is itself again a characteristic cone with vertex at the point $x_0', x_1', \cdots, x_{2k+1}'$ but extending in the direction of increasing values of $x_0'$, i.e., the upper portion of the complete characteristic cone. Indeed, the point $x_0, x_1, \cdots, x_{2k+1}$ and the point $x_0', x_1', \cdots, x_{2k+1}'$ by construction must lie on a common bicharacteristic. Consequently, the set of all points $x_0, x_1, \cdots, x_{2k+1}$ for which $x_0', x_1', \cdots, x_{2k+1}'$ appears on the surface of the characteristic cone directed downward, coincides with the set of all points of all bicharacteristics passing through $x_0', x_1', \cdots, x_{2k+1}'$ and this set is the upper part of the characteristic cone, as was to be shown.

We set:

$$N_0 \frac{D(x_0, x_1, \ldots, x_{2k+1}, \rho, \vartheta_1, \ldots, \vartheta_{2k})}{D(x_0', x_1', \ldots, x_{2k+1}', \rho, \vartheta_1, \ldots, \vartheta_{2k})} = N_0^*$$

and estimate the size of the Jacobian:

$$\frac{D(x_0, x_1, \ldots, x_{2k+1}, \rho, \ldots, \vartheta_{2k})}{D(x_0', x_1', \ldots, x_{2k+1}', \rho, \ldots, \vartheta_{2k})}.$$

This determinant will be bounded. Indeed, we may for its calculation first of all change coordinates from the variables $x_0', x_1', \cdots, x_{2k+1}'$ to the variables $y_0, y_1, \cdots, y_{2k+1}$ and calculate the function

$$\begin{array}{l} y_0(x_0, x_1, \ldots, x_{2k+1}, x_0', x_1', \ldots, x_{2k+1}'); \\ y_1(x_0, x_1, \ldots, x_{2k+1}, x_0', x_1', \ldots, x_{2k+1}'); \\ \cdots\cdots\cdots\cdots\cdots\cdots\cdots\cdots \\ y_{2k+1}(x_0, x_1, \ldots, x_{2k+1}, x_0', x_1', \ldots, x_{2k+1}'), \end{array}$$

where $x_0, x_1, \cdots, x_{2k+1}$ are the coordinates of the vertex of the cone with respect to which the substitution is carried out, and $x_0', x_1', \cdots, x_{2k+1}'$ are the running coordinates related to $y_0, y_1, \cdots, y_{2k+1}$ by the formulas (19.23) investigated above.

It is not difficult to see from general theorems on ordinary differential equations that for fixed $s, p_1^{(0)}, \cdots, p_{2k+1}^{(0)}$ in the equations (19.9), $x_0', x_1', \cdots, x_{2k+1}'$ will be continuous and sufficiently often differentiable functions of the initial data $x_0, x_1, \cdots, x_{2k+1}$ with a functional determinant different from zero (the initial value of this determinant being equal to 1).

As a result:

$$\left.\begin{array}{l} x_0'(x_0, x_1, \ldots, x_{2k+1}, y_0, \ldots, y_{2k+1}), \\ \cdots\cdots\cdots\cdots\cdots\cdots \\ x_{2k+1}'(x_0, x_1, \ldots, x_{2k+1}, y_0, \ldots, y_{2k+1}) \end{array}\right\} \qquad (20.35)$$

will be continuous functions of $x_0, x_1, \cdots, x_{2k+1}$ with functional determinant different from zero. But the functions $x_0'(x_0, \cdots, x_{2k+1}, \rho, \vartheta_1, \cdots, \vartheta_{2k})$ are the same functions (20.35), where we set $y_0 = \rho = \left(\sum_{i=1}^{2k+1} y_i^2\right)^{\frac{1}{2}}$. The boundedness of the determinant

$$D = \frac{D(x_0, x_1, \ldots, x_{2k+1})}{D(x_0', x_1', \ldots, x_{2k+1}')}$$

is proved.

It follows from this without difficulty that the function $N_0^* = N_0 D$ satisfies the inequality

$$|N_0^*| \leqslant M\rho^{-2k}.$$

We shall show that the function $N_0^*$, just like $N_0$, satisfies the integral inequality

$$\left.\begin{array}{r} \int \cdots \int |N_0^*| \rho^{2k} \, d\rho \, dS \leqslant K(T_2 - T_1), \\ x_0 \leqslant T_1 < T_2 \end{array}\right\} \qquad (20.36)$$

For this, in distinction from the previous case, $T_1 > x_0$, since the cone on which $N_0^*$ is defined extends in the direction of increasing $x_0$. From this, as earlier, will follow the existence of an inverse operator $B^*$ in the space of functions vanishing for $x_0 > T$, where $T$ is an arbitrary constant. (In the case of $B$, we had to do with the space of functions vanishing for $x_0 < 0$.) In order to establish (20.36), we need as before to replace $d\rho$ under the integral sign by $dx_0$ for fixed $x_0', \cdots, x_{2k+1}', \vartheta_1, \cdots, \vartheta_{2k}$.

Along the line $\vartheta_1 = \text{const.}, \vartheta_2 = \text{const.}, \cdots, \vartheta_{2k} = \text{const.}$, we have

$$x_0' = \varphi(x_0, x_1, \ldots, x_{2k+1}, \rho, \vartheta_1, \ldots, \vartheta_{2k+1}).$$

Considering this as an implicit equation for $x_0$ for given $x_0'$ and $\rho$, we obtain:

$$0=\frac{\partial\varphi}{\partial x_0}\frac{\partial x_0}{\partial\rho}+\frac{\partial\varphi}{\partial\rho}$$

or

$$\frac{dx_0}{d\rho}=-\frac{\dfrac{\partial x_0'}{\partial x_0}}{\dfrac{\partial x_0'}{\partial\rho}}.$$

From this, there follows the boundedness to be proved.

Our theorem is established.

7. THE ADJOINT INTEGRAL OPERATOR $S^*$. Just as we constructed the operator $B^*$ adjoint to $B$, we may construct the operator $S^*$ adjoint to $S$. We show that such an operator may be defined for all sufficiently smooth functions vanishing for $x_0>T_0$ and $|x_i|>T_0$, where $T_0$ is any constant.

In order to construct the operator $S^*$, we use the previous method. We have:

$$Su=\overbrace{\int\cdots\int}^{2k+1}\sum_{s=0}^{k-1}\sigma_{k-s}\frac{\partial^s u(x_0',\ldots,x_{2k+1})}{\partial y_0^s}\rho^{2k}\,d\rho\,dS=$$

$$=\overbrace{\int\cdots\int}^{2k+1}\sum_{s=0}^{k-1}\sigma_{k-s}\sum_{\Sigma\beta=s}\frac{\partial^s u}{\partial x_0^{\beta_0}\partial x_1^{\beta_1}\cdots\partial x_{k+1}^{\beta_{2k+1}}}\times$$

$$\times\psi^{(s)}_{\beta_0,\beta_1,\ldots,\beta_{2k+1}}\rho^{2k}\,d\rho\,dS. \tag{20.37}$$

Here $\psi^{(s)}_{\beta_0,\cdots\beta_{2k+1}}$ is the coefficient in the representation of the derivative $\partial^s u/\partial y_0^s$ in terms of derivatives with respect to $x_0, x_1,\cdots,x_{2k+1}$, and the integral is over the part of the characteristic cone lying above the plane $x_0=0$. It may be considered as extended over the whole surface of the cone, since $u\equiv 0$ for $x_0<0$.

We form the integral

$$J=\int\cdots\int v(x_0,\ldots,x_{2k+1})\,Su\,dx_0\,dx_1\ldots dx_{2k+1}.$$

Transforming it as above, we obtain:

$$J = \overbrace{\int \cdots \int}^{2k+2} \Big\{ \sum_{\Sigma\beta \leqslant k-1} \frac{\partial^{\beta} u}{\partial x_0^{\beta_0} \partial x_1^{\beta_1} \cdots \partial x_{2k+1}^{\beta_{2k+1}}} (x_0', \ldots, x_{2k+1}') \times$$

$$\times \overbrace{\int \cdots \int}^{2k+1} \sum_{s=0}^{k-1} \sigma_{k-s} (x_0 (x_0', \ldots x_{2k+1}' \rho, \vartheta_1, \ldots, \vartheta_{2k}),$$

$$x_1 (\;) \ldots x_{2k+1} (\ldots)) \cdot \psi^{(s)}_{\beta_0, \beta_1, \ldots, \beta_{2k+1}} \cdot v (x_0 (x_0', \ldots, x_{2k+1}',$$

$$\rho, \vartheta_1, \ldots, \vartheta_{2k}) x_1 (\ldots). \ldots, x_{2k+1} (\ldots)) \rho^{2k} \varkappa \sin^{2k-1} \vartheta_{2k}, \ldots,$$

$$\sin \vartheta_2 \, d\rho \, d\vartheta_1 \, d\vartheta_{2k} \} \, dx_0' \, dx_1' \ldots dx_{2k+1}'. \qquad (20.38)$$

If we take into account the fact that $u$ vanishes for $x_0 < 0$ while the function $v$ vanishes for $x_0 > T$, we may integrate by parts in formula (20.38). Thereby we will have:

$$J = \overbrace{\int \cdots \int}^{2k+2} u (x_0', \ldots, x_{2k+1}') \Big\{ \sum_{\Sigma\beta \leqslant k-1} \frac{\partial^{\beta}}{\partial x_0^{\beta_0} \partial x_1^{\beta_1} \cdots \partial x_{2k+1}^{\beta_{2k+1}}} \times$$

$$\times \overbrace{\int \cdots \int}^{2k+1} \sum_{s=0}^{k-1} \sigma_{k-s} (x_0 (x_0', \ldots, \vartheta_{2k}), x_1 (\ldots), \ldots, x_{2k+1} (\ldots)) \times$$

$$\times \psi^{(s)}_{\beta_0, \beta_1, \ldots, \beta_{2k+1}} \cdot v (x_0 (\ldots), x_1 (\ldots), \ldots x_{2k+1} (\ldots)) \times$$

$$\times \rho^{2k} \varkappa \, d\rho \, d\vartheta, \ldots, d\vartheta_{2k} \Big\} dx_0', \ldots, dx_{2k+1}'.$$

It is obvious from this that the operator $S^*$ has the form:

$$S^* v = \sum_{\Sigma\beta \leqslant k-1} \frac{\partial^{\beta}}{\partial x_0^{\beta_0} \cdots \partial x_{2k+1}^{\beta_{2k+1}}} \int \cdots \int \sigma_{k-s} (x_0 (\;), \ldots,$$

$$\ldots, x_{2k+1} (\;), \rho, \ldots, \vartheta_{2k}) \psi^{(s)}_{\beta_0, \beta_1, \ldots, \beta_{2k+1}} \times$$

$$\times v (x_0, x_1, \ldots, x_{2k+1}) \rho^{2k} d\rho \, \varkappa \, d\vartheta_1, \ldots, d\vartheta_{2k}. \qquad (20.39)$$

From this very form of the operator $S^*$, it is evident that it is defined for functions having continuous derivatives of order $k$. We may now proceed to the proof of the existence of the solution of our Cauchy problem.

8. SOLUTION OF THE CAUCHY PROBLEM FOR AN EVEN NUMBER OF VARIABLES. First of all, we reduce our problem to a simpler one. We calculate the value of all derivatives with respect to $x_0$ up to order $k+1$ inclusive

of the function $u$, and we set:

$$u = w + u\Big|_{x_0=0} + x_0 \frac{\partial u}{\partial x_0}\Big|_{x_0=0} + \\ + \dots + \frac{x_0^{k+1}}{(k+1)!} \frac{\partial^{k+1} u}{\partial x_0^{k+1}}\Big|_{x_0=0}. \tag{20.40}$$

For the new unknown function we will now have a homogeneous Cauchy problem and an equation with another inhomogeneous term $F_1$. It is obvious that $F_1$ together with all its derivatives up to order $k$ inclusive will vanish for $x_0 = 0$.

We consider an arbitrary function $\phi(x_0, x_1, \cdots, x_{2k+1})$ which is sufficiently smooth and differs from zero only in some bounded domain $V_\phi$ contained in a band $0 < x_0 \leqq T_0$. We construct the differential operator $L^*$ of second order adjoint to the operator $L$. The operator $L^*$ will have the form:

$$L^* v = \sum_{i=0}^{2k+1} \sum_{j=0}^{2k+1} \frac{\partial^2}{\partial x_i \partial x_j} (A_{ij} v) - \sum_{i=0}^{2k+1} \frac{\partial}{\partial x_i} (B_i v) + C v.$$

We set:

$$L^* \varphi = \psi.$$

For the function $\phi$, we may pose the Cauchy problem in the domain $x_0 < T_0$. The initial data for this will follow from the condition:

$$\varphi \equiv 0 \quad x_0 > T_0.$$

The theory which we have developed above enables us to write for the function $\phi$ an integral identity of the form

$$B_1 \varphi = S_1 \psi, \tag{20.41}$$

from which the solution of the Cauchy problem may be obtained in the

$$\varphi = B_1^{-1} S_1 \psi.$$

The operators $B_1$ and $S_1$ are analogous to the operators $B$ and $S$ and differ from them merely in that the roll of the direct and reverse cones have been interchanged, in connection with the change in the direction from the surface on which the initial data are given to the domain on which the solution is sought. If one considers that the variable $x_0$ denotes time, then the ordinary Cauchy problem consists in seeking a solution for a later moment of time for given initial conditions at the moment $t = 0$, and the adjoint problem in seeking a solution of an equation at an earlier moment than $t = T_0$, for which the initial data are known.

Multiplying both parts of the equality (20.41) by an arbitrary function $\Phi$ which vanishes for $x_0<0$ and integrating over the whole space, we obtain

$$\overbrace{\int\cdots\int}^{2k+2} \Phi B_1 \varphi \, d\Omega = \overbrace{\int\cdots\int}^{2k+2} \Phi S_1 \psi \, d\Omega.$$

Using the concept of adjoint operators, we rewrite this equation in the form:

$$\overbrace{\int\cdots\int}^{2k+2} \varphi B_1^* \Phi \, d\Omega = \overbrace{\int\cdots\int}^{2k+2} \psi S_1^* \Phi \, d\Omega. \tag{20.42}$$

The right side may be transformed in the following way. We substitute in it $\psi=L^*\phi$. We will have, integrating by parts:

$$\overbrace{\int\cdots\int}^{2k+2} L^*\varphi \cdot S_1^* \Phi \, d\Omega = \overbrace{\int\cdots\int}^{2k+2} \varphi L S_1^* \Phi \, d\Omega. \tag{20.43}$$

The unintegrated terms vanish because of the fact that the function $\phi$ and all its derivatives vanish outside of a bounded domain.

Bringing both integrals in (20.42) to one side and combining them, we obtain using (20.43):

$$\overbrace{\int\cdots\int}^{2k+2} \varphi \, [B_1^*\Phi - L S_1^* \Phi] \, d\Omega = 0.$$

This last equality holds for arbitrary $\phi$. From this, there follows the identity

$$B_1^*\Phi - L S_1^* \Phi = 0.$$

We form now the integral equation:

$$B_1^*\Phi = F, \tag{20.44}$$

where $F$ is the right hand side of equation (19.1). The operator $B_1^*$ has an inverse. Therefore, this equation is always solvable, where

$$\Phi = B_1^{*-1} F.$$

Replacing $\Phi$ in equation (20.44) by its value, we will have:

$$F = L S_1^* B_1^{*-1} F. \tag{20.45}$$

This equality says that the function $S_1^* B_1^{*-1} F$ satisfies the equation $Lu=F$, if the function $F$ has continuous derivatives up to order $k$ inclusive and vanishes for $x_0<0$. From this problem by means of the change of unknown function as we described earlier, we may arrive at the general Cauchy problem for sufficiently smooth $F$ and with sufficiently smooth initial data.

Thus, we have shown the existence of a solution for a linear normal hyperbolic partial differential equation with sufficiently smooth coefficients and sufficiently smooth initial conditions in the case when the number of independent variables is even and is equal to $(2k+2)$.

9. THE CAUCHY PROBLEM FOR AN ODD NUMBER OF VARIABLES. The case in which the number of independent variables is odd, as is well known, can be reduced to the preceding case.

Suppose for example that we need to find the solution of the equation

$$\sum_{i=0}^{2k}\sum_{j=0}^{2k} A_{ij}\frac{\partial^2 u}{\partial x_i\,\partial x_j}+\sum_{i=0}^{2k} B_i\frac{\partial u}{\partial x_i}+Cu=F \tag{20.46}$$

under the conditions

$$\left.\begin{aligned} u\big|_{x_0=0}&=u_0(x_1, x_2, \ldots, x_{2k}),\\ \left.\frac{\partial u}{\partial x_0}\right|_{x_0=0}&=u_1(x_1, x_2, \ldots, x_{2k})\end{aligned}\right\} \tag{20.47}$$

and suppose once more that

$$A_{00}>0, \quad A_{ii}<0, \quad i\neq 0. \tag{20.48}$$

We introduce another independent variable $x_{2k+1}$ in such a way that there is no dependence on this variable for the coefficients, the free term, or the unknown function. The equation (20.46) may be rewritten in the form

$$\sum_{i=0}^{2k}\sum_{j=0}^{2k} A_{ij}\frac{\partial^2 u}{\partial x_i\,\partial x_j}-\frac{\partial^2 u}{\partial x_{2k+1}^2}+\sum_{i=0}^{2k} B_i\frac{\partial u}{\partial x_i}+Cu=F. \tag{20.49}$$

The equation (20.49) will again be normal hyperbolic but with an even number of independent variables, while $A_{ij}, B_i, C$, and $F$ do not depend on the variable $x_{2k+1}$.

By what has been shown, this equation has a solution satisfying the conditions (20.47), which may be interpreted as conditions in the space with $(2k+2)$ variables where the right hand sides in this case do not depend upon $x_{2k+1}$.

The solution of this problem gives us an unknown function $u$ which, as is not hard to see, in its turn does not depend upon $x_{2k+1}$. Indeed, the equation which is satisfied by the function $\partial u/\partial x_{2k+1}$ coincides with the equation for $u$ but the initial data for this function are null. Therefore $\partial u/\partial x_{2k+1} \equiv 0$ is a solution of the equation. However, this solution is unique, as we showed earlier. Therefore, the solution $u$ which we obtained has everywhere $\partial u/\partial x_{2k+1} \equiv 0$ and does not depend on $x_{2k+1}$. However, in this case, the function $u$, as a function of the $2k+1$ variables: $x_0, x_1, \cdots, x_{2k+1}$, satisfies equation (20.46) with initial conditions (20.47), as was to be shown.

§21. **Investigation of linear hyperbolic equations with variable coefficients.**
1. SIMPLIFICATION OF THE EQUATION. In the preceding section we established the existence of a unique solution of a normal hyperbolic linear partial differential equation with variable sufficiently smooth coefficients for sufficiently smooth initial data. The methods developed by us in the first chapter permit us to make a significantly more precise estimate of the order of smoothness of the coefficients of the equation and of the initial data which is needed for the existence of a solution.

Suppose we are given the linear differential equation:

$$\sum_{i=0}^{n} \sum_{j=0}^{n} A_{ij} \frac{\partial^2 u}{\partial x_i \partial x_j} + \sum_{i=0}^{n} B_i \frac{\partial u}{\partial x_i} + Cu = F, \tag{21.1}$$

where $A_{ij} = A_{ji}$, $B_i$ are continuous functions of $x_0, x_1, \cdots, x_n$ while $A_{00} \neq 0$.

In the following, there will be presented additional conditions imposed upon the coefficients $A_{ij}$, $B_i$, $C$ and the free term $F$.

This equation may be simplified with the aid of a change of independent variables.

We set $x_0 = t$ and construct a vector field $\vec{l}$ with the help of the equations

$$l_j = \frac{A_{0j}}{A_{00}}. \tag{21.2}$$

The system of equations (21.2) defines the quantity $\vec{l}$.

We consider the system of ordinary differential equations

$$\frac{dx_s}{dt} = l_s \qquad (s = 1, 2, \ldots, n), \tag{21.3}$$

and let $C_i(t, x_1, \cdots, x_n)$ $(i = 1, 2, \cdots, n)$ be integrals of this system.

The system (21.3) represents the conditions in order that the lines

$$C_i(t, x_1, \ldots, x_n) = \text{const.}$$

be transversals in relation to the planes $x_0 = \text{const}$. We set

$$y_i = C_i(t, x_1, \ldots, x_n). \tag{21.4}$$

In the coordinates $t, y_1, y_2, \cdots, y_n$, the equation (21.1) takes an especially simple form. In it the coefficients $\tilde{A}_{0j}$ for the mixed derivatives $\partial^2 u/\partial y_j \partial t$ vanish.

We shall prove this. As is well known, the coefficients for the mixed derivatives $\tilde{A}_{0j}$ in an equation of hyperbolic type after the transformation will be:

$$\tilde{A}_{0j} = \sum_{l=1} A_{0l} \frac{\partial C_j}{\partial x_l} + A_{00} \frac{\partial C_j}{\partial t}.$$

By virtue of the equations of the system

$$\sum_{l=1}^{n} \frac{\partial C_j}{\partial x_l} \frac{dx_l}{dt} + \frac{\partial C_j}{\partial t} = 0$$

or on the basis of the equations (21.3):

$$\sum_{i=1}^{n} \frac{\partial C_j}{\partial x_i} l_i + \frac{\partial C_j}{\partial t} = 0.$$

Taking (21.2) into consideration, we see that $\tilde{A}_{0j}$ vanishes after the transformation.

Dividing the equation by $\tilde{A}_{00} = A_{00}$, we arrive at a new equation of the form:

$$\frac{\partial^2 u}{\partial t^2} - \sum_{i=1}^{n} \sum_{j=1}^{n} \tilde{A}_{ij} \frac{\partial^2 u}{\partial y_i \partial y_j} - \sum_{i=1}^{n} \tilde{B}_i \frac{\partial u}{\partial y_i} + h \frac{\partial u}{\partial t} - Cu = F.$$

We also set:

$$u = e^{-\frac{1}{2}\int_0^t h\, dt_1} v.$$

Then

$$\frac{\partial u}{\partial t} = \frac{\partial v}{\partial t} e^{-\frac{1}{2}\int_0^t h\, dt_1} + v \frac{\partial}{\partial t} e^{-\frac{1}{2}\int_0^t h\, dt_1};$$

$$\frac{\partial^2 u}{\partial t^2} = \frac{\partial^2 v}{\partial t^2} e^{-\frac{1}{2}\int_0^t h\, dt_1} - h \frac{\partial v}{\partial t} e^{-\frac{1}{2}\int_0^t h\, dt_1} + v \frac{\partial^2}{\partial t^2} e^{-\frac{1}{2}\int_0^t h\, dt_1}$$

$$\frac{\partial^2 u}{\partial t^2} + h \frac{\partial u}{\partial t} = \frac{\partial^2 v}{\partial t^2} e^{-\frac{1}{2}\int_0^t h\, dt_1} + v\left(\frac{\partial^2}{\partial t^2} + h \frac{\partial}{\partial t}\right) e^{-\frac{1}{2}\int_0^t h\, dt}.$$

After this change in the equation, the term containing $\partial u/\partial t$ also vanishes. In the following for the investigation of the general linear equation, we shall always assume from the very beginning that terms involving $\partial^2 u/\partial x_i \partial t$ and $\partial u/\partial t$ do not appear in the equation.

We consider the equation:

$$Lu \equiv \sum_{i=1}^{n} \sum_{j=1}^{n} A_{ij} \frac{\partial^2 u}{\partial x_i \partial x_j} - \frac{\partial^2 u}{\partial t^2} = F - \sum_{i=1}^{n} B_i \frac{\partial u}{\partial x_i} - Cu, \qquad (21.5)$$

where $A_{ij}, B_i, C$, and $F$ are given functions of the point $x_1, x_2, \cdots, x_n, t$. Suppose that at every point of space and for every moment of time:

$$A(p) \equiv \sum_{i=1}^{n} \sum_{j=1}^{n} A_{ij} p_i p_j > c \sum_{i=1}^{n} p_i^2; \; A_{ij} = A_{ji}, \qquad (21.6)$$

where $c > 0$ is some constant.

We seek the solution of this equation satisfying the conditions:

$$u\,|_{t=0} = u_0(x_1, x_2, \ldots, x_n); \; \left.\frac{\partial u}{\partial t}\right|_{t=0} = u_1(x_1, x_2, \ldots, x_n). \qquad (21.7)$$

We shall consider two different formulations of this problem.

2. FORMULATION OF THE CAUCHY PROBLEM FOR GENERALIZED SOLUTIONS. Let $\Omega$ be a domain of the $(n+1)$-dimensional space of $t, x_1, \cdots, x_n$. The function $u(t, x_1, \cdots, x_n)$ defined on $\Omega$ and summable on every bounded portion of $\Omega$ is said to be a generalized solution of equation (21.5) if for every twice continuously differentiable function $v$ on the whole space which vanishes outside of some bounded domain $\Omega' \subset \Omega$, we have the equality:

$$\int \cdots \int_{\Omega} uL^* v \, d\Omega = \int \cdots \int_{\Omega} vF \, d\Omega,$$

where $L^* v$ is the adjoint operator defined in §20, item 8.

The first formulation of the problem consists of the following.

To find a generalized solution of equation (21.5) which, for every value of $t$, is an element of the space $W_2^{(1)}$ while $\partial u/\partial t$ for every value of $t$ is an element of the space $L_2 = W_2$.

The trajectory in $W_2^{(1)}$ and $W_2$ defined by the pair of functions $u$ and $\partial u/\partial t$ in this pair of spaces should be continuous and satisfy the initial conditions (21.7).

We impose upon $A_{ij}, B_i, C$, and $F$ restrictions which we call conditions (o), or conditions for the existence in the generalized $s$ Cauchy problem. These conditions are the following.

(1) The coefficients $A_{ij}$ are continuous, have first derivatives, and satisfy the inequalities

$$|A_{ij}|<A; \quad \left|\frac{\partial A_{ij}}{\partial x_k}\right|<A; \quad \left|\frac{\partial A_{ij}}{\partial t}\right|<A. \tag{21.8}$$

(2) The coefficients $B_i$ are continuous and satisfy the conditions

$$|B_i|<A.$$

(3) The first generalized derivatives of the functions $B_i$ exist and satisfy the inequalities

$$\Big\{\overbrace{\int\cdots\int}^{n}_{\Omega(t)}\Big[\sum_{j=1}^{n}\Big(\frac{\partial B_i}{\partial x_j}\Big)^2+\Big(\frac{\partial B_i}{\partial t}\Big)^2\Big]^{\frac{n+\varepsilon}{2}}d\Omega\Big\}^{\frac{2}{n+\varepsilon}}<$$
$$<A(t)<A \quad (i=1,\,2,\,\ldots,\,n). \tag{21.9}$$

(We note that from the Imbedding Theorem and (3) there follows the boundedness of the $B_i$.) The domain $\Omega(t)$ in the plane $x_1, x_2, \cdots, x_n, t=$ const. is given in some way. The form of $A(t)$, it is clear, depends upon $\Omega(t)$.

(4) The coefficient $C$ satisfies the condition

$$\Big\{\overbrace{\int\cdots\int}^{n}_{\Omega(t)}|C|^{n+\varepsilon}d\Omega\Big\}^{\frac{1}{n+\varepsilon}}\leqslant A(t)<A. \tag{21.10}$$

(5) The first generalized derivatives of $C$ exist and satisfy the inequality:

$$\Big\{\overbrace{\int\cdots\int}^{n}_{\Omega(t)}\Big[\sum_{i=1}^{n}\Big|\frac{\partial C}{\partial x_i}\Big|+\Big|\frac{\partial C}{\partial t}\Big|\Big]^{\frac{n+\varepsilon}{2}}d\Omega\Big\}^{\frac{2}{n+\varepsilon}}\leqslant A(t)<A. \tag{21.11}$$

(6) The free term $F$ satisfies the inequality

$$\Big[\overbrace{\int\cdots\int}^{n}|F|^2d\Omega\Big]^{\frac{1}{2}}\leqslant F(t)<F, \tag{21.12}$$

where $F$ is a constant.

(7) The generalized first derivatives of $F$ exist and satisfy the inequalities

$$\Big\{\overbrace{\int\cdots\int}^{n}_{\Omega(t)}\Big[\sum_{i=1}^{n}\Big(\frac{\partial F}{\partial x_i}\Big)^2+\Big(\frac{\partial F}{\partial t}\Big)^2\Big]d\Omega\Big\}^{\frac{1}{2}}\leqslant F(t)<F. \tag{21.13}$$

(8) The function

$$u_0 \in W_2^{(2)}. \tag{21.14}$$

(9) The function

$$u_1 \in W_2^{(1)}. \tag{21.15}$$

To the question of the solution of the Cauchy problem in the first formulation, the following theorem gives the answer:

THEOREM 1. *The fulfillment of the conditions* (o) *is sufficient for the existence of the solution of the problem in the first formulation. The solution in this case depends continuously upon the initial data together with its first derivatives in the spaces* $W_2^{(1)}$ *and* $L_2$.

Before proving this theorem, we need to establish some important inequalities.

3. BASIC INEQUALITIES. Let $w$ be an arbitrary function of the variables $x_1, x_2, \cdots, x_n, t$ which is continuous together with its derivatives of second order.

We introduce some inequalities analogous to the inequality (17.7). We consider a domain $V$ in the space $(x_1, x_2, \cdots, x_n, t)$ with boundary surface $S$. Then from the Gauss-Ostrogradsky formula:

$$\begin{gathered}\int_S \cdots \int \Big\{ \Big[ \sum_{i=1}^{n} \sum_{j=1}^{n} A_{ij} \frac{\partial w}{\partial x_i} \frac{\partial w}{\partial x_j} + \Big(\frac{\partial w}{\partial t}\Big)^2 \Big] \cos \vec{nt} - \\ - 2 \sum_{i=1}^{n} \sum_{j=1}^{n} A_{ij} \frac{\partial w}{\partial x_j} \frac{\partial w}{\partial t} \cos \vec{nx_i} \Big\} dS = \\ = \int_V \cdots \int \Big\{ -2 \frac{\partial w}{\partial t} \Big[ \sum_{i=1}^{n} \sum_{j=1}^{n} A_{ij} \frac{\partial^2 w}{\partial x_i \partial x_j} - \frac{\partial^2 w}{\partial t^2} \Big] + \\ + \sum_{i=1}^{n} \sum_{j=1}^{n} \Big( \frac{\partial A_{ij}}{\partial t} \frac{\partial w}{\partial x_i} \frac{\partial w}{\partial x_j} - 2 \frac{\partial A_{ij}}{\partial x_j} \frac{\partial w}{\partial x_i} \frac{\partial w}{\partial t} \Big) \Big\} dV.\end{gathered} \tag{21.16}$$

Such a transformation is possible because of the differentiability of all the functions appearing in the form.

If on some part of the surface $S$

$$\sum_{i=1}^{n}\sum_{j=1}^{n} A_{ij}\cos \vec{nx_i}\cos \vec{nx_j} - \cos^2 \vec{nt} \leqslant 0,$$

then on it, the sign of the quadratic form:

$$\left[\sum_{i=1}^{n}\sum_{j=1}^{n} A_{ij}\frac{\partial w}{\partial x_i}\frac{\partial w}{\partial x_j}+\left(\frac{\partial w}{\partial t}\right)^2\right]\cos \vec{nt} - 2\sum_{i=1}^{n}\sum_{j=1}^{n} A_{ij}\frac{\partial w}{\partial x_i}\frac{\partial w}{\partial t}\cos \vec{nx_i} =$$

$$=\frac{1}{\cos \vec{nt}}\sum_{i=1}^{n}\sum_{j=1}^{n} A_{ij}\left(\frac{\partial w}{\partial x_i}\cos \vec{nt}-\frac{\partial w}{\partial t}\cos \vec{nx_i}\right)\left(\frac{\partial w}{\partial x_j}\cos \vec{nt}-\frac{\partial w}{\partial t}\cos \vec{nx_j}\right)+$$

$$+\frac{\cos^2 \vec{nt} - \sum_{i=1}^{n}\sum_{j=1}^{n} A_{ij}\cos \vec{nx_i}\cos \vec{nx_j}}{\cos \vec{nt}}\left(\frac{\partial w}{\partial t}\right)^2 \tag{21.17}$$

coincides with the sign of $\cos\vec{nt}$.

We consider a domain $S_1$ in the plane $t=0$ and let $(S_1, [0, T])$ be the cylinder constructed over $S_1$ and bounded by the planes $t=0$ and $t=T>0$. We set $c_1 = n\cdot \max_{i,j,(S,[0,T])} |A_{ij}|$ and construct through the boundary of the domain $S_1$ the right cone $S_3$ (Figure 12) so that on it

$$\cos^2 nt > \frac{c_1}{1+c_1}.$$

The truncated cone lying in the cylinder $(S_1, [0,T])$ is called the fundamental domain and is denoted by $\Omega^*$; the upper base of this cone is denoted by $S_2$.

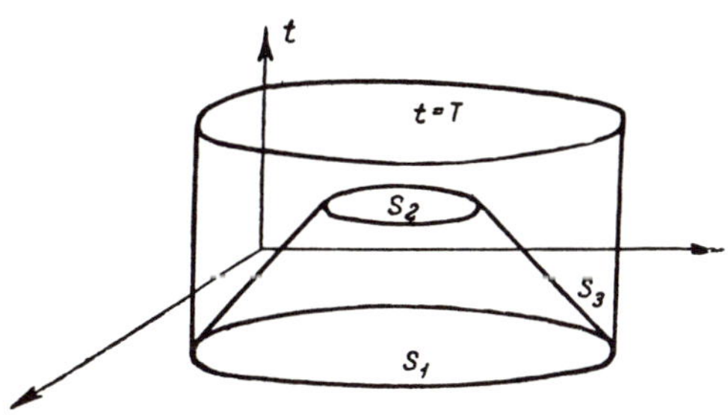


Figure 12

Then on the surface $S_3$, the quadratic form will be positive, since on it $\cos\vec{nt}>0$, and

$$\sum_{i=1}^{n}\sum_{j=1}^{n} A_{ij}\cos \vec{nx_i}\cos \vec{nx_j} - \cos^2 \vec{nt} < c_1\sum_{i=1}^{n}\cos^2 \vec{nx_i} - \cos^2 \vec{nt} =$$

$$=c_1(1-\cos^2 \vec{nt}) - \cos^2 \vec{nt} = c_1-(1+c_1)\cos^2 \vec{nt} < c_1-(1+c_1)\frac{c_1}{1+c_1}=0.$$

From this, we obtain the inequality:

$$\int_{S_2}\cdots\int\left\{\sum_{i=1}^{n}\sum_{j=1}^{n}A_{ij}\frac{\partial w}{\partial x_i}\frac{\partial w}{\partial x_j}+\left(\frac{\partial w}{\partial t}\right)^2\right\}dS\leqslant$$

$$\leqslant\int_{S_1}\cdots\int\left\{\sum_{i=1}^{n}\sum_{j=1}^{n}A_{ij}\frac{\partial w}{\partial x_i}\frac{\partial w}{\partial x_j}+\left(\frac{\partial w}{\partial t}\right)^2\right\}dS+$$

$$+\int_{\Omega^*}\cdots\int\left\{\sum_{i=1}^{n}\sum_{j=1}^{n}\frac{\partial A_{ij}}{\partial t}\frac{\partial w}{\partial x_i}\frac{\partial w}{\partial x_j}-2\sum_{i=1}^{n}\sum_{j=1}^{n}\frac{\partial A_{ij}}{\partial x_i}\frac{\partial w}{\partial x_j}\frac{\partial w}{\partial t}-\right.$$

$$\left.-2\frac{\partial w}{\partial t}Lw\right\}d\Omega, \tag{21.18}$$

which is analogous to the inequality (17.7) obtained earlier.

We denote by $\Omega(t)$ the intersection of the cone $\Omega^*$ with the plane $t=\text{const}$. Suppose that for the function $w$ we have the inequality

$$\int_{\Omega(t)}\cdots\int(Lw)^2\,dx_1\,\ldots\,dx_n=L(t)<L. \tag{21.19}$$

Then we may estimate the integral over $S_2$ in the equality (21.18) in the following way:

$$\left|\int_{\Omega^*}\cdots\int\left\{\sum_{i=1}^{n}\sum_{j=1}^{n}\frac{\partial A_{ij}}{\partial t}\frac{\partial w}{\partial x_i}\frac{\partial w}{\partial x_j}-2\sum_{i=1}^{n}\sum_{j=1}^{n}\frac{\partial A_{ij}}{\partial x_i}\frac{\partial w}{\partial x_j}\frac{\partial w}{\partial t}-\right.\right.$$

$$\left.\left.-2\frac{\partial w}{\partial t}Lw\right\}d\Omega\right|\leqslant\int_0^T\left\{A\int_{\Omega(t_1)}\cdots\int\left[\sum_{i=1}^{n}\sum_{j=1}^{n}\left|\frac{\partial w}{\partial x_i}\right|\left|\frac{\partial w}{\partial x_j}\right|+\right.\right.$$

$$\left.\left.+2n\sum_{i=1}^{n}\left|\frac{\partial w}{\partial x_j}\right|\left|\frac{\partial w}{\partial t}\right|+\frac{2}{A}\left|\frac{\partial w}{\partial t}\right||Lw|\right]dx_1\,dx_2\,\ldots\,dx_n\right\}dt_1\leqslant$$

$$\leqslant C_1\int_0^T\left\{\int_{\Omega(t_1)}\cdots\int\left[\sum_{i=1}^{n}\left(\frac{\partial w}{\partial x_i}\right)^2+\left(\frac{\partial w}{\partial t}\right)^2+2\left|\frac{\partial w}{\partial t}\right||Lw|\right]\times\right.$$

$$\left.\times\,dx_1\,\ldots\,dx_n\right\}dt_1.$$

We set:

$$\int_{\Omega(t)}\cdots\int\left[\sum_{i=1}^{n}\left(\frac{\partial w}{\partial x_i}\right)^2+\left(\frac{\partial w}{\partial t}\right)^2\right]d\Omega=K_1(t\,|\,w). \tag{21.20}$$

Obviously, from (21.6) and the condition $|A_{ij}| < c_1/n$, we have:

$$c_2 K_1(t\,|\,w) \leqslant \left| \int_{\Omega(t)} \cdots \int \left\{ \sum_{i=1}^{n} \sum_{j=1}^{n} A_{ij} \frac{\partial w}{\partial x_i} \frac{\partial w}{\partial x_j} + \left(\frac{\partial w}{\partial t}\right)^2 \right\} dx_1 \ldots dx_n \right| \leqslant$$

$$\leqslant C_2 K_1(t\,|\,w). \tag{21.21}$$

If we also take into account that

$$\int_{\Omega(t)} \cdots \int \left| \frac{\partial w}{\partial t} \right| |Lw| \, dx_1 \ldots dx_n \leqslant$$

$$\leqslant \left( \int_{\Omega(t)} \cdots \int \left| \frac{\partial w}{\partial t} \right|^2 dx_1 \ldots dx_n \right)^{\frac{1}{2}} \left( \int_{\Omega(t)} \cdots \int [Lw]^2 \, dx_1 \ldots dx_n \right)^{\frac{1}{2}} \leqslant$$

$$\leqslant L(t)^{\frac{1}{2}} K_1(t\,|\,w)^{\frac{1}{2}}$$

we obtain from (21.18)

$$c_2 K_1(t\,|\,w) \leqslant C_2 K_1(0\,|\,w) + C_1 \int_0^t [K_1(t_1\,|\,w) +$$

$$+ K_1(t_1\,|\,w)^{\frac{1}{2}} L(t_1)^{\frac{1}{2}}] \, dt_1. \tag{21.22}$$

For our purposes it is necessary to derive another inequality. We set

$$\int_{\Omega(t)} \cdots \int w^2 \, dx_1 \ldots dx_n = K_0(t\,|\,w). \tag{21.23}$$

It is obvious that

$$K_0(t_1, w) \leqslant \int_{\Omega(t)} \cdots \int w^2(t_1, x_1, \ldots, x_n) \, dx_1 \ldots dx_n,$$

if $t_1 > t$. Therefore

$$\left. \frac{dK_0(t_1\,|\,w)}{dt_1} \right|_{t_1=t} \leqslant \left[ \frac{d}{dt_1} \int_{\Omega(t)} \cdots \int w^2(t_1, x_1, \ldots, x_n) \times \right.$$

$$\left. \times \, dx_1 \ldots dx_n \right]_{t_1=t} = 2 \int_{\Omega(t)} \cdots \int w \frac{\partial w}{\partial t} \, dx_1 \ldots dx_n \leqslant$$

$$\leqslant 2 \left[ \int_{\Omega(t)} \cdots \int w^2 \, dx_1 \ldots dx_n \right]^{\frac{1}{2}} \left[ \int_{\Omega(t)} \cdots \int \left( \frac{\partial w}{\partial t} \right)^2 dx_1 \ldots dx_n \right]^{\frac{1}{2}} \leqslant$$

$$\leqslant 2 [K_0(t\,|\,w)]^{\frac{1}{2}} [K_1(t\,|\,w)]^{\frac{1}{2}}.$$

Then:

$$K_0(t\,|\,w) = \int_0^t \frac{d}{dt_1} K_0(t_1\,|\,w)\,dt_1 + K_0(0\,|\,w) \leqslant$$

$$\leqslant K_0(0\,|\,w) + 2\int_0^t [K_0(t_1\,|\,w)]^{\frac{1}{2}} [K_1(t_1\,|\,w)]^{\frac{1}{2}} dt_1. \qquad (21.24)$$

The inequalities (21.22) and (21.24) are basic for the following.

4. A LEMMA ON ESTIMATES FOR APPROXIMATING SOLUTIONS. We now prove a basic lemma. We consider an equation satisfying the conditions (o).

We replace in it the functions $A_{ij}$, $B_i$, $C$, and $F$ by their averaged values $A_{ijh}$, $B_{ih}$, $C_h$, and $F_h$ with respect to $x_1, x_2, \cdots, x_n, t$, and the initial data $u_0$ and $u_1$ by their averaged values $u_{0h}$ and $u_{1h}$ with respect to $x_1, x_2, \cdots, x_n$. The new equation will be called the averaged equation.

The new equations with the new conditions have solutions $u_h$, as shown above.

We form for these solutions the integrals

$$K_0(t\,|\,u_h), \quad K_1(t\,|\,u_h)$$

and

$$K_2(t\,|\,u_h) = \int \cdots \int_{\Omega(t)} \left[\sum_{i=1}^n \sum_{j=1}^n \left(\frac{\partial^2 u_h}{\partial x_i \partial x_j}\right)^2 + \sum_{i=1}^n \left(\frac{\partial^2 u_h}{\partial x_i \partial t}\right)^2 + \left(\frac{\partial^2 u_h}{\partial t^2}\right)^2\right] d\Omega.$$

Then for the functions $K_0(t|, u_h)$, $K_1(t|, u_h)$ and $K_2(t|, u_h)$, we have the inequalities

$$|K_i(t\,|\,u_h)| < y_i(t) \quad i = 0, 1, 2, \ldots \qquad (21.25)$$

where $y_i(t)$ are the solutions of the system of equations:

$$\left.\begin{aligned} \frac{dy_0^{\frac{1}{2}}}{dt} &= y_1^{\frac{1}{2}}; \\ \frac{dy_1^{\frac{1}{2}}}{dt} &= M(y_0^{\frac{1}{2}} + y_1^{\frac{1}{2}} + F(t)); \\ \frac{dy_2^{\frac{1}{2}}}{dt} &= M(y_0^{\frac{1}{2}} + y_1^{\frac{1}{2}} + y_2^{\frac{1}{2}} + \dot{F}(t)) \end{aligned}\right\} \qquad (21.26)$$

for the conditions

$$\left.\begin{aligned} y_0\big|_{t=0} &= CK_0(0\,|\,u); \\ y_1\big|_{t=0} &= CK_1(0\,|\,u); \\ y_2\big|_{t=0} &= CK_2(0\,|\,u), \end{aligned}\right\} \tag{21.27}$$

while the constant $M$ and the function $F(t)$ do not depend upon the radius of averaging $h$. Thus, $K_0(t|, u_h)$, $K_1(t|u_h)$ and $K_2(t|u_h)$ have estimates not depending upon $h$.

For the proof of the lemma, we return to the inequality (21.22) and set:

$$Lw = F(x_1, x_2, \ldots, x_n, t) - \sum_{i=1}^{n} B_i \frac{\partial w}{\partial x_i} - Cw, \tag{21.28}$$

where the $B_i$ are continuous and bounded

$$|B_i| < A, \tag{21.29}$$

and the function $C$ satisfies the inequality

$$\left[\int \cdots \int |C|^{n+\varepsilon} d\Omega\right]^{\frac{1}{n+\varepsilon}} \leqslant A(t) \leqslant A, \tag{21.30}$$

and suppose in addition that:

$$\left[\int_{\Omega(t)} \cdots \int F^2(x_1, x_2, \ldots, x_n, t)\, dx_1 \ldots dx_n\right]^{\frac{1}{2}} = F(t) < F. \tag{21.31}$$

In that case, the quantity $Lw$ may be estimated without difficulty. We shall have in the domain

$$\|Lw\|_{L_2} \leqslant \|F\|_{L_2} + \left\|\sum_{i=1}^{n} B_i \frac{\partial w}{\partial x_i}\right\|_{L_2} + \|Cw\|_{L_2}. \tag{21.32}$$

It is obvious for this that:

$$\|F\|_{L_2} = F(t) < F; \quad \left\|\sum_{i=1}^{n} B_i \frac{\partial w}{\partial x_i}\right\|_{L_2} \leqslant A\,\|w\|_{L_2^{(1)}}.$$

For the estimate of $\|Cw\|_{L_2}$, we use Hölder's inequality and Minkowski's theorem. We have

$$\left[\int \underset{\Omega(t)}{\cdots} \int [Cw]^2 d\Omega\right]^{\frac{1}{2}} \leqslant \left[\int \underset{\Omega(t)}{\cdots} \int |C|^{n+\varepsilon} d\Omega\right]^{\frac{1}{n+\varepsilon}} \times$$

$$\times \left[\int \underset{\Omega(t)}{\cdots} \int |w|^{\frac{1}{\frac{1}{2}-\frac{1}{n+\varepsilon}}} d\Omega\right]^{\frac{1}{2}-\frac{1}{n+\varepsilon}} \leqslant A(t)\,\|w\|_{W_2^{(1)}} \leqslant \tag{21.33}$$

$$\leqslant A\,[K_1(t|w)^{\frac{1}{2}} + K_0(t|w)^{\frac{1}{2}}].$$

Or finally

$$L(t) \leqslant C_1\,[K_1(t|w)^{\frac{1}{2}} + K_0(t|w)^{\frac{1}{2}} + F(t)], \tag{21.34}$$

where $C$ is a constant not depending on $w$.

Setting this estimate in the inequality (21.22), we obtain:

$$K_1(t|w) \leqslant L_1 K_1(0|w) + L_2 \int_0^t K_1(t_1|w)^{\frac{1}{2}} [K_0(t_1|w)^{\frac{1}{2}} +$$

$$+ K_1(t_1|w)^{\frac{1}{2}} + F(t_1)]\,dt. \tag{21.35}$$

Together with (21.24), we obtain a system of inequalities from which it is easy to estimate the functions $K_0(t|w)$ and $K_1(t|w)$.

We denote by $y_0(t)$ and $y_1(t)$ the functions satisfying the system of equations:

$$\left.\begin{aligned} y_1(t) &= L_1 K_1(0|w) + L_2 \int_0^t y_1^{\frac{1}{2}} [y_0^{\frac{1}{2}} + y_1^{\frac{1}{2}} + F]\,dt_1 \\ y_0(t) &= K_0(0|w) + 2\int_0^t y_0^{\frac{1}{2}} y_1^{\frac{1}{2}}\,dt_1. \end{aligned}\right\} \tag{21.36}$$

It is not hard to prove that it is always true that:

$$y_0(t) > K_0(t|w), \quad y_1(t) > K_1(t|w).$$

Indeed the functions $y_0$ and $y_1$ may be found as the limits of the increasing sequences: $y_0^{(n)}$ and $y_1^{(n)}$ defined by the formulas:

$$y_1^{(n)} = L_1 K_1(0|w) + L_2 \int_0^t y_1^{(n-1)\frac{1}{2}} (y_0^{(n-1)\frac{1}{2}} + y_1^{(n-1)\frac{1}{2}} + F)\,dt_1;$$

$$y_0^{(n)} = K_0(0|w) + 2\int_0^t y_1^{(n-1)\frac{1}{2}} y_0^{(n-1)\frac{1}{2}}\,dt,$$

if for the first terms of these sequences we take $K_0(t|w)$ and $K_1(t|w)$. Our estimate is proved. The functions $y_0$ and $y_1$ are solutions of the system of differential equations

$$\left.\begin{aligned} y_0' &= 2y_0^{\frac{1}{2}} y_1^{\frac{1}{2}} \\ y_1' &= L_2 y_1^{\frac{1}{2}} (y_0^{\frac{1}{2}} + y_1^{\frac{1}{2}} + F). \end{aligned}\right\} \tag{21.37}$$

Taking for new variables $y_0^{1/2} = z_0$ and $y_1^{1/2} = z_1$, we obtain:

$$\left.\begin{aligned} z_0' &= z_1; \\ z_1' &= \frac{L_2}{2}(z_0 + z_1 + F). \end{aligned}\right\} \tag{21.38}$$

It remains for us to estimate $K_2(t|w)$.

We form the new equations for the functions $u = \partial u_h/\partial x_l$, $w = \partial u_h/\partial t$. Differentiating the averaged equation (21.28) with respect to $x_l$ and $t$, we have:

$$\left.\begin{aligned} L\frac{\partial u_h}{\partial x_l} &= \frac{\partial F_h}{\partial x_l} - \sum_{i=1}^{n}\sum_{j=1}^{n} \frac{\partial A_{ijh}}{\partial x_l}\frac{\partial^2 u_h}{\partial x_i \partial x_j} - \sum_{i=1}^{n} B_{ih}\frac{\partial^2 u_h}{\partial x_i \partial x_l} - \\ &\quad - \sum_{i=1}^{n}\frac{\partial B_{ih}}{\partial x_l}\frac{\partial u}{\partial x_i} - C_h\frac{\partial u}{\partial x_l} - u_h\frac{\partial C_h}{\partial x_l}. \\ L\frac{\partial u_h}{\partial t} &= \frac{\partial F_h}{\partial t} - \sum_{i=1}^{n}\sum_{j=1}^{n}\frac{\partial A_{ijh}}{\partial t}\frac{\partial^2 u_h}{\partial x_i\partial x_j} - \sum_{i=1}^{n} B_{ih}\frac{\partial^2 u_1}{\partial x_i \partial t} - \\ &\quad - \sum_{i=1}^{n}\frac{\partial B_{ih}}{\partial t}\frac{\partial u_h}{\partial x_i} - C_h\frac{\partial u_h}{\partial t} - u_h\frac{\partial C_h}{\partial t}. \end{aligned}\right\} \tag{21.39}$$

We now set:

$$K_2(t|w) = \int_{\Omega(t)} \cdots \int \Big[\sum_{i=1}^{n}\sum_{j=1}^{n}\Big(\frac{\partial^2 w}{\partial x_i \partial x_j}\Big)^2 + \sum_{i=1}^{n}\Big(\frac{\partial^2 w}{\partial x_i \partial t}\Big)^2 + \Big(\frac{\partial^2 w}{\partial t^2}\Big)^2\Big]\, d\Omega. \tag{21.40}$$

We may easily estimate the quantity $K_2(E|w)$ with the aid of an application of the inequality (21.22). Applying this separately to each derivative and noting that

$$\sum_{i=1}^{n} K_1\left(t \,\Big|\, \frac{\partial w}{\partial x_i}\right) + K_1\left(t \,\Big|\, \frac{\partial w}{\partial t}\right) = K_2(t \,|\, w),$$

we obtain

$$c_2 K_2(t \,|\, u_h) \leqslant C_2 K_2(0 \,|\, u_2) + C_1 \int_0^t \Big\{ K_2(t_1 \,|\, w) +$$

$$+ K_2(t_1 \,|\, w)^{\frac{1}{2}} \Big[ \sum_{i=1}^{n} [L_l(t_1)]^{\frac{1}{2}} + L_0(t_1)^{\frac{1}{2}} \Big] \Big\} \, dt_1,$$

where by $L_l(t)$ we denote $\| L(\partial u_h / \partial x_l) \|_{L_2}$, and $L_0(t) = \| L(\partial u / \partial t) \|_{L_2}$. We obtain:

$$\| L_l(t) \| \leqslant \left\| \frac{\partial F_h}{\partial x_l} \right\|_{L_2} + A \sum_{i=1}^{n} \sum_{j=1}^{n} \left\| \frac{\partial^2 u_h}{\partial x_i \, \partial x_j} \right\|_{L_2} + A \sum_{i=1}^{n} \left\| \frac{\partial^2 u_h}{\partial x_l \, \partial x_i} \right\|_{L_2} +$$

$$+ \sum_{i=1}^{n} \left\| \frac{\partial B_{ih}}{\partial x_l} \frac{\partial u_h}{\partial x_i} \right\|_{L_2} + \left\| C_h \frac{\partial u_h}{\partial x_l} \right\|_{L_2} + \left\| \frac{\partial C_h}{\partial x_l} u_h \right\|_{L_2}.$$

Estimating $\| \partial B_i / \partial x_l \; \partial u_h / \partial x_i \|$ and $\| C(\partial u_h / \partial x_l) \|$ as above, we obtain:

$$\left\| \frac{\partial B_i}{\partial x_l} \frac{\partial u_h}{\partial x_i} \right\| \leqslant CA(t) \, [K_2(t \,|\, u_h)^{\frac{1}{2}} + K_1(t \,|\, u_h)^{\frac{1}{2}}]; \tag{21.41}$$

$$\left\| C_h \frac{\partial u_h}{\partial x_i} \right\| \leqslant CA(t) \, [K_2(t \,|\, u_h)^{\frac{1}{2}} + K_1(t \,|\, u_h)^{\frac{1}{2}}]. \tag{21.42}$$

Finally, for the estimate of $\| (\partial C_h / \partial x_l) \, u_h \|$, we again use the Imbedding Theorem. We will have

$$\left\| \frac{\partial C_h}{\partial x_l} u_h \right\|_{L_2} \leqslant \Big[ \int_{\Omega(t)} \cdots \int \left( \frac{\partial C_h}{\partial x_l} \right)^{\frac{n+\varepsilon}{2}} d\Omega \Big]^{\frac{2}{n+\varepsilon}} \times$$

$$\times \Big[ \int_{\Omega(t)} \cdots \int | u_h |^{\frac{1}{\frac{1}{2} - \frac{2}{n+\varepsilon}}} d\Omega \Big]^{\frac{1}{2} - \frac{2}{n+\varepsilon}} \leqslant A(t) \, \| u_h \|_{W_2^{(2)}} \leqslant$$

$$\leqslant A(t) \, [K_0(t \,|\, u_h)^{\frac{1}{2}} + K_1(t \,|\, u_h)^{\frac{1}{2}} + K_2(t \,|\, u_h)^{\frac{1}{2}}]. \tag{21.43}$$

Gathering together all the above, we obtain

$$K_2(t\,|\,u_h) \leqslant M_1\Big\{K_2(0\,|\,u_h) + \int_0^t [K_0(t_1\,|\,u_h)^{\frac{1}{2}} + K_1(t_1\,|\,u_h)^{\frac{1}{2}} +$$

$$+ K_2(t_1\,|\,u_h)^{\frac{1}{2}}]\, K_2(t_1\,|\,u_h)^{\frac{1}{2}} dt_1. \qquad (21.44)$$

We set again:

$$y_2(t) = M_1\Big[K_2(0\,|\,u_h) + \int_0^t y_2(t_1)^{\frac{1}{2}}[y_0(t_1)^{\frac{1}{2}} + y_1(t_1)^{\frac{1}{2}} +$$

$$+ y_2(t_1)^{\frac{1}{2}} + F(t_1)\Big]dt. \qquad (21.45)$$

Then

$$K_2(t\,|\,u_h) < y_2(t). \qquad (21.46)$$

For $y_2(t)$ we have the equation:

$$y_2'(t) = M_1 y_2(t)^{\frac{1}{2}}[y_0^{\frac{1}{2}} + y_1^{\frac{1}{2}} + y_2^{\frac{1}{2}} + F(t)]. \qquad (21.47)$$

Setting $y_i^{1/2} = z_i$ $(i = 0, 1, 2)$, we will have

$$z_2 = \frac{M_1}{2}(z_0 + z_1 + z_2 + F), \qquad (21.48)$$

from which we conclude the uniform boundedness of the function $z_2$ satisfying equation (21.48).

The lemma is proved.

5. SOLUTION OF THE GENERALIZED PROBLEM. We pass to the proof of the basic theorem.

The sequence $u_h$ by our estimates will have a bounded integral for the sum of the squares of all second derivatives in $(n+1)$ variables. In other words $u_h \in L_2^{(2)}$, while $\| u_h \|_{L_2^{(2)}} < M$, where $M$ is a constant not depending on $h$. From the uniform boundedness of $\| u_h \|_{L_2^{(2)}}$ and $\| u_h \|_{L_{2_2}}$ we conclude:

$$\| u_h \|_{W_2^{(2)}} < M_1. \qquad (21.49)$$

By the Imbedding Theorem, $u_h$ and its first derivatives are compact in $W_2^{(1)}$ on an arbitrary $n$-dimensional set $t=$ const., while the values of these functions considered as elements of $W_2^{(1)}$ and depending upon $t$ yield an equicontinuous family of continuous trajectories in $W_2^{(1)}$.

Consequently, from the $u_h$ we may choose a subsequence converging uniformly to a trajectory, continuous in $W_2^{(1)}$, together with its first derivatives. Passing to the limit in the inequalities (21.25) and using the theorem on the existence of generalized derivatives, we obtain

$$K_0(t\,|\,u) \leqslant y_0(t); \quad K_1(t\,|\,u) \leqslant y_1(t); \quad K_2(t\,|\,u) \leqslant y_2(t). \tag{21.50}$$

We denote the limit function by $u$. The subsequence $u_h$ obviously satisfies the integral identity

$$\underset{\substack{\Omega(t)\\0<t<T}}{\int \cdots \int} u_h \overset{*}{L_h} v \, d\Omega \, dt = \underset{\substack{\Omega(t)\\0<t<T}}{\int \cdots \int} v F_h \, d\Omega \, dt, \tag{21.51}$$

where $v$ is an arbitrary function of the variables $x_1, x_2, \cdots, x_n, t$ continuous, together with its derivatives up to third order inclusive, and vanishing outside of some domain $V_v$ lying completely in the interior of the domain $0<t<T$; $\Omega(t)$.

By $L_h^* v$ we denote the operator adjoint to the operator $L_h$.

Passing to the limit as $h \to 0$, we obtain:

$$\underset{\substack{\Omega(t)\\0<t<T}}{\int \cdots \int} u L^* v \, d\Omega \, dt = \underset{\substack{\Omega(t)\\0<t<T}}{\int \cdots \int} v F \, d\Omega \, dt. \tag{21.52}$$

This equality, by the definition, says that the function is a generalized solution of equation (21.1).

It is easy to see that this function satisfies the initial conditions.

The continuous dependence of the solutions upon the initial data is also proved without difficulty.

If $F(t)$ is sufficiently small and $K_0(0\,|\,u)$ and $K_1(0\,|\,u)$ are small in their turn, then the functions $y_0(t)$ and $y_1(t)$ will be small. By the inequality (21.25), there follows the smallness of $K_0(t_1\,|\,u)$ and $K_1(t\,|\,u)$.

The theorem is proved.

The Cauchy problem is solved in the first formulation. We pass to the second formulation.

6. FORMULATION OF THE CLASSICAL CAUCHY PROBLEM. We shall seek a solution of equation (19.1) with conditions (19.3), having everywhere in our domain continuous derivatives of second order.

On the coefficients $A_{ij}$, $B_i$, $C$, and $F$, as well as upon the initial conditions, we shall impose some further conditions, which we shall call the conditions (n).

The conditions (n) are the following:

(1) The coefficients $A_{ij}$ and their first derivatives $\partial A_{ij}/\partial x_k$ and $\partial A_{ij}/\partial t$ are bounded:

$$|A_{ij}| < A; \quad \left|\frac{\partial A_{ij}}{\partial t}\right| < A; \quad \left|\frac{\partial A_{ij}}{\partial x_k}\right| < A. \tag{21.53}$$

(2) The derivatives of $A_{ij}$ of order $2, 3, \cdots, k_1$ satisfy the inequalities:

$$\left\{\int \cdots \int_{\Omega(t)} \left[\sum \frac{l!}{\alpha_0! \ldots \alpha_n!} \times \right.\right.$$
$$\left.\left. \times \left(\frac{\partial^l A_{ij}}{\partial t^{\alpha_0} \partial x_1^{\alpha_1} \ldots \partial x_n^{\alpha_n}}\right)^2\right]^{\frac{n+\varepsilon}{l-1}} d\Omega\right\}^{\frac{l-1}{n+\varepsilon}} \leqslant A. \tag{21.54}$$
$$l = 2, 3, \ldots k_1.$$

(3) The coefficients $B_i$ are bounded functions

$$|B_i| \leqslant A. \tag{21.55}$$

(4) The derivatives of $B_i$ of order $1, 2, \cdots, k_1 - 1$ satisfy the inequalities:

$$\left\{\int \cdots \int_{\Omega(t)} \left[\sum \frac{l!}{\alpha_0!\alpha_1! \ldots \alpha_n!} \times \right.\right.$$
$$\left.\left. \times \left(\frac{\partial^l B_i}{\partial t^{\alpha_0} \partial x_1^{\alpha_1} \ldots \partial x_n^{\alpha_n}}\right)^2\right]^{\frac{n+\varepsilon}{l}} d\Omega\right\}^{\frac{l}{n+\varepsilon}} \leqslant A. \tag{21.56}$$

(5) The derivatives of $C$ or orders: $0, 1, 2, \cdots, k_1 - 1$ satisfy the inequalities:

$$\left\{\int \cdots \int_{\Omega(t)} \left[\sum \frac{l!}{\alpha_0!\alpha_1! \ldots \alpha_n!} \times \right.\right.$$
$$\left.\left. \times \left(\frac{\partial^l C}{\partial t^{\alpha_0} \partial x_1^{\alpha_1} \ldots \partial x_n^{\alpha_n}}\right)^{\frac{n+\varepsilon}{l+1}}\right\}^{\frac{l+1}{n+\varepsilon}} \leqslant A. \tag{21.57}$$

(6) The derivatives of $F$ of orders $0, 1, \cdots, k_1 - 1$ satisfy the inequalities:

$$\left\{\int \cdots \int_{\Omega(t)} \left[\sum \frac{l!}{\alpha_0! \ldots \alpha_n!} \left(\frac{\partial^l F}{\partial t^{\alpha_0} \partial x_1^{\alpha_1} \ldots \partial x_n^{\alpha_n}}\right)^2\right] d\Omega\right\}^{\frac{1}{2}} \leqslant$$
$$\leqslant F(t) \leqslant F, \tag{21.58}$$

where $F$ is a constant.

(7) $$u_0 \in W_2^{(k_1)}. \tag{21.59}$$

(8)
$$u_1 \in W_2^{(k_1-1)}. \tag{21.60}$$

We may now state the basic theorem.

THEOREM 2. *In order that the Cauchy problem should have solutions which are continuous with all derivatives up to order* $m$ *it suffices that the conditions* (n) *should be satisfied for* $2k_1 > 2m+n$, *i.e., for*

$$k_1 \geqslant m+1+\left[\frac{n}{2}\right].$$

The proof of this theorem is based upon the application of the Imbedding Theorem and the use of the same inequalities which enabled us to prove Theorem 1.

For $m=2$, we obtain the solution of the Cauchy problem in the second formulation.

First of all, before passing to the proof, we make some remarks.

It is convenient, just as we did in the proof of Theorem 1, to construct the averaged functions for $A_{ij}$, $B_i$, $C$, $F$, $u_0$, and $u_1$ and form the new equation

$$L_h u_h = F_h - \sum_{i=1}^{n} B_{ih} \frac{\partial u_h}{\partial x_i} - C_h u_h \tag{21.61}$$

for which we seek solutions under the conditions:

$$\left. \begin{aligned} u_h \big|_{t=0} &= u_{0h}, \\ \frac{\partial u_h}{\partial t}\bigg|_{t=0} &= u_{1h}. \end{aligned} \right\} \tag{21.62}$$

Together with the equations (21.61), we consider all possible equations for the functions:

$$v_{\alpha_0, \alpha_1, \ldots, \alpha_n h} = \frac{\partial^l u_h}{\partial t^{\alpha_0} \partial x_1^{\alpha_1} \ldots \partial x_n^{\alpha_n}}. \tag{21.63}$$

We will obviously have:

$$L_h v_{\alpha_0, \alpha_1, \ldots \alpha_n h} = \frac{\partial^l F_h}{\partial t^{\alpha_0} \partial x_1^{\alpha_1} \ldots \partial x_n^{\alpha_n}} -$$

$$- \sum_{\Sigma \beta \leqslant \alpha} C_{\alpha_0, \alpha_1, \ldots \alpha_n}^{\beta_0, \beta_1, \ldots \beta_n} \left\{ \sum_{i=1}^{n} \sum_{j=1}^{n} \frac{\partial^\beta A_{ijh}}{\partial t^{\beta_0} \partial x_1^{\beta_1} \ldots \partial x_n^{\beta_n}} \times \right.$$

$$\times \frac{\partial^{l-\beta+2} u_h}{\partial t^{\alpha_0 - \beta_0 + 2} \partial x_1^{\alpha_1 - \beta_1} \ldots \partial x_n^{\alpha_n - \beta_n} \partial x_i \partial x_j} + \sum_{i=1}^{n} \frac{\partial^\beta B_{ih}}{\partial t^{\beta_0} \partial x_1^{\beta_1} \ldots \partial x_n^{\beta_n}} \times$$

$$\times \frac{\partial^{l-\beta+1} u_n}{\partial t^{\alpha_0 - \beta_0 + 1} \partial x_1^{\alpha_1 - \beta_1} \ldots \partial x_n^{\alpha_n - \beta_n} \partial x_i} + \frac{\partial^\beta C_h}{\partial t^{\beta_0} \partial x_1^{\beta_1} \ldots \partial x_n^{\beta_n}} \times$$

$$\times \frac{\partial^{l-\beta} u_n}{\partial t^{\beta_0 - \alpha_0} \partial x_1^{\beta_1 - \alpha_1} \ldots \partial x_n^{\beta_n - \alpha_n}} \equiv \Phi_{\alpha_0, \ldots, \alpha_n}^{(l)} \qquad (21.64)$$

where $C_{\alpha_0, \alpha_1, \ldots, \alpha_n}^{\beta_0, \beta_1, \ldots, \beta_n}$ are binomial coefficients.

We introduce some notation. Let

$$K_\rho (t \mid w) = \int_{\Omega(t)} \ldots \int \sum_{\Sigma \gamma = \rho} \frac{\rho!}{\gamma_0! \gamma_1! \ldots \gamma_n!} \left[ \frac{\partial^\rho w}{\partial t^{\gamma_0} \partial x_1^{\gamma_1} \ldots \partial x_n^{\gamma_n}} \right]^2 d\Omega. \qquad (21.65)$$

7. A LEMMA ON ESTIMATES FOR DERIVATIVES. *Because of the fulfillment of the conditions* (n), *the functions $u_h$ satisfy the inequalities*

$$K_\rho (t \mid u_h) \leqslant y_\rho \quad \rho = 1, 2, \ldots k_1, \qquad (21.66)$$

*where*

$$\frac{d (y_\rho^{\frac{1}{2}})}{dt} = M [y_0^{\frac{1}{2}} + y_1^{\frac{1}{2}} + \ldots + y_\rho^{\frac{1}{2}} + F(t)]; \qquad (21.67)$$

*M is a constant and F(t) is the function standing on the right side of the sixth condition.*

For $\rho = 0,1$, this estimate was established earlier. We pass to the proof of the lemma for the remaining $\rho$.

First of all, we estimate $\| \Phi_{\alpha_0, \alpha_1, \cdots, \alpha_n}^{(l)} \|_{L_2}$.

We will have, using Hölder's inequality:

$$J_C=\left[\int_{\Omega(t)}\cdots\int\left(\frac{\partial^{\beta} C}{\partial t^{\beta_0}\partial x_1^{\beta_1}\ldots\partial x_n^{\beta_n}}\right)^2\left(\frac{\partial^{l-\beta} u_n}{\partial t^{\alpha_0-\beta_0}\partial x_1^{\alpha_1-\beta_1}\ldots\partial x_n^{\alpha_n-\beta_n}}\right)^2 d\Omega\right]^{\frac{1}{2}}\leqslant$$

$$\leqslant\left\{\int_{\Omega(t)}\cdots\int\left|\frac{\partial^{\beta} C}{\partial t^{\beta_0}\partial x_1^{\beta_1}\ldots\partial x_n^{\beta_n}}\right|^{\frac{n+\varepsilon}{\beta+1}}d\Omega\right\}^{\frac{\beta+1}{n+\varepsilon}}\times$$

$$\times\left\{\int_{\Omega(t)}\cdots\int\left|\frac{\partial^{l-\beta} u_n}{\partial t^{\alpha_0-\beta_0}\partial x_1^{\alpha_1-\beta_1}\ldots\partial x_n^{\alpha_n-\beta_n}}\right|^{\frac{1}{\frac{1}{2}-\frac{\beta+1}{n+\varepsilon}}}d\Omega\right\}^{\frac{1}{2}-\frac{\beta+1}{n+\varepsilon}}\leqslant$$

$$\leqslant A\left\{\int_{\Omega(t)}\cdots\int\left|\frac{\partial^{l-\beta} u_n}{\partial t^{\alpha_0-\beta_0}\partial x_1^{\alpha_1-\beta_1}\ldots\partial x_n^{\alpha_n-\beta_n}}\right|^{\frac{1}{\frac{1}{2}-\frac{\beta+1}{n+\varepsilon}}}d\Omega\right\}^{\frac{1}{2}-\frac{\beta+1}{n+\varepsilon}} \tag{21.68}$$

On the basis of the Imbedding Theorem, we may estimate the last integral by the norms of derivatived of higher order. We will have:

$$J_C\leqslant A\,[K_0(t\,|\,u_h)^{\frac{1}{2}}+K_1(t\,|\,u_h)^{\frac{1}{2}}+\ldots+K_l(t\,|\,u_h)^{\frac{1}{2}}]. \tag{21.69}$$

Analogously, we obtain the estimates:

$$J_A=\left[\int_{\Omega(t)}\cdots\int\left(\frac{\partial^{\beta} A_{ij}}{\partial t^{\beta_0}\partial x_1^{\beta_1}\ldots\partial x_n^{\beta_n}}\right)^2\times\right.$$

$$\left.\times\left(\frac{\partial^{l-\beta+2} u_h}{\partial t^{\alpha_0-\beta_0+2}\partial x_1^{\alpha_1-\beta_1}\ldots\partial x_n^{\alpha_n-\beta_n}}\right)^2 d\Omega\right]^{\frac{1}{2}}\leqslant$$

$$\leqslant A\sum_{i=0}^{l}K_i(t\,|\,u_h)^{\frac{1}{2}}. \tag{21.70}$$

$$J_\beta=\left[\int_{\Omega(t)}\cdots\int\left(\frac{\partial^{\beta} B_i}{\partial t^{\beta_0}\partial x_1^{\beta_1}\ldots\partial x_n^{\beta_n}}\right)^2\times\right.$$

$$\left.\times\left(\frac{\partial^{l-\beta+1} u_h}{\partial t^{\alpha_0-\beta_0+1}\partial x_1^{\alpha_1-\beta_1}\ldots\partial x_n^{\alpha_n-\beta_n}}\right)^2\right]^{\frac{1}{2}}\leqslant A\sum_{i=0}^{l}K_i(t|u_h)^{\frac{1}{2}}. \tag{21.71}$$

Gathering these estimates together, we obtain:

$$\|\Phi^{(l)}_{\alpha_0,\alpha_1,\ldots,\alpha_n}\|_{L_2}\leqslant K_1 A\sum_{i=0}^{l}K_i(t/u_h)^{\frac{1}{2}}, \tag{21.72}$$

where $K_1$ is some constant.

Returning now to the inequality (21.22), we set in it $w=v_{\alpha_0,\cdots,\alpha_n}$ and sum over all derivatives of order $l$. We get:

$$c_2 \sum_{\Sigma\alpha=l} K_1(t\,|\,v_{\alpha_0,\ldots,\alpha_n}) \leqslant C_2 \sum_{\Sigma\alpha=l} K_1(0\,|\,v_{\alpha_0,\alpha_1,\ldots,\alpha_n}) +$$

$$+ C_1 \int_0^t \left\{K_1(t_1\,|\,v_{\alpha_0,\alpha_1,\ldots,\alpha_n}) + K_1(t_1\,|\,v_{\alpha_0,\ldots,\alpha_n})^{\frac{1}{2}} \|\Phi^{(l)}_{\alpha_0,\ldots,\alpha_n}\|_{L_2}\right\} dt_1 .$$

If we use the estimates for $\|\Phi^{(l)}\|$ and the fact that

$$\sum_{\Sigma\alpha=l} K_1(t\,|\,v_{\alpha_0,\ldots,\alpha_n}) = K_{l+1}(t\,|\,u), \tag{21.73}$$

we will have, finally:

$$K_{l+1}(t|u_h) \leqslant M\left[K_{l+1}(0|u_h) + \int_0^t K_{l+1}(t_1|u_h)^{\frac{1}{2}} \left\{\sum_{i=1}^{l+1} K_i(t_1\,|\,u_h)^{\frac{1}{2}} + \right.\right.$$

$$\left.\left. + F(t_1)\right\} dt_1\right]$$

$$l = 1, 2, \ldots, k_1 - 1, \tag{21.74}$$

where $M$ is some constant.

From this, as in the preceding, setting

$$y_{l+1}(t) = M\left[y_{l+1}(0) + \int_0^t y_{l+1}(t_1)^{\frac{1}{2}} \left\{\sum_{i=1}^{l+1} y_i(t_1)^{\frac{1}{2}} + F(t_1)\right\} dt_1\right],$$

we may prove the inequalities

$$K_{l+1}(t\,|\,u_h) \leqslant y_{l+1}(t). \tag{21.75}$$

For the functions $y$, we obtain

$$\frac{dy_{l+1}^{\frac{1}{2}}}{dt} = \frac{M}{2}\left[y_0^{\frac{1}{2}} + y_1^{\frac{1}{2}} + \ldots + y_{l+1}^{\frac{1}{2}} + F(t)\right]. \tag{21.76}$$

The lemma is proved.

8. SOLUTION OF THE CLASSICAL CAUCHY PROBLEM. We may now prove Theorem 2.

We consider the sequence of solutions $u_h$ of the averaged equations with averaged initial conditions.

These functions will have bounded integrals

$$K_\rho(t|u_h), \qquad \rho=0,1,\cdots, m+1+[n/2],$$

and consequently, bounded norms in the spaces $W_2^{(\rho)}$, from which there follows their uniform boundedness and equicontinuity as functions of the variables $x_1, x_2, \cdots, x_n$ for fixed $t$ together with their derivatives of order up to $m$ inclusive.

The derivatives of order $m+1$ on the basis of the Imbedding Theorem will, as functions of $x_1, \cdots, x_n$, lie in the spaces

$$L_{q^*}, \text{ где } \frac{1}{q^*} > \frac{1}{2} - \frac{\left[\frac{n}{2}\right]}{n};$$

$$\left\| \frac{\partial^{m+1} u_h}{\partial t^{\alpha_0} \partial x_1^{\alpha_1} \ldots \partial x_n^{\alpha_n}} \right\|_{L_{q^*}} < A,$$

where $A$ does not depend on $h$. However

$$\frac{\left[\frac{n}{2}\right]}{n} = \begin{cases} \frac{1}{2}, & \text{if } n \text{ is even} \\ \frac{n-1}{2n} = \frac{1}{2} - \frac{1}{2n}, & \text{if } n \text{ is odd} \end{cases}$$

and as a result

$$\frac{1}{q^*} > \frac{1}{2n},$$

which means that for $n \geqq 2$, we may choose $q^*$ so that we have

$$\frac{1}{q^*} < \frac{1}{n+1}.$$

Thereby it is evident that the derivatives $\partial^{m+1} u_h / \partial t^{\alpha_0} \partial x_1^{\alpha_1} \cdots \partial x_n^{\alpha_n}$ will have bounded norms in $L_q^*$ in the space of dimension $n+1$.

Consequently, it is obvious from the Imbedding Theorem that the derivatives of order $m$ of $u_h$ will form a compact subset of $C$ not only in the $n$-dimensional but in the $(n+1)$-dimensional space.

We choose from this set a subsequence converging uniformly together with all derivatives of order $m$, and passing to the limit on the averaged equations

$$L_h u_h = F_h - \sum B_{ih} \frac{\partial u_h}{\partial x_i} - C u_h, \qquad (21.61)$$

we obtain our theorem.

It is useful to note that the inequality (21.75) proved by us for $u_h$ remains true also for the solution $u$. This follows from the theorem on the existence of generalized derivatives (cf. §5, item 2).

## § 22. Quasi-linear equations.

1. FORMULATION OF FUNCTIONS OF FUNCTIONS. Let $A_i(t, x_1, \cdots, x_n)$, $B_i(t, x_1, \cdots, x_n)$ $(i=1,2,\cdots, m)$ be $2m$ continuous functions defined on some domain $\Omega$ of the $(n+1)$-dimensional space of $(t, x_1, \cdots, x_n)$ with $A_i<B_i$. By $D$ we denote the domain of $(n+m+1)$-dimensional space $(t, x_1, \cdots, x_n, y_1, \cdots, y_m)$ where $(t, x_1, \cdots, x_n)\in\Omega$ and $A_i\leqq y_i\leqq B_i$.

We shall say that the family of $m$ continuous functions $y_i=\eta_i(t, x_1, \cdots, x_n)$ $(i=1,\cdots,m)$, defined on $\Omega$ belongs to $D_y$ if $A_i\leqq\eta_i\leqq B_i$.

In the following by $\Omega(t)$ we shall denote the domain of the $n$-dimensional space $(x_1, \cdots, x_n)$ which is the intersection of $\Omega$ with the plane $t=\text{const.}$

We consider some bounded function

$$\Phi(t, x_1, \ldots, x_n; y_1, \ldots, y_m), \tag{22.1}$$

defined on the domain $D$. Suppose the function $\Phi$ is continuous and has continuous derivatives up to order $l$ with respect to $y_1, y_2, \cdots, y_m$:

$$\Phi_{\beta_1,\ldots,\beta_m}=\frac{\partial^\beta\Phi}{\partial y_1^{\beta_1}\partial y_2^{\beta_2}\cdots\partial y_m^{\beta_m}}, \quad (0\leqslant\beta=\beta_1+\beta_2+\cdots+\beta_m\leqslant l), \tag{22.2}$$

while the functions $\Phi_{\beta_1,\cdots,\beta_n}$ have generalized derivatives of order with respect to $t, x_1, \cdots, x_n$ for each fixed choice of values of $y_1, y_2, \cdots, y_m$.

Suppose now that $\Omega$ is a bounded domain. We shall say that the function $\Phi$ has property $T$ (or $\Phi\in T$) if:

1. There exists a number $p>1$, $p>n/l$ such that the result of substituting in the function

$$\frac{\partial^\alpha\Phi_{\beta_1,\ldots,\beta_m}}{\partial t^{\alpha_0}\partial x_1^{\alpha_1}\cdots\partial x_n^{\alpha_n}}=\Phi^{\alpha_0,\alpha_1,\ldots,\alpha_n}_{\beta_1,\ldots,\beta_m}, \quad (\alpha\leqslant l)$$

an arbitrary system from $D_y$ in place of $y_1, y_2, \cdots, y_m$ is a composite function of $(t, x_1, \cdots, x_n)$ such that for each fixed $t$, we have $[\Phi^{\alpha_0,\alpha_1,\cdots,\alpha_n}_{\beta_1,\cdots,\beta_n}]_{y_i=\eta_i}\in L_{((1/p)-((l-\alpha)/n))-1}$ on $\Omega(t)$ if $l\geqq\alpha>l-(n/p)$

$$\left[\Phi^{\alpha_0,\alpha_1,\ldots,\alpha_n}_{\beta_1,\ldots,\beta_m}\right]_{y_i=\eta_i}\in C, \quad \alpha<l-\frac{n}{p}. \tag{22.3}$$

2. There exists a constant $A_\Omega$ such that

$$\int_{\Omega(t)} \cdots \int \left| \left[ \Phi^{\alpha_0, \alpha_1, \ldots, \alpha_n}_{\beta_1, \ldots, \beta_m} \right]_{y_i = \eta_i} \right|^{\frac{1}{\frac{1}{p} - \frac{l-\alpha}{n}}} dx_1 \ldots dx_n \leqslant A_\Omega^{\frac{1}{\frac{1}{p} - \frac{l-\alpha}{n}}},$$

$$\text{if} \quad l \geqslant \alpha > l - \frac{n}{p} \tag{22.4}$$

$$\left| \left[ \Phi^{\alpha_0, \alpha_1, \ldots, \alpha_n}_{\beta_1, \ldots, \beta_m} \right]_{y_i = \eta_i} \right| < A_\Omega, \quad \text{if} \quad \alpha < l - \frac{n}{p},$$

where the constant $A_\Omega$ depends only on the choice of the domain $\Omega$ and does not depend on $\alpha$, $\beta$, or on the choice of the family $\{\eta_i\} \in D_y$.

In the case $\alpha = l - (n/p) [\Phi^{\alpha_0, \alpha_1, \ldots, \alpha_n}_{\beta_1, \ldots, \beta_n}]_{y_i = \eta_i} \in L_q$, where $q > 1$ is an arbitrary number and the constant $A_\Omega$ depends upon the choice of the number $q$. In the following, this case as well as the case $\alpha < l - (n/p)$ will not be specifically mentioned, which will lead to no misunderstanding if wherever $((1/p) - ((l-\alpha)/n))^{-1}$ appears it is replaced in these cases by an arbitrary $q > 1$. Obviously, if $\Phi$ has the property $T$ on $\Omega$, it has that property for every $\Omega_1 \subset \Omega$.

The simplest example of a function $\Phi \in T$ is given by a polynomial in the variables $y_1, y_2, \cdots, y_m$ with coefficients depending on $t, x_1, \cdots, x_n$

$$\Phi(t, x_1, \ldots, x_n; y_1, \ldots, y_m) =$$
$$= \sum_{\Sigma \gamma_i \leqslant N} a_{\gamma_1 \gamma_2 \ldots \gamma_m}(t, x_1, \ldots, x_n) y_1^{\gamma_1} y_2^{\gamma_2}, \ldots, y_m^{\gamma_m},$$

if the coefficients $a_{\gamma_1 \gamma_2 \cdots \gamma_m}$ are such that

$$\frac{\partial^{\alpha_0} a_{\gamma_1 \ldots \gamma_m}}{\partial t^{\alpha_0}} \in W^{(l-\alpha_0)}_{\frac{1}{\frac{1}{p} - \frac{l-\alpha_0}{n}}} \quad \text{in } \Omega(t).$$

We now consider a $(n+k+1)$-dimensional domain $D_1$ of the space $(t, x_1, \cdots, x_n; z_1, z_2, \cdots, z_k)$ analogous to the domain $D$, where $(t, x_1, x_2, \cdots, x_n) \in \Omega$. Let

$$y_i = \varphi_i(t, x_1, \ldots, x_n; z_1, \ldots, z_k) \quad (i = 1, 2, \ldots, m) \tag{22.5}$$

and assume that $y_i \in D_y$ if $\{z_i\} \in D_z$. Then, setting (22.5) in (22.1), we obtain the new function

$$\Psi(t, x_1, \ldots, x_n; z_1, \ldots, z_k) = \Phi(t, x_1, \ldots, x_n; \varphi_1, \ldots, \varphi_m).$$

THEOREM. *If the functions* (22.5) *have property* $T$, *then the result of substituting them in the function* $\Phi$ *having property* $T$, *is again a function having that property.*

This theorem is evident if $\Phi$ and the $\phi_i$ are polynomials in $y$ and $z$. It is not hard to prove that if we consider a polynomial in $y_1,\cdots,y_m$ and replace the $y_j$ by the polynomials in $z_s$, with coefficients depending upon $x_0,x_1,\cdots\cdots,x_n$ in the fashion indicated above, we obtain a new polynomial of the same form.

We prove this theorem in its general form by estimating a norm for all derivatives of $\Psi(x_0, x_1,\cdots,x_n, z_1,\cdots,z_k)$ with respect to $z$ in the corresponding space $L$ of the variables $x_1, x_2,\cdots,x_n$. We turn to the proof of the theorem.

PROOF. Since the $\phi_i$ have property $T$, for $z_i$ in $D_z$ the inequalities

$$\overbrace{\int\cdots\int_{\Omega(t)}}^{n}\left[\frac{\partial^{\alpha}}{\partial x_0^{\alpha_0}\partial x_1^{\alpha_1}\cdots\partial x_n^{\alpha_n}}\left(\frac{\partial^{\beta}\varphi_i}{\partial z_1^{\beta_1}\partial z_2^{\beta_2}\cdots\partial z_n^{\beta_n}}\right)\right]^{\frac{1}{\frac{1}{p}-\frac{l-\alpha}{n}}}\times$$

$$\times\, dx_1 dx_2\,\ldots\, dx_n \leqslant B_{\Omega}^{\frac{1}{\frac{1}{p}-\frac{l-\alpha}{n}}}. \tag{22.6}$$

hold. We set

$$\Psi_{\beta_1,\beta_2,\ldots\beta_k}=\frac{\partial^{\beta}\Psi}{\partial z_1^{\beta_1}\partial z_2^{\beta_2}\cdots\partial z_k^{\beta_k}}=\sum_{\gamma\leqslant\beta}\frac{\partial^{\gamma}\Phi}{\partial y_1^{\gamma_1}\partial y_2^{\gamma_2}\cdots\partial y_m^{\gamma_m}}A_{\gamma}(\varphi_i), \tag{22.7}$$

where $A_\gamma(\phi_i)$ are polynomials of degree $\gamma$ in the derivatives of the functions $\phi_i$ with respect to the $z_j$ of order $\beta-\gamma$. We establish the validity of the inequality:

$$\left[\overbrace{\int\cdots\int_{\Omega(t)}}^{n}\left|\frac{\partial^{\alpha}\Psi_{\beta_1,\ldots,\beta_k}}{\partial x_0^{\alpha_0}\partial x_1^{\alpha_1}\cdots\partial x_n^{\alpha_n}}\right|^{\frac{1}{\frac{1}{p}-\frac{l-\alpha}{n}}}d\Omega_x\right]^{\frac{1}{p}-\frac{l-\alpha}{n}}\leqslant$$

$$\leqslant N_1 A_{\Omega}(B_{\Omega}^{\alpha}+B_{\Omega}^{\alpha-1}+\ldots+1)\leqslant N_2 A_{\Omega}(1+B_{\Omega}^{\alpha}), \tag{22.8}$$

where $N_1$ and $N_2$ are constants not depending upon $\phi_i$.

For the proof of (22.8), we introduce the concept of the inverse index. We shall say that the function $\psi(x_1, x_2,\cdots,x_n)$ defined on some bounded domain $D_x$ of the space $(x_1, x_2,\cdots,x_n)$ has the index $\lambda$ $(0<\lambda<1)$, with the bound $M$ on $D_x$ if

$$\left[\int\cdots\int_{D_x}|\psi|^{\frac{1}{\lambda}}d\Omega_x\right]^{\lambda}\leqslant M.$$

The following properties of the inverse index are obvious:

(1) The sum of two functions $\psi_1$ and $\psi_2$ with the index $\lambda$ has the index $\lambda$ with a bound equal to the sum of the bounds of both functions (which follows from the Minkowski inequality).

(2) If $\psi$ has the index $\lambda$ with bound $M$, then it has every index $\lambda_1>\lambda$ with bound $KM$, where the constant $K$ depends only upon the domain $D_x$.

Indeed, from Hölder's inequality,

$$\int_{D_x}\cdots\int |\psi|^{\frac{1}{\lambda_1}} d\Omega_x \leqslant \left[\int_{D_x}\cdots\int |\psi|^{\frac{1}{\lambda}} d\Omega_x\right]^{\frac{\lambda}{\lambda_1}}\left[\int_{D_x}\cdots\int d\Omega_x\right]^{\frac{1}{1-\frac{\lambda}{\lambda_1}}} \leqslant KM^{\frac{1}{\lambda_1}}.$$

(3) Multiplying a function by a bounded multiplier does not disturb the index, while the bound is changed only by multiplying it by the maximum absolute value of the multiplier.

(4) If we have the functions $\psi_1,\psi_2,\cdots,\psi_l$ with the indices $\lambda_1,\cdots,\lambda_l$, where $\lambda_1+\lambda_2+\cdots+\lambda_l<1$, then the product $\psi_1\psi_2\cdots\psi_l$ has the index equal to the sum of the indices with a bound equal to the product of the bounds of the factors.

The proof follows from Hölder's inequality:

$$\int_{D_x}\cdots\int (\psi_1\ \psi_2\ \ldots\ \psi_l)^{\frac{1}{\lambda_1+\lambda_2+\cdots+\lambda_l}} d\Omega_x \leqslant$$

$$\leqslant \left[\int_{D_x}\cdots\int \psi_1^{\frac{1}{\lambda_1}} d\Omega_x\right]^{\frac{\lambda_1}{\lambda_1+\lambda_2+\cdots+\lambda_l}}\cdots\left[\int_{D_x}\cdots\int |\psi_l|^{\frac{1}{\lambda_l}} d\Omega_x\right]^{\frac{\lambda_l}{\lambda_1+\lambda_2+\cdots+\lambda_l}}.$$

From (22.6) and (22.7) it follows that for every differentiation with respect to $x_i$ and $t$, the index of the functions $\Phi_{\beta_1,\cdots,\beta_m}$ and $\phi_{i,\beta_1,\beta_2,\cdots,\beta_m}= \partial^\beta\phi_i/\partial z_1^{\beta 1}\cdots\partial z_k^{\beta k}$ increases by $1/n$.

With the aid of the concept of the index, we may, without special difficulty, prove the inequality (22.8) for continuous functions. Indeed, suppose that for derivatives of order $\beta$ the calculation of the indices and bounds for the polynomials (22.7) shows the correctness of (22.8). We differentiate (22.7) with respect to $x_i$. In doing this, we differentiate either $\Phi_{\beta_1,\cdots,\beta_m}$ or one of the derivatives of $\phi_i$. Each such differentiation increases the index by $1/n$, while the bound is multiplied by a constant not depending respectively on $B_D$ or $CB_D$, if the differentiation of $\Phi_{\beta_1\cdots\beta_m}$ is carried through by means of the auxiliary functions. By virtue of the fact that the bound of the sum does not exceed the sum of the bounds of the summands, we obtain the desired estimate. Thus (22.8) is established for continuous functions.

We now assume that the derivatives of $\Phi_{\beta_1,\beta_2,\cdots,\beta_m}$ and $\phi_{i\beta_1,\cdots,\beta_m}$ exist only in the generalized sense. We construct the averaged functions with respect to $x_1, x_2, \cdots, x_n$ for the functions $\Phi$ and $\phi_i$. Let these averaged functions be $\Phi_h$ and $\phi_{ih}$. For the functions $\Phi_h$ and $\phi_{ih}$, the estimates (22.8) are valid. Obviously we have the formula

$$\lim \frac{\partial^\beta \Phi_h}{\partial y_1^{\gamma_1} \partial y_2^{\gamma_2} \cdots \partial y_n^{\gamma_n}}\bigg|_{y_i=\varphi_{ih}} = \frac{\partial^\beta \Phi}{\partial y_1^{\gamma_1} \cdots \partial y_n^{\gamma_n}}\bigg|_{y_i=\varphi_i}. \tag{22.9}$$

Applying the theorem on the criterion for the existence of generalized derivatives, we see that for the function $\Phi$, the theorem and the inequality (22.8) remain valid.

2. BASIC INEQUALITIES. We consider the quasi-linear equation of second order

$$\sum_{i=1}^{n} \sum_{j=1}^{n} A_{ij}\left(t, x_1, \ldots, x_n; u, \frac{\partial u}{\partial t}, \frac{\partial u}{\partial x_1}, \ldots, \frac{\partial u}{\partial x_n}\right) \frac{\partial^2 u}{\partial x_i \partial x_j} - \frac{\partial^2 u}{\partial t^2} =$$
$$= F\left(t, x_1, \ldots, x_n; u, \frac{\partial u}{\partial t}, \frac{\partial u}{\partial x_1}, \ldots, \frac{\partial u}{\partial x_n}\right), \tag{22.10}$$

where the $A_{ij}$ are assumed to be continuous functions of all their arguments and $A_{ij} = A_{ji}$.

We set

$$y = u, \quad y_0 = \frac{\partial u}{\partial t}, \; y_1 = \frac{\partial u}{\partial x_1}, \ldots, y_n = \frac{\partial u}{\partial x_n}. \tag{22.11}$$

We consider the quadratic form

$$\sum_{i=1}^{n} \sum_{j=1}^{n} A_{ij}(t, x_1, \ldots, x_n; y, y_0, y_1, \ldots, y_n) \xi_i \xi_j.$$

The equation (22.10) is said to be a quasilinear hyperbolic equation on some domain $D^*$ of the $(2n+3)$-dimensional space $(t, x_1, \cdots, x_n; y, y_0, y_1, \cdots \cdots, y_n)$ if the quadratic form is positive definite when $(t, x_1, \cdots, x_n; y, y_0, y_1, \cdots, y_n) \in D^*$.

Then for every domain $D$, lying together with its boundary in $D^*$, we may find $c > 0$ such that

$$\sum\sum A_{ij} \xi_i \xi_j > c \sum \xi_i^2.$$

We assume that the domain $D$ contains some domain $D^\circ$ of the plane $t = 0$.

Let $u_0(x_1, \cdots, x_n)$ be a continuously differentiable function, $u_1(x_1, \cdots \cdots, x_n)$ a continuous function, both on some bounded domain $D_x$ of the space $(x_1,\cdots,x_n)$. If for all points of this domain, we have

$$\left(x_1,\ x_2,\ \ldots,\ x_n;\quad u_0,\ u_1,\ \frac{\partial u_0}{\partial x_1},\ \ldots,\ \frac{\partial u_0}{\partial x_n}\right)\in D^\circ$$

and if, in addition, there exist $\delta_1>0$ and $\delta>0$ such that

$$(t,\ x_1,\ \ldots,\ x_n;\ y,\ y_0,\ y_1,\ \ldots,\ y_n)\in D,$$

if

$$0\leqslant t\leqslant\delta_1,\quad (x_1,\ \ldots,\ x_n)\in D_x$$

and

$$|y-u_0|<\delta;\quad |y_0-u_1|<\delta;$$
$$\left|y_1-\frac{\partial u_0}{\partial x_1}\right|<\delta;\ \ldots,\ \left|y_n-\frac{\partial u_0}{\partial x_n}\right|<\delta, \tag{22.12}$$

we shall say that $u_0(x_1,\cdots,x_n)$ and $u_1(x_1,\cdots,x_n)$ are permissible boundary data. The inequality (22.12) defines the set $D_y$.

Let $c_1>0$ be such that $\max|A_{ij}|<c_1/n$ if $0\leqq t\leqq\delta_1$, $(x_1,\cdots,x_n)\in D_x$, and $(y_1,y_0,y_1,\cdots,y_n)\in D_y$. Then just as in §21, item 3, we may construct a fundamental conical domain $\Omega^*$, the construction of which is determined by the choice of $c_1$. We pose the problem: To find a twice continuously differentiable solution in $\Omega^*$ of the equation (22.10), satisfying the initial conditions

$$u\,|_{t=0}=u_0,\quad \left.\frac{\partial u}{\partial t}\right|_{t=0}=u_1. \tag{22.13}$$

THEOREM. *If the functions $A_{ij}$ and $F$, as functions of the variables $t,x_1,\cdots,x_n$ with parameters $y,y_0,y_1,\cdots,y_n$, satisfy the condition $T$ on the domain $\Omega^*$ for $p=2$ and $l=[n/2]+2$, while the permissible initial data satisfy the conditions*

$$u_0\in W_2^{(l+1)},\quad u_1\in W_2^{(l)},$$

*then we may find $\delta_2>0$ $(0<\delta_2\leqq\delta_1)$ such that on that part of $\Omega^*$ where $0\leqq t\leqq\delta_2$, there exists a unique solution $u(t,x_1,\cdots,x_n)$ of the equation (22.10), continuous with its derivatives of second order inclusive, satisfying the initial conditions (22.13). In addition, $u\in W_2^{(l+1)}$ on each plane $t=\text{const.}$ $(0\leqq t\leqq\delta_2)$ and $(u,\partial u/\partial t,\partial u/\partial x_1,\cdots,\partial u/\partial x_n)\in D_y$. The number $\delta_2$ depends upon the choice of $u_0$ and $u_1$.*

We pass to the proof of the theorem.

By virtue of our conditions, $A_{ij}$ and $F$ will be continuous functions of the variables $t, x_1, \cdots, x_n$, taken from $\Omega^*$ with parameters $y, y_0, y_1, \cdots, y_n$ taken from $D_y$ and having continuous derivatives up to order $[n/2]+2=l$ with respect to the parameters $y, y_0, \cdots, y_n$. Let $\Omega(t)$ be the intersection of $\Omega^*$ with the plane $t=\text{const.}$ $(0 \leqq t \leqq \delta_1)$.

Set

$$A_{ij}^{\alpha_*, \alpha_0, \ldots, \alpha_n} = \frac{\partial^\alpha A_{ij}}{\partial u^{\alpha_*} \partial \left(\frac{\partial u}{\partial t}\right)^{\alpha_0} \partial \left(\frac{\partial u}{\partial x_1}\right)^{\alpha_1} \cdots \partial \left(\frac{\partial u}{\partial x_n}\right)^{\alpha_n}}$$

$$F^{\alpha_*, \alpha_0, \ldots, \alpha_n} = \frac{\partial^\alpha F}{\partial u^{\alpha_*} \partial \left(\frac{\partial u}{\partial t}\right)^{\alpha_0} \partial \left(\frac{\partial u}{\partial x_1}\right)^{\alpha_1} \cdots \partial \left(\frac{\partial u}{\partial x_n}\right)^{\alpha_n}}.$$

Then, after a substitution in $A_{ij}^{\alpha_*, \alpha_0, \cdots, \alpha_n}$ and $F^{\alpha_*, \alpha_0, \cdots, \alpha_n}$ of arbitrary functions from $D_y$, we have the inequalities:

$$\left\{ \int_{\Omega(t)} \cdots \int \left[ \frac{\partial^\gamma A_{ij}^{\alpha_*, \alpha_0, \ldots, \alpha_n}}{\partial t^{\gamma_0} \partial x_1^{\gamma_1} \ldots \partial x_n^{\gamma_n}} \right]^{\frac{1}{\frac{1}{2} - \frac{[\frac{n}{2}]+2-\gamma}{n}}} \times d\Omega_x \right\}^{\frac{1}{2} - \frac{[\frac{n}{2}]+2-\gamma}{n}} \leqslant B; \tag{22.14}$$

$$\left\{ \int_{\Omega(t)} \cdots \int \left[ \frac{\partial^\gamma F^{\alpha_*, \alpha_0, \ldots, \alpha_n}}{\partial t^{\gamma_0} \partial x_1^{\gamma_1} \ldots \partial x_n^{\gamma_n}} \right]^{\frac{1}{\frac{1}{2} - \frac{[\frac{n}{2}]+2-\gamma}{n}}} \times d\Omega_x \right\}^{\frac{1}{2} - \frac{[\frac{n}{2}]+2-\gamma}{n}} \leqslant C. \tag{22.15}$$

If in $A_{ij}$ and $F$, in place of $y, y_0, y_1, \cdots, y_n$ we set functions $y_i(t, x_1, \cdots, x_n, z_1, \cdots, z_k)$ having derivatives of arbitrary order $\gamma \leqq [n/2]+2$, integrable to the power $((1/2)-([n/2]+2-\gamma)/n)^{-1}$, then by means of (22.8) from the theorem on functions satisfying condition $T$, we obtain:

$$A(t) = \max\left\{ \int_{\Omega(t)} \cdots \int \left[ \frac{\partial^\gamma A_{ij}^{\alpha_*, \alpha_0, \alpha_1, \ldots, \alpha_n}}{\partial t^{\gamma_0} \partial x_1^{\gamma_1} \ldots \partial x_n^{\gamma_n}} \right]^{\frac{1}{\frac{1}{2} - \frac{\left[\frac{n}{2}\right]+2-\gamma}{n}}} d\Omega \right\}^{\frac{1}{2} - \frac{\left[\frac{n}{2}\right]+2-\gamma}{n}} \leqslant$$

$$\leqslant MB\,[1 + L(t)^\gamma]; \tag{22.16}$$

$$F(t) = \max\left\{ \int_{\Omega(t)} \cdots \int \left[ \frac{\partial^\gamma F^{\alpha_*, \alpha_0, \alpha_1, \ldots, \alpha_n}}{\partial t^{\gamma_0} \partial x_1^{\gamma_1} \ldots \partial x_n^{\gamma_n}} \right]^{\frac{1}{\frac{1}{2} - \frac{\left[\frac{n}{2}\right]+2-\gamma}{n}}} d\Omega \right\}^{\frac{1}{2} - \frac{\left[\frac{n}{2}\right]+2-\gamma}{n}} \leqslant$$

$$\leqslant MC\,[1 + L(t)^\gamma], \tag{22.17}$$

where $d\Omega = dx_1 \cdots dx_n$;

$$L(t) = \sum_{\substack{i \\ 0 \leqslant \sum \alpha_k = \alpha \leqslant \left[\frac{n}{2}\right]+2}} \left[ \int_{\Omega(t)} \cdots \int \left( \frac{\partial^\alpha y_i}{\partial t^{\alpha_0} \partial x_1^{\alpha_1} \ldots dx_n^{\alpha_n}} \right)^2 d\Omega \right]^{\frac{1}{2}}. \tag{22.18}$$

3. Petrovsky's functional equation. We replace our quasilinear equation by a new functional equation.

$$\sum_{i=1}^{n} \sum_{j=1}^{n} A_{ij}\left(u(t-\eta), \frac{\partial u}{\partial t}(t-\eta), \ldots, \frac{\partial u}{\partial x_n}(t-\eta)\right) \frac{\partial^2 u}{\partial x_i \partial x_j} - \frac{\partial^2 u}{\partial t^2} =$$

$$= F\left(u(t-\eta), \frac{\partial u}{\partial t}(t-\eta), \ldots, \frac{\partial u}{\partial x_n}(t-\eta)\right). \tag{22.19}$$

This equation is obtained from (22.10) by setting in the coefficients in place of the values of $u$ and its derivatives at the given point, their values at $x_1, \cdots, x_n, t-\eta$. The first use of such functional equations is due to Academician I. G. Petrovsky.

Then for the coefficients of the new equation (22.19), we obtain:

$$L(t) \leqslant \sum_{\alpha=0}^{\left[\frac{n}{2}\right]+3} K_\alpha (t-\eta)^{\frac{1}{2}}, \tag{22.20}$$

and the inequalities (22.16) and (22.17) may be rewritten as:

$$\left.\begin{aligned} A(t) &\leqslant MB\left\{1+\left[\sum_{\rho=0}^{\left[\frac{n}{2}\right]+3} K_\rho (t-\eta)^{\frac{1}{2}}\right]^{\left[\frac{n}{2}\right]+2}\right\} \\ F(t) \leqslant F_1(t) &\leqslant MC\left\{1+\left[\sum_{\rho=0}^{\left[\frac{n}{2}\right]+3} K_\rho (t-\eta)^{\frac{1}{2}}\right]^{\left[\frac{n}{2}\right]+2}\right\}. \end{aligned}\right\} \tag{22.21}$$

Suppose that the function $F(t, x_1, x_2, \cdots, x_n, u, \partial u/\partial t, \cdots, \partial u/\partial x_n)$ after the substitution

$$u \equiv \frac{\partial u}{\partial t} \equiv \frac{du}{dx_1} \equiv \ldots \equiv \frac{\partial u}{\partial x_n} = 0$$

vanishes identically. Then we may strengthen the second of the inequalities (22.21).

Suppose that $y, y_0, y_1, \cdots, y_n$ depend only on $t, x_1, \cdots, x_n$ (and do not depend on the $z_k$); then, applying Minkowski's inequality, we obtain:

$$\left\{\int_{\Omega(t)}\!\!\cdots\int\left[\frac{d^\alpha F}{dt^{\alpha_0}\, dx_1^{\alpha_1}\ldots dx_n^{\alpha_n}}\right]^2 d\Omega\right\}^{\frac{1}{2}} \leqslant$$

$$\leqslant\left\{\int_{\Omega}\!\!\cdots\int\left(\frac{\partial^\alpha F}{\partial t^{\alpha_0}\,\partial x_1^{\alpha_1}\ldots \partial x_n^{\alpha_n}}\right)^2 d\Omega\right\}^{\frac{1}{2}} + MC_1\sum_{\gamma=1}^{\alpha} L(t)^\gamma \tag{22.22}$$

(the notation of total derivatives is understood without explanation).

We note that

$$\sum_{\gamma=i}^{\alpha} L(t)^\gamma \leqslant N\{L(t) + L(t)^\alpha\},$$

and shall estimate the first term on the right side of (22.22).

Set

$$\begin{aligned} F_\lambda(t,\, x_1,\, x_2,\, \ldots,\, x_n,\, y,\, y_0,\, y_1,\, \ldots,\, y_n) &= \\ = F(t,\, x_1,\, \ldots,\, x_n,\, \lambda y,\, \lambda y_0,\, \ldots,\, \lambda y_n). \end{aligned}$$

Then we have:

$$\begin{aligned} &F_\lambda(t,\, x_1,\, \ldots,\, x_n;\, y,\, y_0,\, y_1,\, \ldots,\, y_n) = \\ &= \int_0^1 (F_\lambda^{1,0,0,\ldots,0} y + F_\lambda^{0,1,0,\ldots,0} y_0 + \ldots + F_\lambda^{0,0,\ldots,1} y_n)\, d\lambda, \end{aligned}$$

where, by $F_\lambda^{1,0,0,\cdots,0}$ we denote as above, the derivatives of $F_\lambda$ with respect to $y, y_0, \cdots, y_n$.

Consider $t, x_1, \cdots, x_n, y, y_0, y_1, \cdots, y_n$ as independent variables and form the expression for the derivatives

$$\frac{\partial^\alpha F(t, x_1, x_2, \ldots, x_n, y, y_0, y_1, \ldots, y_n)}{\partial t^{\alpha_0} \partial x_1^{\alpha_1} \partial x_2^{\alpha_2} \ldots \partial x_n^{\alpha_n}}.$$

Using the definition of generalized derivatives we have:

$$\int \cdots \int_V \frac{\partial^\alpha \psi}{\partial t^{\alpha_0} \partial x_1^{\alpha_1} \ldots \partial x_n^{\alpha_n}} F(t, x_1, x_2, \ldots, x_n)\, d\Omega\, dt =$$

$$= \int \cdots \int_V \int_0^1 \frac{\partial^\alpha \psi}{\partial t^{\alpha_0} \partial x_1^{\alpha_1} \ldots \partial x_n^{\alpha_n}} (F_\lambda^{1,0,0,\ldots,0} y + F_\lambda^{0,1,0,\ldots,0} y_0 +$$

$$+ F_\lambda^{0,0,1,\ldots,0} y_1 + \ldots + F_\lambda^{0,0,0,\ldots,1} y_n)\, d\lambda\, d\Omega\, dt =$$

$$= \int_0^1 \left\{ \int \cdots \int_V \frac{\partial^\alpha \psi}{\partial t^{\alpha_0} \partial x_1^{\alpha_1} \ldots \partial x_n^{\alpha_n}} (F_\lambda^{1,0,0,\ldots,0} y + F_\lambda^{0,1,0,\ldots,0} y_0 + \right.$$

$$\left. + F_\lambda^{0,0,1,\ldots,0} y_1 + \ldots + F_\lambda^{0,0,0,\ldots,1} y_n)\, d\Omega\, dt \right\} d\lambda =$$

$$= \int_0^1 \left\{ \int \cdots \int_V (-1)^\alpha \psi(x_1, x_2, \ldots, x_n, t) \left[ y \frac{\partial^\alpha F_\lambda^{1,0,0,\ldots,0}}{\partial t^{\alpha_0} \partial x_1^{\alpha_1} \ldots \partial x_n^{\alpha_n}} + \right. \right.$$

$$\left. \left. + \ldots + y_n \frac{\partial^\alpha F_\lambda^{0,0,0,\ldots,1}}{\partial t^{\alpha_0} \partial x_1^{\alpha_1} \ldots \partial x_n^{\alpha_n}} \right] d\Omega\, dt \right\} d\lambda =$$

$$= \int \cdots \int_V (-1)^\alpha \psi(x_1, x_2, \ldots, x_n, t) \left\{ \int_0^1 \left[ y \frac{\partial^\alpha F_\lambda^{1,0,0,\ldots,0}}{\partial t^{\alpha_0} \partial x_1^{\alpha_1} \ldots \partial x_n^{\alpha_n}} + \right. \right.$$

$$\left. \left. + \ldots + y_n \frac{\partial^\alpha F_\lambda^{0,0,0,\ldots,1}}{\partial t^{\alpha_0} \partial x_1^{\alpha_1} \ldots \partial x_n^{\alpha_n}} \right] d\lambda \right\} d\Omega\, dt.$$

From this we obtain:

$$\frac{\partial^\alpha F(t, x_1, x_2, \ldots, x_n, y, y_0, y_1, \ldots, y_n)}{\partial t^{\alpha_0} \partial x_1^{\alpha_1} \partial x_2^{\alpha_2} \ldots \partial x_n^{\alpha_n}} =$$

$$= \int_0^1 \left[ y \frac{\partial^\alpha F_\lambda^{1, 0, 0, \ldots, 0}}{\partial t^{\alpha_0} \partial x_1^{\alpha_1} \ldots \partial x_n^{\alpha_n}} + \ldots + y_n \frac{\partial^\alpha F_\lambda^{0, 0, 0, \ldots, 1}}{\partial t^{\alpha_0} \partial x_1^{\alpha_1} \ldots \partial x_n^{\alpha_n}} \right] d\lambda.$$

We use this formula for the estimation of the first summand on the right side of (22.22) to obtain:

$$\int_{\Omega(t)} \cdots \int \left( \frac{\partial^\alpha F}{\partial t^{\alpha_0} \partial x_1^{\alpha_1} \ldots \partial x_n^{\alpha_n}} \right)^2 d\Omega =$$

$$= \int_{\Omega(t)} \cdots \int \left[ \int_0^1 \left( y \frac{\partial^\alpha F_\lambda^{1, 0, 0, \ldots, 0}}{\partial t^{\alpha_0} \partial x_1^{\alpha_1} \ldots \partial x_n^{\alpha_n}} + \ldots + \right.\right.$$

$$\left.\left. + y_n \frac{\partial^\alpha F_\lambda^{0, 0, 0, \ldots, 1}}{\partial t^{\alpha_0} \ldots \partial x_n^{\alpha_n}} \right) d\lambda \right]^2 d\Omega.$$

Estimating with the aid of the Minkowski and Bunjakovsky inequalities, we have:

$$\left\{ \int_{\Omega(t)} \cdots \int \left( \frac{\partial^\alpha F}{\partial t^{\alpha_0} \partial x_1^{\alpha_1} \ldots \partial x_n^{\alpha_n}} \right)^2 d\Omega \right\}^{\frac{1}{2}} \leqslant$$

$$\leqslant \sum \left\{ \int_{\Omega(t)} \cdots \int \left( \int_0^1 y_i^2 \, d\lambda \right)^{\frac{1}{2}} \left[ \int_0^1 \left( \frac{\partial^\alpha F^{0, 0, \ldots, 1, 0, \ldots, 0}}{\partial t^{\alpha_0} \partial x_1^{\alpha_1} \ldots \partial x_n^{\alpha_n}} \right)^2 d\lambda \right]^{\frac{1}{2}} d\Omega \right\} \leqslant$$

$$\leqslant \max |y_i| \sum \left[ \int_{\Omega(t)} \cdots \int \int_0^1 \left( \frac{\partial^\alpha F^{0, 0, 0, \ldots, 1, 0, \ldots, 0}}{\partial t^{\alpha_0} \partial x_1^{\alpha_1} \ldots \partial x_n^{\alpha_n}} \right)^2 d\lambda \, d\Omega_x \right]^{\frac{1}{2}},$$

and since for positive $x_1, x_2, \cdots, x_n$

$$(x_1 + x_2 + x_3 + \ldots + x_n)^2 \leqslant n (x_1^2 + x_2^2 + \ldots + x_n^2)$$

and $\max |y_i|^2 \leqslant C_2 L(t)^2$, then

$$\left[\int_{\Omega(t)}\cdots\int\left(\frac{\partial^\alpha F}{\partial t^{\alpha_0}\partial x_1^{\alpha_1}\dots\partial x_n^{\alpha_n}}\right)^2 d\Omega\right]^{\frac{1}{2}}\leqslant CC_2 L(t). \qquad (22.23)$$

From (22.22) and (22.23) follows:

$$F(t)\leqslant N\{L(t)+[L(t)]^{\left[\frac{n}{2}\right]+2}\}. \qquad (22.24)$$

This is the desired strengthening of the inequality (22.21).

Set

$$\left.\begin{aligned}\sum_{\rho=0}^{\left[\frac{n}{2}\right]+3} K_\rho(t)=Z(t),\\ \left[\frac{n}{2}\right]+2=k.\end{aligned}\right\} \qquad (22.25)$$

We note that

$$K_\rho(t)^{\frac{1}{2}}<Z(t)^{\frac{1}{2}}.$$

In an analogous way to that by which we obtained the inequalities (21.74), we find

$$K_{\rho+1}(t)\leqslant M\left\{K_{\rho+1}(0)+\int_0^t\left[A(t_1)K_{\rho+1}(t_1)^{\frac{1}{2}}\sum_{s=1}^{\rho+1}K_s(t_1)^{\frac{1}{2}}+\right.\right.$$

$$\left.\left.+F(t_1)K_{\rho+1}(t_1)^{\frac{1}{2}}\right]dt_1\right\}.$$

We combine this inequality and a similar inequality for $K_0(t)$. Then, taking a sufficiently large $C_3$, we obtain:

$$Z(t)\leqslant C_3\left\{Z(0)+\int_0^t\left[A(t_1)Z(t_1)+F(t_1)Z(t_1)^{\frac{1}{2}}\right]dt_1\right\}. \qquad (22.26)$$

In addition, (22.21) may be rewritten as:

$$\left.\begin{aligned}A(t)\leqslant C_3B\left\{1+Z(t-\eta)^{\frac{k}{2}}\right\},\\ F(t)\leqslant C_3C\left\{1+Z(t-\eta)^{\frac{k}{2}}\right\}.\end{aligned}\right\} \qquad (22.27)$$

Setting (22.27) in (22.26), we obtain:

$$Z(t) \leqslant C_4 \left\{ Z(0) + \int_0^t \left[ B[1 + Z(t_1 - \eta)^{\frac{k}{2}}] Z(t_1) + \right.\right.$$

$$\left.\left. + C[1 + Z(t_1 - \eta)^{\frac{k}{2}}] Z(t_1)^{\frac{1}{2}} \right] dt_1 \right\}. \tag{22.28}$$

If the free term of equation (22.10) vanishes for $u = \partial u/\partial t = \partial u/\partial x_1 = \cdots \cdots = \partial u/\partial x_n = 0$, then using (22.24) we obtain

$$F(t) \leqslant C_5 C \left\{ Z(t-\eta)^{\frac{1}{2}} + Z(t-\eta)^{\frac{k}{2}} \right\}, \tag{22.29}$$

and then obtain in place of (22.28):

$$Z(t) \leqslant C_5 \left\{ Z(0) + \int_0^t \left[ BZ(t_1) [1 + Z(t_1 - \eta)^{\frac{k}{2}}] + \right.\right.$$

$$\left.\left. + CZ(t_1)^{\frac{1}{2}} [Z(t_1 - \eta)^{\frac{1}{2}} + Z(t_1 - \eta)^{\frac{k}{2}}] \right] dt_1 \right\}. \tag{22.30}$$

4. THE CAUCHY PROBLEM WITH HOMOGENEOUS INITIAL CONDITIONS. We shall first establish our theorem in a special case. Suppose that the functions $u_0$ and $u_1$ are equal to zero and that the free term of the equation vanishes identically on the segment $-h \leqq t \leqq 0$ after the substitution into it of $u \equiv \partial u/\partial t \equiv \partial u/\partial x_1 \equiv \cdots \equiv \partial u/\partial x_n = 0$. Then on this segment, the equation (22.10) has the solution

$$u \equiv 0,$$

satisfying the initial conditions (22.13).

We now consider the functional equation (22.19), setting $\eta < h$. For $-h \leqq t \leqq 0$ it will have the solution $u = 0$. We continue this solution with respect to $t$.

For this purpose, we decompose the interval of values of $t$ into pieces by points $\alpha_0, \alpha_1, \cdots, \alpha_n$ such that $\alpha_{i+1} - \alpha_i < \eta$. Then the values of the coefficients (22.19) on the $i$th segment depend only on the values of $u$ and its derivatives on the $(i-1)$th segment, and the solution of the quasilinear equation thus reduces to the solution of a linear equation. If the coefficients of this equation have generalized derivatives up to order $[n/2]+2$ and the initial data up to order $[n/2]+3$, and if that part of the fundamental domain $\Omega^*$ which falls within our segment remains fundamental, then for the linear equation we may construct a solution according to §20. Thus, we may calculate $u$ by going from one segment to the next.

We now show that we can choose $\Omega^*$ so that it remains fundamental within the region of regularity of the conoid. Indeed, by the Imbedding Theorem on that domain, when the integrals of the sum of the squares of the derivatives of $u$ up to order $[n/2]+3$ are bounded by a number $Q$, we may find $\delta_1(Q)$ so that on the segment $0 \leqq t \leqq \delta_1(Q)$ the values of $u$, $\partial u/\partial t$, $\partial u/\partial x_1, \cdots, \partial u/\partial x_n$ differ from their initial values (i.e., from zero) by not more than $\Delta$. Then if one can give estimates for $Z(t)$ not depending upon the character of the solution, one can find a domain $0<t<\delta_2(Q)$ where $Z(t)<Q$. Then taking $\delta=\min(\delta_1(Q), \delta_2(Q))$, we obtain a domain on which all the inequalities are valid and, therefore, equation (22.10) will be solvable.

We now estimate $Z(t)$. We use the inequality $x^\alpha \leqq A(1+x^\beta)$ for $x>0$ and $0 \leqq \alpha \leqq \beta$ arbitrary ($A$ does not depend on $x$). Then it follows from (22.28) that:

$$Z(t) \leqslant C_6 \int_0^t [1+Z(t_1)^2+Z(t_1-\eta)^k]\, dt_1 \tag{22.31}$$

and analogously from (22.30):

$$Z(t) \leqslant C_6 \int_0^t [Z(t_1)+Z(t_1)^2+Z(t_1-\eta)+Z(t_1-\eta)^k]\, dt_1. \tag{22.32}$$

We consider (22.31) first. We construct the sequence of functions:

$$v_0(t),\ v_1(t),\ \ldots,\ v_m(t)$$

by the rule

$$v_0 = M,$$

$$v_1(t) = C_6 \int_0^t [1+v_{i-1}(t)^2+v_{i-1}(t)^k]\, dt_1. \tag{22.33}$$

According to the theory of differential equations, $v_i$ converges uniformly to $v(t)$, the unique solution of the equation

$$v'(t) = C_5(1+v^2+v^k); \quad v|_{t=0}=0 \tag{22.34}$$

wherever this solution exists.

From this, if

$$Z(t) < v_i(t),$$

then, since

$$v_i(t) > v_i(t-\eta),$$

it follows from (22.31) that

$$Z(t) < v_{i+1}(t),$$

and then obviously,

$$Z(t) < v(t). \tag{22.35}$$

From these estimates there follows the existence of solutions to all the equations (22.19), while for the family of these solutions we have uniform bounds to the integrals over the domain $\Omega(t)$ of the sum of the squares of the derivatives for arbitrary $\eta \to 0$. From the Imbedding Theorem, using Arzela's theorem, we may choose from this family a uniformly convergent subsequence. Passing to the limit in (22.19) for $\eta \to 0$, we find that the limit of such functions will be a solution of the quasilinear equation (22.10).

We show the uniqueness of the solution. Let $u_1$ be the constructed solution of the equation (22.10) for $0 \leqq t \leqq h$. We introduce a new unknown function $v = u - u_1$. Then the new quasilinear equation will have as one of its solutions $v \equiv 0$, and therefore, the free term for

$$v \equiv 0, \quad \frac{\partial v}{\partial t} \equiv 0, \quad \frac{\partial v}{\partial x_1} \equiv 0, \quad \ldots, \quad \frac{\partial v}{\partial x_n} \equiv 0$$

will vanish. We shall show that there does not exist any other solution than $v = 0$. In our case $Z(t)$ satisfies the condition (22.30). Setting $\eta = 0$, we obtain:

$$Z(t) \leqslant C_5 \int_0^t [2Z(t_1) + Z(t_1)^2 + Z(t_1)^k]\, dt_1 \tag{22.36}$$

and if $Z(t)$ is bounded, then it is no larger than the unique solution of the equation

$$y' = C_5 (2y + y^2 + y^k),$$

vanishing for $t = 0$. Obviously, $y \equiv 0$, and therefore $Z(t) \leqq 0$, which proves the uniqueness, since from $Z(t) = 0$ follows $v = 0$. Thus in the special case, our theorem is proved.

5. PROPERTIES OF AVERAGED FUNCTIONS. Before going over to the consideration of the general case, it is necessary for us to establish some properties of averaged functions.

Suppose the function $\mu(x_1', \cdots, x_n')$ has generalized derivatives up to order $l$ and suppose there exists $\epsilon < 1$ such that the derivatives of $\mu$ of order $s$, where

$$l \geqslant s > -n\varepsilon$$

are summable to the power $(\epsilon + s/n)^{-1}$ on every bounded portion of the space.

If $\epsilon<0$, then by the Imbedding Theorem all the derivatives of order $s<-n\epsilon$ are continuous.

Suppose that $F(\xi)=0$ for $\xi>1$ and for $\xi<\frac{1}{2}$ and that $F(\xi)$ is continuous together with all its derivatives. Suppose also

$$\int_0^1 \frac{F(\xi)}{\xi}\,d\xi=\frac{1}{\varkappa_n},$$

where $u_n$ is the surface of the sphere of unit radius.

We consider the averaged potential:

$$U(t, x_1, \ldots, x_n)=\int \cdots \int \frac{F\left(\frac{r}{t}\right)}{r^n}\,\mu(x_1', \ldots, x_n')\,d\Omega. \quad (22.37)$$

By the theorems of §§2, 3, for $t\to 0$ we have $\lim_{t\to 0} U=\mu(x_1,\cdots,x_n)$ and for $t>0$, $U$ has continuous derivatives of arbitrary order.

THEOREM. *The derivatives of the potential $U$ with respect to the variables $t, x_1,\cdots,x_n$ of order $s\le l$ are integrable on every bounded domain $\Omega_x$ to the power $(\epsilon+(s/n))^{-1}$ and have almost everywhere for $t\to 0$ a limit in the sense of $L_{(\epsilon+(s/n))^{-1}}$. The derivatives of $U$ of order $s=l+\alpha$ multiplied by $t^\alpha$ have uniformly bounded integrals for their $(\epsilon+(s/n))^{-1}$ power on every bounded domain $\Omega_x$ and have almost everywhere for $t\to 0$ a limit in the sense of*

$$L_{\frac{1}{\varepsilon+\frac{s}{n}}}.$$

PROOF. (1) $s\le l$. Differentiating (22.37) in the ordinary way with respect to $x_1, x_2,\cdots,x_n$, we obtain:

$$\frac{\partial^{s_1} U}{\partial x_1^{\alpha_1}\partial x_2^{\alpha_2}\cdots\partial x_n^{\alpha_n}}=$$

$$=\int\cdots\int \mu(x_1' \ldots x_n')\frac{\partial^{s_1}}{\partial x_1^{\alpha_1}\cdots\partial x_n^{\alpha_n}}\left[\frac{1}{r^n}F\left(\frac{r}{t}\right)\right]d\Omega'$$

and since $(1/r^n)F(r/t)$ depends only on the difference $x_i-x_i'$,

$$\frac{\partial^{s_1}}{\partial x_1^{\alpha_1}\,\partial x_2^{\alpha_2}\cdots\partial x_n^{\alpha_n}}\left[\frac{1}{r^n}F\left(\frac{r}{t}\right)\right]=$$

$$=(-1)^{s_1}\frac{\partial^{s_1}}{\partial x_1'^{\alpha_1}\partial x_2'^{\alpha_2}\cdots\partial x_n'^{\alpha_n}}\left[\frac{1}{r^n}F\left(\frac{r}{t}\right)\right],$$

and therefore

$$\frac{\partial^{s_1} U}{\partial x_1^{\alpha_1} \dots \partial x_n^{\alpha_n}} =$$

$$= \int \cdots \int \mu(x_1', \dots, x_n') (-1)^{s_1} \frac{\partial^{s_1}}{\partial x_1'^{\alpha_1} \partial x_2'^{\alpha_2} \dots \partial x_n'^{\alpha_n}} \left[ \frac{1}{r^n} F\left(\frac{r}{t}\right) \right] d\Omega' =$$

$$= \int \cdots \int \frac{1}{r^n} F\left(\frac{r}{t}\right) \frac{\partial^{s_1} \mu(x_1', \dots, x_n')}{\partial x_1'^{\alpha_1} \partial x_2'^{\alpha_2} \dots \partial x_n'^{\alpha_n}} d\Omega' =$$

$$= \int \cdots \int \frac{1}{r} F\left(\frac{r}{t}\right) \frac{\partial^{s_1} \mu}{\partial x_1'^{\alpha_1} \partial x_2'^{\alpha_2} \dots \partial x_n'^{\alpha_n}} dr\, dS_1. \tag{22.38}$$

We differentiate now $\alpha_0$ times with respect to $t$ and use the equality:

$$\frac{\partial}{\partial t} \left[ \frac{1}{r} F\left(\frac{r}{t}\right) \right] = -\frac{1}{t^2} F'\left(\frac{r}{t}\right) = \frac{\partial}{\partial r} \left[ -\frac{1}{t} F\left(\frac{r}{t}\right) \right] =$$

$$= \frac{\partial}{\partial r} \left[ \frac{1}{r} \left( -\frac{r}{t} \right) F\left(\frac{r}{t}\right) \right]$$

and further

$$\frac{\partial^k}{\partial t^k} \left[ \frac{1}{r} F\left(\frac{r}{t}\right) \right] = \frac{\partial^k}{\partial r^k} \left[ \frac{1}{r} \left( -\frac{r}{t} \right)^k F\left(\frac{r}{t}\right) \right]. \tag{22.39}$$

Using these equalities, we obtain:

$$\frac{\partial^{s_1} U}{\partial t^{\alpha_0} \partial x_1^{\alpha_1} \dots \partial x_n^{\alpha_n}} =$$

$$= \int \cdots \int \frac{\partial^{s_1} \mu}{\partial x_1'^{\alpha_1} \dots \partial x_n'^{\alpha_n}} \frac{\partial^{\alpha_0}}{\partial r^{\alpha_0}} \left[ \frac{1}{r} \left( -\frac{r}{t} \right)^{\alpha_0} F\left(\frac{r}{t}\right) \right] dr\, dS_1 =$$

$$= (-1)^{\alpha_0} \int \cdots \int \frac{\partial^{s_1} \mu}{\partial x_1'^{\alpha_1} \dots \partial x_n'^{\alpha_n}} \frac{\partial^{\alpha_0}}{\partial r^{\alpha_0}} \left[ \frac{1}{r} \left( \frac{r}{t} \right)^{\alpha_0} F\left(\frac{r}{t}\right) \right] dr\, dS_1 =$$

$$= \int \cdots \int \sum \frac{\partial^{s} \mu}{\partial x_1'^{\beta_1} \dots \partial x_n'^{\beta_n}} \Lambda_{\beta_1 \dots \beta_n} (x_i - x_i') \times \tag{22.40}$$

$$\times \frac{1}{t^{\alpha_0} r^{n-\alpha_0+1}} F\left(\frac{r}{t}\right) d\Omega'.$$

This result may be briefly formulated in the following way. The derivatives of the potential may be put in the form of a linear combination of averaged functions of the derivatives $\partial^s\mu/\partial x_1'^{\beta_1}\cdots\partial x_n'^{\beta_n}$ with regular kernels. By virtue of the fact that the derivatives $\partial^s\mu/\partial x_1'^{\beta_1}\cdots\partial x_n'^{\beta_n}$ are summable to the power $(\epsilon+(s/n))^{-1}$, we immediately show the first part of the theorem by a simple application of theorems on averaged functions (cf. §5, item 2).

(2) $s=l+\alpha$. Carrying through a transformation as in the preceding discussion, we obtain:

$$t^\alpha \frac{\partial^{l+\alpha} U}{\partial t^{\alpha_0}\partial x_1^{\alpha_1}\dots\partial x_n^{\alpha_n}} =$$

$$= \int\cdots\int \frac{\partial^l \mu}{\partial x_1'^{\beta_1}\partial x_2'^{\beta_2}\dots\partial x_n'^{\beta_n}}\, t^\alpha \frac{\partial^\alpha}{\partial t^{\alpha_0}\partial x_1^{\gamma_1}\dots\partial x_n^{\gamma_n}} \times$$

$$\times \left\{\frac{1}{r} F\left(\frac{r}{t}\right)\right\} d\Omega', \tag{22.41}$$

if

$$\alpha_1+\alpha_2+\dots+\alpha_n > l$$

or

$$t^\alpha \frac{\partial^{l+\alpha} U}{\partial t^{\alpha_0}\partial x_1^{\alpha_1}\dots\partial x_n^{\alpha_n}} = \tag{22.42}$$

$$= \int\cdots\int \sum \frac{\partial^l \mu}{\partial x_1'^{\beta_1}\dots\partial x_n'^{\beta_n}}\, t^\alpha \frac{\partial^\alpha}{\partial r^\alpha}\left[\frac{1}{r}\left(-\frac{r}{t}\right)^\alpha F\left(\frac{r}{t}\right)\right]\frac{d\Omega'}{r^{n-1}},$$

if $\alpha_1+\alpha_2+\cdots+\alpha_n\leqq l$. In both cases, the kernels of the averaged functions are homogeneous functions of $r$ and $t$ of degree $(-n)$. By the theorem on averaged functions, we obtain the final result.

COROLLARIES. (1) *The derivatives of $t^\alpha U$ of order $s=0,1,\cdots,l+\alpha$ have bounded integrals on every domain $\Omega$ to their $(\epsilon+((s-\alpha)/n))^{-1}$ powers. The derivatives of order $s<\alpha-n\epsilon$ will be simply continuous functions.*

This follows from the Imbedding Theorem.

(2) *The derivatives of order $s$ have a limit in $L_{(\epsilon+((s-\alpha)/n))^{-1}}$ for $t\to 0$ and that limit for $s<\alpha$ will equal zero, while*

$$\lim_{t\to 0} \frac{\partial^\alpha (t^\alpha U)}{\partial t^\alpha} = \alpha!\,\mu(x_1, x_2, \ldots, x_n).$$

This corollary follows from the properties of averaged functions.

We may now prove an important theorem.

6. TRANSFORMATION OF THE INITIAL CONDITIONS.

THEOREM. *Suppose we are given* $k+1$ *functions such that*

$$\left.\begin{array}{l} \int\cdots\int_\Omega |u_k|^p\,d\Omega \leqslant A_\Omega^p; \\ \int\cdots\int_\Omega |u_{k-1}|^{\frac{1}{\frac{1}{p}-\frac{1}{n}}}\,d\Omega \leqslant A_\Omega^{\frac{1}{\frac{1}{p}-\frac{1}{n}}}; \\ \int\cdots\int_\Omega \left|\frac{\partial u_{k-1}}{\partial x_j}\right|^p d\Omega_x \leqslant A_\Omega^p \\ \cdots\cdots\cdots\cdots\cdots \\ \int\cdots\int_\Omega |u_0|^{\frac{1}{\frac{1}{p}-\frac{k}{n}}}\,d\Omega \leqslant A_\Omega^{\frac{1}{\frac{1}{p}-\frac{k}{n}}} \\ \cdots\cdots\cdots\cdots\cdots \\ \int\cdots\int_\Omega \left|\frac{\partial^k u_0}{\partial x_1^{\alpha_1}\cdots\partial x_n^{\alpha_n}}\right|^p d\Omega \leqslant A_\Omega^p, \end{array}\right\} \qquad (22.43)$$

*if* $\Omega$ *is an arbitrary bounded domain in the space* $x_1, x_2, \cdots, x_n$.

*If*

$$k > \left[\frac{n}{p}\right],$$

*then from* (22.43) *it follows that the functions*

$$u_0, \frac{\partial u_0}{\partial x_i}, \ldots, \frac{\partial^{k-[\frac{n}{p}]} u_0}{\partial x_1^{\gamma_1}\ldots\partial x_n^{\gamma_n}} \cdot u_1, \frac{\partial u_1}{\partial x_j}, \ldots, u_{k-[\frac{n}{p}]}$$

*will be bounded and continuous.*

*Then for* $t \geqq 0$, *we may construct a function* $v(t, x_1, \cdots, x_n)$ *for which the functions* $u_i$ *will coincide with the initial values of its time derivatives. The function* $v$ *will have for* $t>0$ *continuous derivatives of arbitrary order and*

$$\lim_{t\to 0}\left[\int\cdots\int\left(\frac{\partial^\alpha v}{\partial t^\alpha}-u_\alpha\right)^{\frac{1}{\frac{1}{p}-\frac{k-\alpha}{n}}} d\Omega\right]=0, \quad \left(\alpha>k-\left[\frac{n}{p}\right]\right), \tag{22.44}$$

*where the integrals* (22.44) *have a sense and we also have amost everywhere in* $\Omega$:

$$\lim_{t\to 0}\frac{\partial^\alpha v}{\partial t^\alpha}=u_\alpha. \tag{22.45}$$

*The derivatives up to order* $k-[n/p]$ *will be continuous for* $t=0$.

*In addition, we will have the inequalities*

$$\int_\Omega\cdots\int\left|\frac{\partial^\alpha v}{\partial t^{\alpha_0}\partial x_1^{\alpha_1}\cdots\partial x_n^{\alpha_n}}\right|^{\frac{1}{\frac{1}{p}-\frac{k-\alpha}{n}}} d\Omega < B_\Omega; \quad \left(\alpha>k-\left[\frac{n}{p}\right]\right)$$

*on an arbitrary bounded domain* $\Omega$ *for arbitrary* $t>0$, *where* $B_\Omega$ *is some new constant.*

To prove this theorem we seek the function $v(t, x_1,\cdots, x_n)$ in the form:

$$v=u_0+\frac{t}{1!}U_1+\frac{t^2}{2!}U_2+\ldots+\frac{t^k}{k!}U_k, \tag{22.46}$$

where

$$U_j=\int\cdots\int\frac{1}{r^n}F\left(\frac{r}{t}\right)\mu_j(x_1', \ldots, x_n')\,d\Omega'. \tag{22.47}$$

We choose the $\mu_i$ by completeness, using the conditions (22.45). We have:

$$\lim_{t\to 0} v=u_0; \tag{22.48}$$

$$\lim_{t\to 0}\frac{\partial v}{\partial t}=\lim_{t\to 0}U_1=\mu_1=u_1, \tag{22.49}$$

$$\lim_{t\to 9}\frac{\partial^2 v}{\partial t^2}=2\lim_{t\to 0}\frac{\partial u_1}{\partial t}+\lim_{t\to 0}U_2=\sum_{j=1}^n \alpha_j\frac{\partial \mu_1}{\partial x_j}+\mu_2=u_2,$$

from which

$$\mu_2 = u_2 - \sum_{j=1}^{n} \alpha_i \frac{\partial u_1}{\partial x_j}. \tag{22.50}$$

If we suppose now

$$\mu_{i-1} = \sum_{\beta=1}^{i-1} \alpha^{(i-1)}_{\beta_1 \beta_2, \ldots, \beta_n} \frac{\partial^{i-\beta-1} u}{\partial x_1^{\beta_1} \partial x_2^{\beta_2} \ldots \partial x_n^{\beta_n}},$$

we show that the $\mu_i$ have the same form. Indeed, differentiating (22.46) $i$ times with respect to $t$, we obtain:

$$\frac{\partial^i U}{\partial t^i} = i \frac{\partial^{i-1} U_1}{\partial t^{i-1}} + t \frac{\partial^i U_1}{\partial t^i} + (i-1) \frac{\partial^{i-2} U_2}{\partial t^{i-2}} + \ldots + \frac{t^k}{k!} \frac{\partial^i U_k}{\partial t^i}$$

$$\lim \frac{\partial^i v}{\partial t^i} = i \lim_{t \to 0} \frac{\partial^{i-1} U_i}{\partial t^{i-1}} + (i-1) \lim_{t \to 0} \frac{\partial^{i-2} U_2}{\partial t^{i-2}} + \ldots + \lim_{t \to 0} U_1 =$$

$$= \sum \gamma^{(1)}_{\beta_1, \beta_2, \ldots, \beta_n} \frac{\partial^{i-1} \mu_i}{\partial x_1^{\beta_1} \ldots \partial x_n^{\beta_n}} +$$

$$+ \sum \gamma^{(2)}_{\beta_1, \beta_2, \ldots, \beta_n} \frac{\partial^{i-2} \mu^2}{\partial x_1^{\gamma_1} \ldots \partial x_n^{\gamma_n}} t + \ldots + \mu_i = u_i,$$

As a result,

$$\mu_i = \sum_{\beta=1}^{i} \alpha^{(i)}_{\beta_1, \beta_2, \ldots, \beta_n} \frac{\partial^{i-\beta} u_\beta}{\partial x_1^{\beta_1} \ldots \partial x_n^{\beta_n}}. \tag{22.51}$$

Using (22.51), we shall now easily establish with the help of the Minkowski inequality that

$$\int_\Omega \cdots \int \left| \frac{\partial^{k-i} \mu_i}{\partial x_1^{\alpha_1} \partial x_2^{\alpha_2} \ldots dx_n^{\alpha_n}} \right|^p d\Omega \leqslant M A_\Omega^p.$$

On the basis of the theorem on potentials, we obtain our theorem.

7. THE GENERAL CASE FOR THE CAUCHY PROBLEM FOR QUASILINEAR EQUATIONS. In the general case, we have to seek a solution of the equation

$$\sum_{i=1}^{n} \sum_{j=1}^{n} A_{ij} \frac{\partial^2 u}{\partial x_i \partial x_j} - \frac{\partial^2 u}{\partial t^2} = F \tag{22.52}$$

for the conditions:

$$u \Big|_{t=0} = u_0; \quad \frac{\partial v}{\partial t} \Big|_{t=0} = u_1, \tag{22.53}$$

where

$$\left.\begin{aligned}
&\int\limits_{\Omega}\cdots\int\left[\frac{\partial^{\rho} u_0}{\partial x_1^{\alpha_1}\partial x_2^{\alpha_2}\cdots\partial x_n^{\alpha_n}}\right]^2 d\Omega \leqslant U,\\
&\qquad \rho = 0, 1, 2, \ldots, \left[\frac{n}{2}\right]+3\\
&\int\limits_{\Omega}\cdots\int\left[\frac{\partial^{\rho} u_1}{\partial x_1^{\alpha_1}\partial x_2^{\alpha_2}\cdots\partial x_n^{\alpha_n}}\right]^2 d\Omega \leqslant U,\\
&\qquad \rho = 0, 1, \ldots, \left[\frac{n}{2}\right]+2.
\end{aligned}\right\} \tag{22.54}$$

On the initial hyperplane $t=0$, using (22.53) and (22.52) we may calculate the values of all derivatives of $u$ up to order $[n/2]+3$ inclusive, for which each function

$$\left.\frac{\partial^s u}{\partial t^{\alpha_0}\partial x_1^{\alpha_1}\cdots\partial x_n^{\alpha_n}}\right|_{t=0}$$

for

$$s = 3, 4, \ldots, \left[\frac{n}{2}\right]+3$$

will be summable to the power

$$\frac{1}{\dfrac{1}{2}-\dfrac{\left[\frac{n}{2}\right]+3-s}{2}},$$

and for $s=0,1,2,$ will be continuous.

We prove this by induction. Suppose that the assertion is true for all derivatives for which the order of differentiation in $t$ is less than $\alpha_0$ (if the differentiation is carried out with respect to $x_i$, then the assertion is a consequence of the Imbedding Theorem).

We consider $\partial^s u/\partial t^{\alpha_0}\partial x_1^{\alpha_1}\cdots\partial x_n^{\alpha_n}$. As a consequence of (22.52), we have:

$$\frac{\partial^s u}{\partial t^{\alpha_0}\partial x_1^{\alpha_1}\cdots\partial x_n^{\alpha_n}} = -\frac{\partial^{s-2}F}{\partial t^{\alpha_0-2}\partial x_1^{\alpha_1}\cdots\partial x_n^{\alpha_n}}+$$

$$+\frac{\partial^{s-2}}{\partial t^{\alpha_0-2}\partial x_1^{\alpha_1}\cdots\partial x_n^{\alpha_n}}\sum_{i=1}^{n}\sum_{j=1}^{n}A_{ij}\frac{\partial^2 u}{\partial x_i\partial x_j}.$$

We use the concept of inverse index. The inverse index for the initial values of the derivatives of order $s$ of $u$ will be:

$$\left.\begin{array}{ll} \dfrac{1+2(s-3)}{2n}; & n=2m \\ \dfrac{1+(s-3)}{n}; & n=2m+1 \end{array}\right\} \quad s=3,4,\ldots,\left[\frac{n}{2}\right]+3.$$

Reading off the index for the right side and noting that each differentiation in $t$ and substitution of $t=0$ increases the index by not more than $1/n$, we obtain our assertion.

We introduce for $t>0$, the new function $w=u-v$, where $v$ is constructed as indicated above and takes on the initial hyperplane the same values as $u$. After this substitution, the equation remains quasilinear with coeffiicients having property $T$.

We extend the coefficients of the equation to the halfspace $t<0$. For this purpose, the $A_{ij}$ are extended arbitrarily and

$$F(t, x_1, \ldots, x_n, u, \ldots)|_{t<0} = F(0, x_1, \ldots, x_n, 0, \ldots) = 0.$$

Thereby the problem is reduced to the special case considered above, and for $t<0$, the solutions of (22.52) will be identically zero.

Thus, our theorem is proved in the general case.